OXYGEN TRANSPORT TO TISSUE IX

ADVANCES IN EXPERIMENTAL MEDICINE AND BIOLOGY

Recent Volumes in this Series

A Continuation Order Plan is available for this series. A continuation order will bring delivery of each new volume immediately upon publication. Volumes are billed only upon actual shipment. For further information please contact the publisher.

OXYGEN TRANSPORT TO TISSUE IX

Edited by

I. A. Silver
and A. Silver

The Medical School
University of Bristol
Bristol, United Kingdom

PLENUM PRESS • NEW YORK AND LONDON

Library of Congress Cataloging in Publication Data

Oxygen transport to tissue IX.

  Based on the ISOTT meeting, held July 27–30, 1986, at Churchill College, Cambridge
United Kingdom.
  Includes bibliographies and index.
  1. Oxygen transport (Physiology) — Congresses. 2. Oxygen in the body — Congresses.
I. Silver, I. A. II. Silver, A. (Ann) III. International Society on Oxygen Transport to
Tissue. IV. Title: Oxygen transport to tissue 9. [DNLM: 1. Biological Transport —
congresses. 2. Oxygen — blood — congresses. 3. Oxygen — metabolism — congresses.
4. Oxygen Consumption — congresses. QV 312 0983 1986]
QP99.3.090939 1987   612′.01524   87-3329
ISBN 978-1-4684-7435-0   ISBN 978-1-4684-7433-6 (eBook)
DOI 10.1007/978-1-4684-7433-6

Proceedings of an ISOTT meeting on Oxygen Transport to Tissue,
held July 27–30, 1986, at Churchill College, Cambridge, United Kingdom

© 1987 Plenum Press, New York
Softcover reprint of the hardcover 1st edition 1987

A Division of Plenum Publishing Corporation
233 Spring Street, New York, N.Y. 10013

PREFACE

These papers stem from the ISOTT Meeting held at Churchill College, Cambridge, from July 27th to 30th, 1986. Although the sun did not shine so brightly as during the Cambridge meeting in 1977, the communications and discussions were as lively and informative and some heat, as well as light, was generated in the presentation of differing views. The meeting was conducted in a generally informal way which allowed maximum time for discussion but the relatively unstructured nature of the debates made them unsuitable for publication. The amount of editing necessary meant that the printed version of the exchanges would bear little resemblance to the original, hence their omission.

All the papers presented here have been scrutinized and retyped in a standard format. However, the diverse interests of ISOTT's members, reflected in the wide spectrum of the material submitted, made total editorial uniformity an unrealistic goal. Complete consistency in the use of symbols, abbreviations and units seemed less important than speed of publication.

We are very grateful to all those who contributed to the smooth running of the Meeting itself and to the production of this book. In particular we would like to thank Fiona and Angus Silver, John Powter, Charles Drown and Judy Deas for their work before and during the Symposium and Mrs Avril Lear for her efforts in dealing with the correspondence and for the initial retyping of the papers. Finally we would like to thank the staff of the Department of Pathology at Bristol for their tolerance and help.

I.A. Silver
Ann Silver

January 1987

CONTENTS

METHODS

BLOOD SUBSTITUTES

MODELLING

PHYSIOLOGY

OXYGEN TOXICITY

TUMOURS

MORPHOMETRIC ANALYSIS OF SPARSE CAPILLARY NETWORKS

S. EGGINTON, Z. TUREK and L. HOOFD

Department of Physiology, University of Birmingham
Medical School, Birmingham, B15 2TJ, England, and
Department of Physiology, Faculty of Medicine, Catholic
University of Nijmegen, 6500 HB Nijmegen, The Netherlands

SUMMARY

Two methods were used to assess the heterogeneity of capillary supply to muscles of widely differing metabolic capacity and fibre size. Using the method of capillary domains (DOM; Hoofd et al., 1985) and the closest-individual method (CI; Kayar et al., 1981) radii of Kroghian cylinders (R) can be calculated, and the heterogeneity of their lognormal distribution represented by the logarithmic standard deviation (Log SD). Both methods yield similar values for mean R in a tissue. DOM is more direct and quicker than CI, and may be particularly useful in the analysis of capillary oxygen supply during functional hypertrophy and in muscle regeneration where a broad distribution of fibre areas may be found.

Despite a 500-fold range of capillary density, to a minimum of 20 capillaries mm^{-2}, heterogeneity of capillary supply was similar in all muscles, indicating a functionally homologous spatial distribution. The relationship between number of fibres overlapped by a capillary domain, and domain area has zero correlation in most tissues but shows a negative trend in fish fast muscle, reflecting hyperplastic and hypertrophic growth. Capillary/fibre ratio is inappropriate for sparse networks whereas the cumulative fraction of domains vs fibre area shows a strong correlation, suggesting that maximal oxygen supply to muscle fibres is not restricted to contiguous capillaries, but also involves those remote from the fibre surface.

INTRODUCTION

Direct estimates of oxygen release from capillaries to, and its consumption in, tissue have numerous methodological problems leading many workers into the field of mathematical analysis and modelling of tissue oxygen supply. Unfortunately, the complexity of the problem often requires gross simplifications to be employed, invoking assumptions that lack experimental justification. Analytical problems can be minimized by examining a highly structured, or anisotropic tissue such as skeletal muscle. The modelling approach often fails to allow for capillary functions other than oxygen delivery, and is usually restricted to cases of high capillary density (CD) and geometrically regular networks. Most tissue does not fall into this category having either low CD, irregular capillary supply, or both. Morphometric data, on the other hand, reveal empirical anatomical relationships which may then profitably be correlated with diverse physiological or metabolic characteristics of the tissue in question. We present here a comparison of analytical methods using data from widely differing muscle types. Mammalian slow and fast skeletal muscle are tissues of high capillary density. Fish skeletal muscle is separable into functionally discrete and anatomically distinct tissues containing a single fibre type, of relatively low CD.

Two basic approaches have been adopted in describing capillary oxygen supply to skeletal muscle using either eccentric diffusion from a capillary to surrounding tissue, or concentric diffusion from neighbouring capillaries to a fibre or tissue mass. In cardiac muscle a number of analytical methods have been shown to produce similar estimates for the derived Kroghian radius, R, although the apparent heterogeneity of capillary supply is dependent on the model used (Turek, Hoofd and Rakusan, 1987). This study considers the utility of computerized analyses for use in a range of striated muscles, and suggests a number of factors that need to be incorporated into future models of the capillary supply.

MATERIALS AND METHODS

Tissue preparation

Female Sprague-Dawley rats, around 220 g body weight, used in a previous study (Egginton, 1986) were killed by anaesthetic overdose and tissue immediately excised and placed on ice. Leg muscles sampled were extensor digitorum longus and soleus, representing typical mammalian fast and slow skeletal muscle, respectively. Steaks of tissue from the mid region of the muscle were mounted in an inert medium (OCT compound; Lamb) on cork and rapidly frozen in isopentane (2-methyl butane) cooled in liquid nitrogen. Cryostat sections (12 μm, cut at -20°C) were stained for alkaline phosphatase activity to localize capillary endothelium.

Conger eels, Conger conger, were used in a previous study to determine the degree of anisotropy in the capillary bed of fish skeletal muscle (Egginton and Johnston, 1984). Skin was removed, and bundles of superficial slow muscle and deep fast muscle carefully dissected. These were fixed at resting length by pinning to strengthened cork strips and immersed in 3% gluteraldehyde, 0.15 M phosphate buffer (pH 7.4 at 20°C) for 8-12 hours at 4°C. Samples were subsequently post-fixed in 1% OsO_4, dehydrated in a series of alcohols, cleared in propylene oxide and embedded in Araldite CY212 resin (EM Scope; Trent). Semi-thin (0.5 μm) sections were stained with toluidine blue (1% in 1% sodium tetraborate). Capillaries were identified using morphological criteria since alkaline phosphatase activity is negligible in these vessels.

Analysis

Traces of fibre cross-sections and location of capillary centres were made directly from slides, using a microscope drawing arm and a final magnification of x400 for all skeletal muscles samples, except for fish fast muscle where the low capillary density and large fibre area necessitated a much lower magnification of x100. Where oblique profiles of capillaries were encountered one coordinate was registered at the mid-point. Areas of tissue for analysis were selected at random and delineated by a square counting frame, measuring 10 x 10 cm. The size of tissue cross-section available for analysis (whole muscle, or representative subsample) and CD determined the degree of replication required (see below).

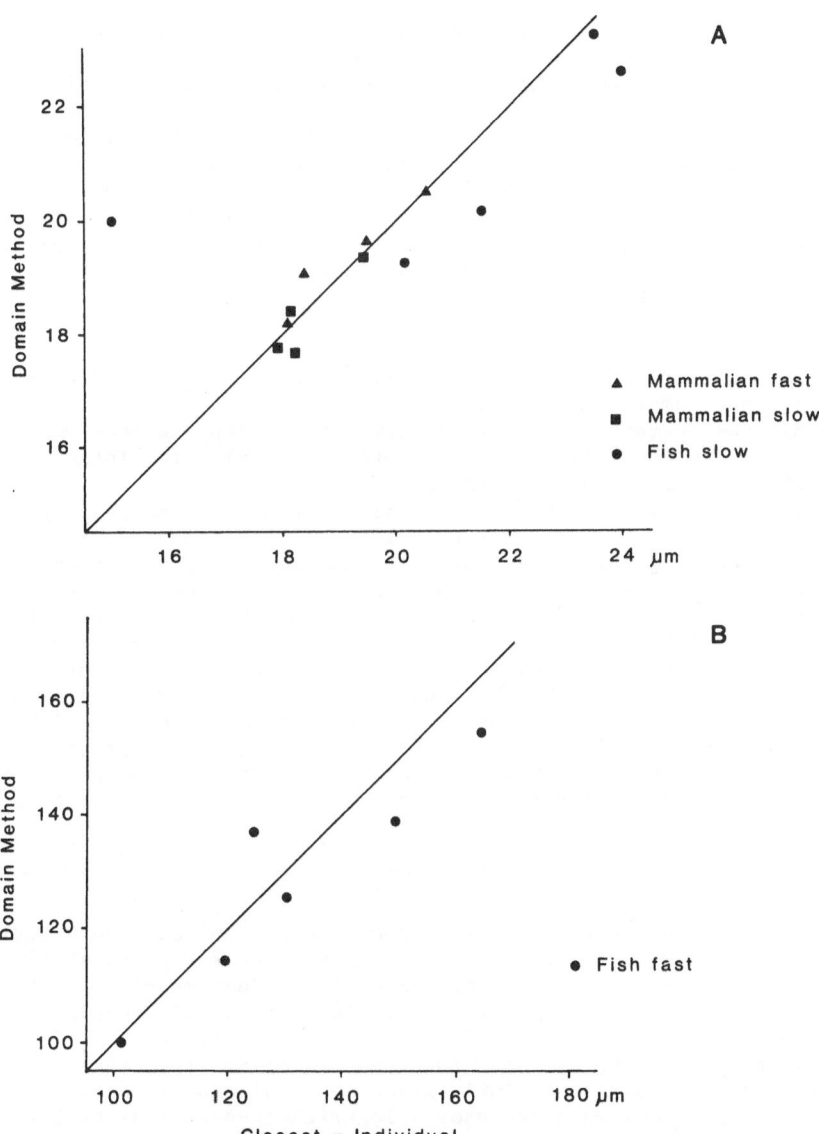

Figure 1. Comparative estimates of mean Kroghian cylinder radii, R, using the DOM and CI methods for skeletal muscle with relatively high capillary densities (A) and with a sparse capillary network (B).

X, Y coordinates of capillary centres from traces were registered, and their spatial distribution subsequently analysed using in-house software (Hoofd et al., 1985; Turek, Hoofd and Rakusan, 1986). In the CI method a continuous distribution of the distances of tissue points to the nearest capillary is obtained by means of a grid overlay, and a fitting routine used to calculate the mean radius of the Kroghian tissue cylinder, R. The DOM method constructs polygons around each capillary by bisecting lines between adjacent capillaries, and connecting extra- polated lines drawn normal to the original line, and the resultant area calculated. The equivalent R is then calculated as:

$$\sqrt{(mean\ domain\ area/pi)}.$$

Capillary density is calculated per unit area of tissue sample, and is only approximated by the inverse of mean domain area due to the exclusion of incomplete (peripheral) domains from the analysis.

In addition a new method for the evaluation of the local capillary supply was derived from the overlapping of domains and muscle fibres, as previously outlined (Hoofd and Turek, 1985). A further step was required so that muscle fibre outlines could also be read into the computer as contour coordinates. Fibre cross-sections and domains were then superimposed and the extent of their overlap calculated. Three indices can then be derived:

a) the number of muscle fibres that overlap each domain, usually the number of fibres surrounding one capillary. This is plotted against domain area;
b) the number of capillary domains overlapping each muscle fibre cross-section, giving the number of capillaries influencing each fibre. This is plotted against fibre area;
c) the sum of the fractional area of domains overlapping each fibre cross-section was calculated. This index can be thought of as the sum of fractions of capillaries interacting with an individual fibre. Unlike the integer values obtained with (a) and (b), when plotted against fibre area this index shows that many fibres interact with a number of 'complete' capillaries plus a fraction of an additional one. In reality, of course, a fibre will interact with fractions of many capillaries as a capillary domain usually overlaps (is 'in contact' with) more fibres than indicated by index (a). This last index, then, provides a local capillary-to-fibre ratio (C:F), as distinct from the average C:F obtained from numerical counts of capillaries and fibres in a section.

Statistics
Group means were derived from either 4 animals and 6 replicates (mammalian slow and fast muscle), 6 animals and 2 replicates (fish slow muscle) or 7 animals and 2 replicates (fish fast muscle). Sufficiency of sample size was determined by inspection of cumulative means (Egginton and Johnston, 1982). Intra-group variability of normally-distributed parameters, and inter-group variability, were assessed by coefficient of variance (CV) given as 100[standard deviation/mean]. Both R's and domain areas show a logarithmic-normal distribution, and the heterogeneity can therefore be represented by the logarithmic standard deviation, Log SD (Turek et al., 1986). Scatter plots of related parameters were overlain with regression lines, and correlation coefficients calculated by linear regression using least-squares best fit.

4

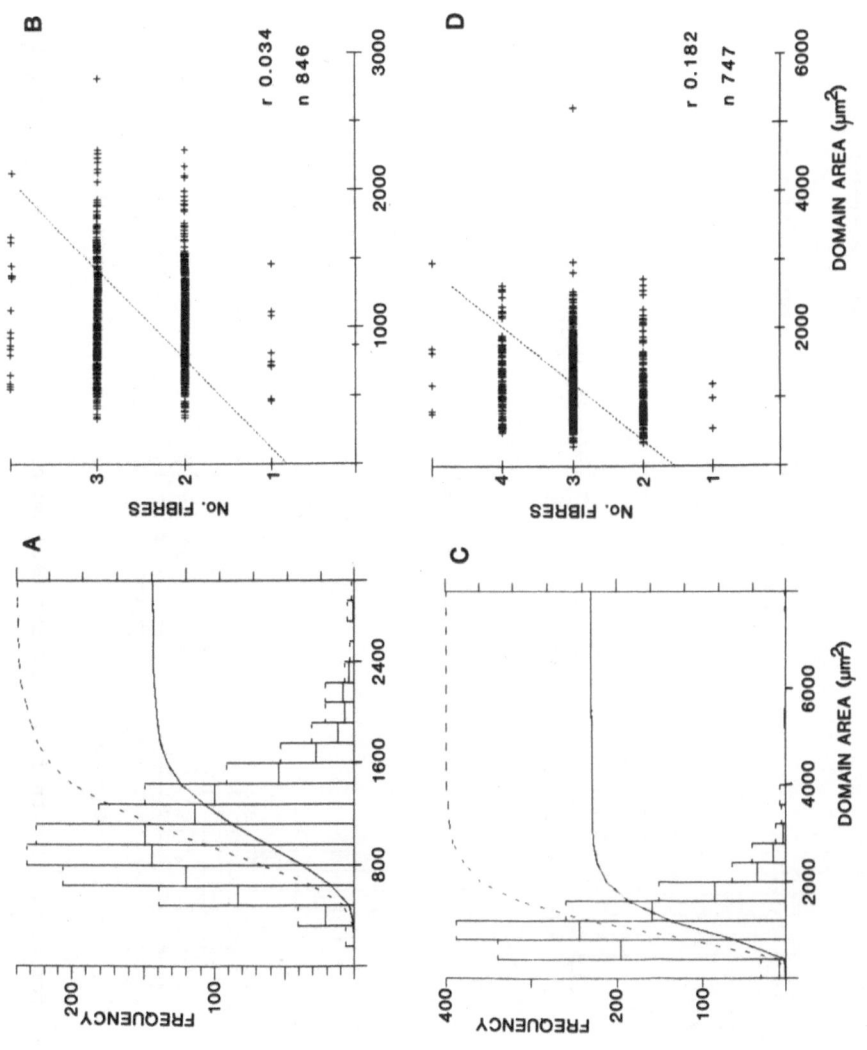

Figure 2. Mammalian slow (top) and fast (bottom) skeletal muscle. A,C: Frequency distribution of capillary domain cross-sectional areas using all samples (broken line) and only those domains free from error due to proximity of the grid edge (solid line) B,D: Numerical coincidence of muscle fibres and capillary domains as a function of domain area.

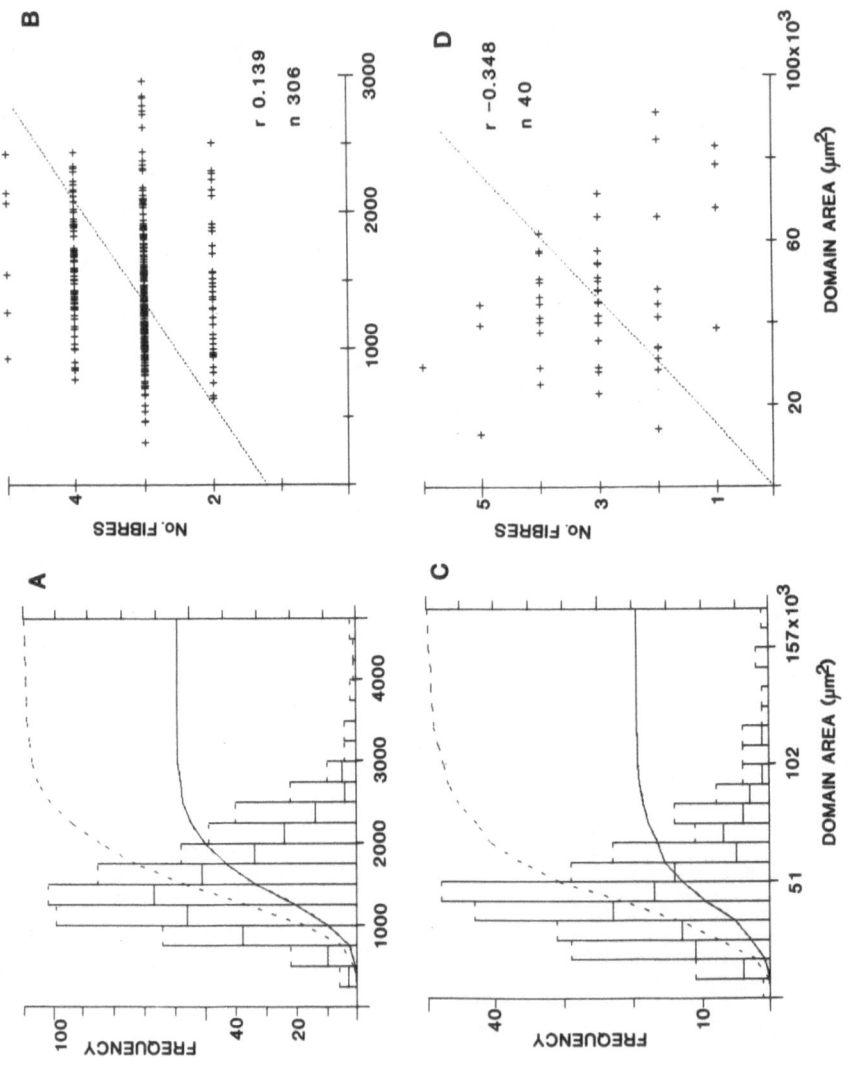

Figure 3. As for Figure 2 with fish slow skeletal muscle (top) and fast skeletal muscle (bottom).

RESULTS

Heterogeneity of capillary distribution in cross-sections of a variety of muscles can successfully be evaluated by both the domain (DOM) and closest-individual (CI) methods. Both provide similar estimates of the mean Kroghian radius, R (Fig. 1). Capillary density shows an inverse progression with increasing fibre areas of all tissues sampled, while the complementary trend is clearly seen for mean R (Table 1).

Table 1.   Estimates of capillary supply by capillary domains method

	Equiv. Kroghian Radius (μm)			Capillary Density (mm^{-2})	Mean Fibre Area (μm^2)
	Mean	Median	Log SD		
Mamm. slow (N=4; n=23)	18.30	17.76	0.077	934 $\pm$ 32.5	2373 $\pm$ 64.8**
Mamm. fast (N=4; n=24)	19.25	18.33	0.096	825 $\pm$ 59.0	1301 $\pm$ 36.2
Fish slow (N=7; n=14)	21.63	21.00	0.077	679 $\pm$ 51.1**	952 $\pm$ 53.7**
Fish fast (N=6; n=13)	121.7	116.1	0.097	19.5 $\pm$ 2.32	3789 $\pm$ 204.3

Note: Mamm., Mammalian; CD and fibre area given as mean $\pm$ S.E.M.; sample size is given as N=animals, n=fields.   ** P< 0.001.

Inter-animal variability in estimates of R is inversely proportional to CD with CV being 4.1, 5.0, 8.6 and 13.6% for mammalian slow and fast muscle, and fish slow and fast muscle, respectively. As with cardiac tissue, Log SD is larger in CI estimates, probably reflecting an asymmetry of the capillary domains since DOM is known to under-estimate the heterogeneity with respect to CI (Turek et al., 1986). In all tissues the cross-sectional areas of capillary domains show a skewed distribution (Figs 2 and 3). The DOM method is more convenient and faster than the CI method when used in conjunction with existing software.

Correlation between the constituent number of fibres overlapping a capillary domain, and domain area is around zero for most muscles but shows a negative correlation in fish fast muscle due to the growth pattern of this tissue which results in many young (smaller) fibres in a sample. Although mean domain area tends to increase directly with mean fibre area, within a sample clustering of small fibres appears more common within the smaller capillary domains. Combination of DOM with imaging of muscle fibre profiles suggests that the correlation between number of capillaries around a fibre vs fibre cross-sectional area is relatively poor. There is a better correlation and a linear regression of local capillary-to-fibre ratio on fibre size (Figs 4 and 5).

DISCUSSION

It has long been recognized from the Krogh-Erlang formula (Krogh, 1919) that the radius of the tissue cylinder supplied by a capillary, sometimes called the diffusion distance, R, is an important determinant of oxygen supply. A number of methods are available for calculating R

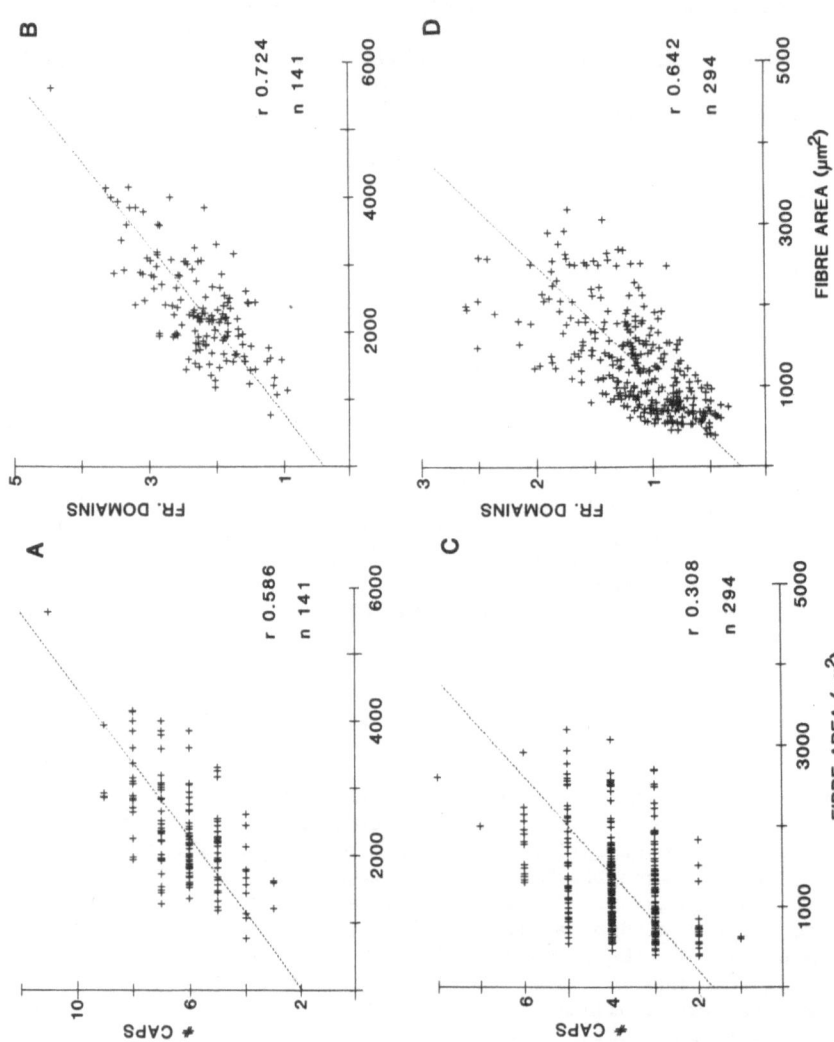

Figure 4. Mammalian slow (top) and fast (bottom) skeletal muscle. A,C: Number of capillaries adjacent to individual muscle fibres as a function of fibre cross-sectional area. B,D: Correlation is improved by plotting the cumulative fraction of overlapping domains (FR. Domains) against fibre area.

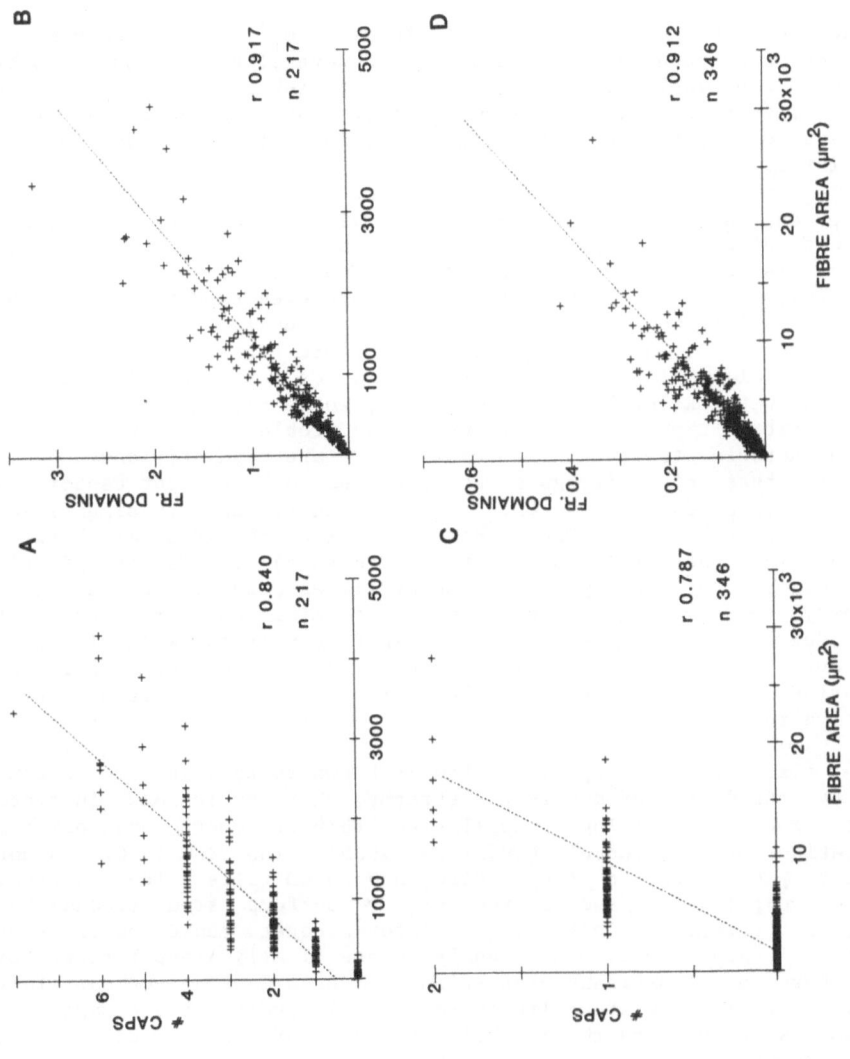

Figure 5. As for Figure 4 with fish slow skeletal muscle (top) and fast skeletal muscle (bottom).

from the average tissue volume supplied by a capillary, but such analyses ignore heterogeneity of capillary spacing which is known to affect both mean tissue PO_2 and the proportion of poorly oxygenated tissue (Turek and Rakusan, 1981). Heterogeneity of capillary spacing can be estimated from tissue cross-sections by a number of methods, of which two are considered here. First, one can estimate the distribution of tissue at various distances from the nearest capillary by CI analysis (Kayar et al., 1982) from which the distribution of R may be derived, although this requires additional computation. Second, one can estimate the area of tissue supplied by each capillary, or DOM, from which R is easily derived (Hoofd et al., 1985). Practical advantages of the DOM method, therefore, seem to be the ease of computation and reduced data variance. However, this has so far only been shown applicable to cardiac tissue (Turek et al., 1987) and in order to investigate further the validity and application of this method, we felt it necessary to perform the analysis on a wide range of muscle tissue.

Most analyses of capillary supply use assumptions of either homogeneous distribution patterns and/or high capillary density, and so integrate tissue-wide variation, unless one adopts more laborious methods (e.g. Gray and Renkin, 1978) although even then the utility of the data is limited. In mixed mammalian muscle this is a serious problem since the influence of different fibre types on the capillary bed is unequal. This is likely to be less of a problem with cardiac muscle, and with skeletal muscle from fish, where tissue is composed of a single fibre type. A more intriguing problem than the situation found in aerobic tissue where capillaries are regularly and densely packed, is that of an irregular capillary network and most especially where this involves a low CD, such as in tissue with significant glycolytic potential. The large metabolic differential between mammalian slow (aerobic) and fast (glycolytic) muscle is greatly accentuated in the homologous systems of fishes, and provides a useful opportunity to investigate an unambiguous microvascular supply. A similar degree of heterogeneity in capillary supply to skeletal muscle exists over a 500-fold range of CD, suggesting that differences in the organization of the capillary bed are quantitative rather than qualitative in nature.

The size of the capillary bed is known to be regulated to some extent by the degree of muscle hypertrophy, where an increase in fibre diameter forces existing capillaries further apart and elicits profileration of new vessels (Hudlicka, 1985). The rise in C:F is not in direct proportion to fibre size, and so CD falls. The numerical ratio of capillaries and fibres either suffers from tissue-wide integration (C:F), or only takes into account those capillaries adjacent to each fibre. Both analyses are clearly inappropriate for tissue where a significant proportion of fibres lack direct capillary contact. It is therefore likely that a relatively distant capillary may also contribute to the metabolic activity of a fibre, albeit to a reduced degree. This is particularly the case in muscle with a heterogeneous distribution of fibre size during hyperplasia. In addition, the influence of distant capillaries is likely to increase with decreasing CD as reflected in the greater correlation between the local capillary-to-fibre ratio (cumulative domain fraction) and fibre area in fish muscle. This suggests that the oxygen supply of working skeletal muscle fibres is not restricted to capillaries contiguous with the fibre surface, but extends to remote capillaries whose domains show some overlap with fibres with which they have no direct contact. Of course it should be understood that such indices only describe the

10

anatomical interaction between capillaries and muscle fibres, and not the functional relationships which depend on additional factors such as oxygen consumption, blood flow etc. This anatomical evidence complements work suggesting that active fibres may actually extract oxygen from adjacent resting fibres, whose myoglobin may act as an oxygen conduit and short term store (Honig et al., 1984), drawing on these remote capillaries.

ACKNOWLEDGEMENTS

This work was in part supported by the Wellcome Trust (S.E.). The technical assistance of B.E.M. Ringnalda is gratefully acknowledged.

REFERENCES

Egginton, S. (1986). Effects of an anabolic hormone on performance and composition of rat skeletal muscle. J. Physiol. 373, 35P.

Egginton, S. and Johnston, I.A. (1982). Suitability of measured parameters and minimum sample size required to quantify capillary supply to fish muscle. Acta Stereologica, 1, 309-319.

Egginton, S. and Johnston, I.A. (1984). An estimate of capillary anisotropy and determination of surface and volume densities of capillaries in skeletal muscles of the conger eel (Conger conger L.). Q. J. Exp. Physiol. 68, 603-617.

Gray, S.D. and Renkin, E.M. (1978). Microvascular supply in relation to fiber metabolic type in mixed skeletal muscles of rabbit. Microvasc. Res. 16, 406-425.

Honig, C.R., Gayeski, T.E.J., Federspiel, W., Clark, A., and Clark, P. (1984). Muscle O_2 gradients from hemoglobin to cytochrome: new concepts, new complexities. In: Oxygen Transport to Tissue-V. Eds Lubbers, D.W., Acker, H., Leniger-Follert, E. and Goldstick, T.K., Plenum Pless, New York and London, (Adv. Exp. Med. Biol. 169, 23-38).

Hoofd, L. and Turek, Z. (1985). Modelling the capillary supply of muscle fibres in the myocardium. Proc. 26th Dutch Federation Meeting. No. 143.

Hoofd, L., Turek, Z., Kubat, K., Ringnalda, B.E.M. and Kazda, S. (1985). Variability of intercapillary distance estimated on histological sections of rat heart. In: Oxygen Transport to Tissue-VII. Eds Kreuzer, F., Cain, S.M., Turek, Z. and Goldstick, T.K., Plenum Press, New York and London, (Adv. Exp. Med. Biol. 191, 239-247).

Hudlicka, O. (1985). Development and adaptability of microvasculature in skeletal muscle. J. Exp. Biol. 115, 215-228.

Kayar, S.R., Archer, P.G., Lechner, A.J. and Banchero, N. (1981). The closest-individual method in the analysis of the distribution of capillaries. Microvasc. Res. 24, 326-341.

Krogh, A. (1919). The number and distribution of capillaries in muscles with calculations of the oxygen pressure head necessary for supplying the tissue. J. Physiol. 52, 409-415.

Turek, Z. and Rakusan, K. (1981). Lognormal distribution of intercapillary distance in normal and hypertrophic rat hearts as estimated by the method of concentric circles: its effect on tissue oxygenation. Pflugers Arch. 391, 17-21.

Turek, Z., Hoofd, L. and Rakusan, K. (1986). Myocardial capillaries and tissue oxygenation. Can. J. Cardiol. 2, 98-103.

Turek, Z., Hoofd, L. and Rakusan, K. (1987). A comparsion of the methods for the assessment of the heterogeneity of myocardial capillary spacing. This volume.

A COMPARISON OF THE METHODS FOR ASSESSMENT OF THE HETEROGENEITY OF MYOCARDIAL CAPILLARY SPACING

Z. TUREK, L. HOOFD*and K. RAKUSAN

*Department of Physiology, Faculty of Medicine, Catholic University, Nijmegen, The Netherlands, and Department of Physiology, School of Medicine, University of Ottawa, Ottawa, Ontario, Canada

INTRODUCTION

In 1982, during the ISOTT Meeting in Dortmund, we presented a paper evaluating the methods for assessment of myocardial capillarity (Rakusan and Turek, 1984). In that paper we compared various morphometric methods for the estimation of the mean intercapillary distance and the heterogeneity of capillary spacing. The triangular method (Renkin et al., 1981) and the method of concentric circles (Loats, Sillau and Banchero, 1978) were used for obtaining the index of the heterogeneity. Both these methods are manual and hence, time consuming. Recently, we have computerized the closest-individual method introduced by Kayar et al. (1982) and have in addition developed our own method, the method of capillary domains (Hoofd et al., 1985). These two computerized methods are much faster and more convenient to handle than manual methods and therefore more attractive to use. However, before introducing the computerized methods, we had already evaluated some results with the old methods (Rakusan et al., 1984; Rakusan et al., 1986). Thus the question arose, whether the results obtained with the old manual methods and with the new computerized methods were comparable. Therefore we decided to compare the various methods through a critical evaluation of how they reflect experimentally-induced changes in the heterogeneity of the myocardial capillary spacing. In addition, we evaluated the methods with respect to their practicability.

METHODS

This study was performed on 7 rats with aortic constriction and on 5 sham-operated control animals. In order to achieve a large cardiomegaly and presumably large differences in capillary spacing, a ligature was placed around the aorta of 5-day-old male Sprague-Dawley rats. The animals were killed 5 months later, at which time the left ventricular mass of rats with aortic constriction was about twice that of the controls. Standard histological techniques were used for the preparation of tissue samples, as described in our previous paper (Rakusan and Turek, 1984). In all animals, the subendocardial region of the left ventricle was investigated. The capillarization of the myocardium was assessed using all four methods as mentioned above: triangular, concentric circles, closest-individual and capillary domains.

The triangular method (Renkin et al., 1981) uses a direct measurement of the distances between capillaries connected so as to obtain a network of non-overlapping triangles. It was shown by Renkin et al. (1981) in skeletal muscle, and confirmed by us in cardiac muscle (Rakusan et al., 1986) that the distribution of the intercapillary distances can be described by a logarithmic-normal distribution. Half the mean intercapillary distance can be taken to represent a mean radius of the Kroghian tissue cylinder (Rm) whereas the logarithmic standard deviation (log SD) serves as a heterogeneity index of the capillary spacing.

In the next method concentric circles are used to estimate the distribution of tissue points at different distances from the nearest capillary (Loats et al., 1978). This is done in the following manner. Four sets consisting of four concentric circles are drawn on a transparent sheet which is moved over the photomicrograph in a random fashion and the location of the capillaries noted. They can be found within the smallest circles or between two of the four circles. If no capillary is located within the whole set, one capillary will be registered as being situated outside the largest circle. In this way the distances from the centres of the circles, each representing a tissue point, to the nearest capillaries are obtained. When repeated often enough this gives a frequency distribution of the distances from tissue points to the nearest capillary. The radii of the circles define the interval limits, so that a distribution of distances, consisting of five classes, will be derived. The distance from the tissue point can be measured either to the centre or to the edge of the lumen of the capillary. The latter measurement was used in this experiment. The method does not estimate directly the intercapillary distance but the average radius of a Kroghian tissue cylinder as well as its log SD can be derived by a fitting procedure, adopting the logarithmic-normal distribution (Turek and Rakusan, 1981; Rakusan and Turek, 1984). As the distance is estimated to the edge of the capillary lumen, the mean capillary radius is also required for the fitting procedure.

The closest-individual method (Kayar et al., 1982) also estimates the distances of tissue points to the nearest capillary. The co-ordinates of all capillaries are transferred into the computer using an image analyser and stored in its memory. Subsequently, a square-array rectangular grid is superimposed upon the coordinates by a computer program and distances from the grid points to the nearest capillaries are calculated. In this case it is more convenient to use the distances between the grid points and centres of the capillaries, which yield their continuous frequency distribution. The same fitting procedure mentioned above can also be used to derive the mean value of the radius of the tissue cylinder and its log SD. As the distances from tissue points are estimated to the centres of the capillaries, the capillary radius is not needed.

The method of capillary domains (Hoofd et al., 1985) can use previously stored coordinates of capillary locations. The computer program calculates the straight lines located in the middle between two neighbouring capillaries. These lines intersect and thus form polygons, or 'capillary domains', around each capillary. The surface area of each domain is calculated and stored in the memory. The frequency distribution of the surfaces also follows a logarithmic normal distribution (Van Haelst, Hoofd and Turek, 1985). Domains can be replaced with circles of identical surface area and their radii will represent the radii of tissue cylinders as defined earlier. Using standard procedures, the average value and the log SD of these radii can be obtained.

In this way, results of all four methods can be transformed into only two indices of the capillary spacing, the average radius of tissue cylinder and the log SD representing the heterogeneity. Thus a comparison among all the methods can be easily performed.

RESULTS

Figure 1 shows the mean radii of tissue cylinders from each rat, using the triangular method, the method of concentric circles and the closest-individual methods, plotted against Rm's obtained by the method of capillary domains. In the same figure the regression lines and correlation coefficients are also shown. Correlation coefficients were all highly significant ($P < 0.01$). Also the further correlations, not displayed here, among results derived by the remaining methods, were all significant ($P < 0.01$). Log SD's obtained by the triangular method, the method of concentric circles and the closest-individual method, plotted against log SD's derived by the domain method, and the corresponding regression lines and correlation coefficients are shown in Figure 2. Here again, the correlations were all significant not only between log SD obtained by the domain method and the three other methods ($P < 0.01$) but also between results of these three methods ($P < 0.05$).

Animals with cardiomegaly had both a larger Rm ($P < 0.01$) and log SD ($P < 0.05$) when compared with control rats, irrespective of which method was used. The traditional index of the capillarization, the capillary density, was also estimated and found to be significantly lower ($P < 0.01$) in rats with aortic constriction (1915 $\pm$ 240 per mm^2, mean value $\pm$ S.D.) than in control animals (2849 $\pm$ 276 per mm^2, mean value $\pm$ S.D.).

DISCUSSION

It has been shown that results of all methods are interrelated with respect to Rm as well as to log SD. In the hearts of rats with aortic constriction all methods demonstrated an increase in both the average intercapillary distance and the degree of heterogeneity, when compared with their controls. However, the absolute values of the mean radius of tissue cylinder and of log SD obtained with different methods were quite different.

The triangular method provided the largest values for the average radius of tissue cylinder and also for its log SD. The method of concentric circles resulted in mean values for the radius of tissue cylinder which were larger than those obtained by both the closest-individual and the domain method, while the log SD did not differ from that of the closest-individual method and was larger than log SD derived by the domain method. There was no difference between the average radius derived by the closest-individual and the domain methods. The domain method consistently yielded the smallest log SD.

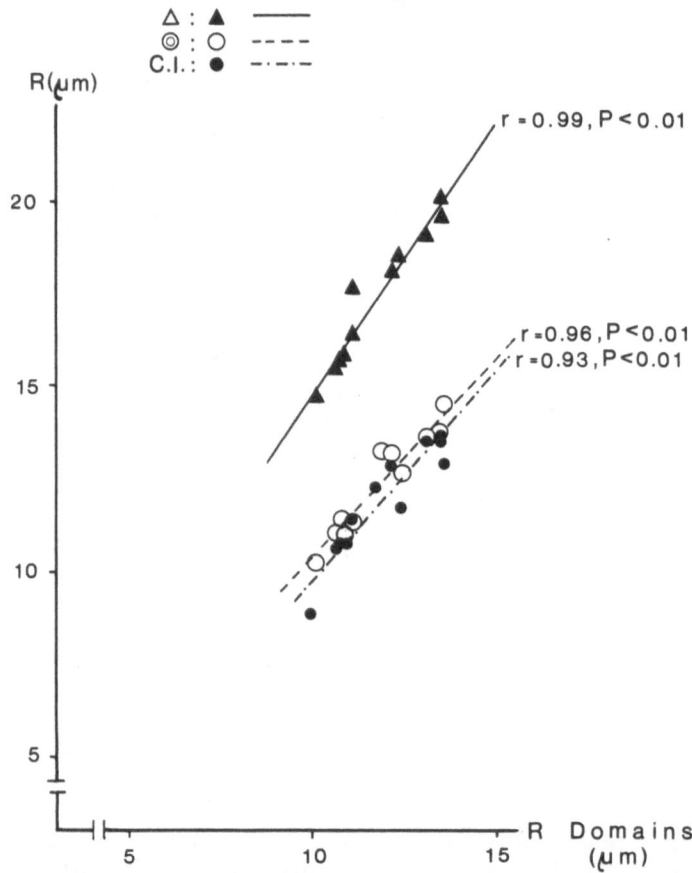

Figure 1. Regression of mean radius of tissue cylinder as obtained by the triangular method (△), the method of concentric circles (◎) and the closest-individual method (C.I.), on the mean radius of tissue cylinder derived by the method of capillary domains.

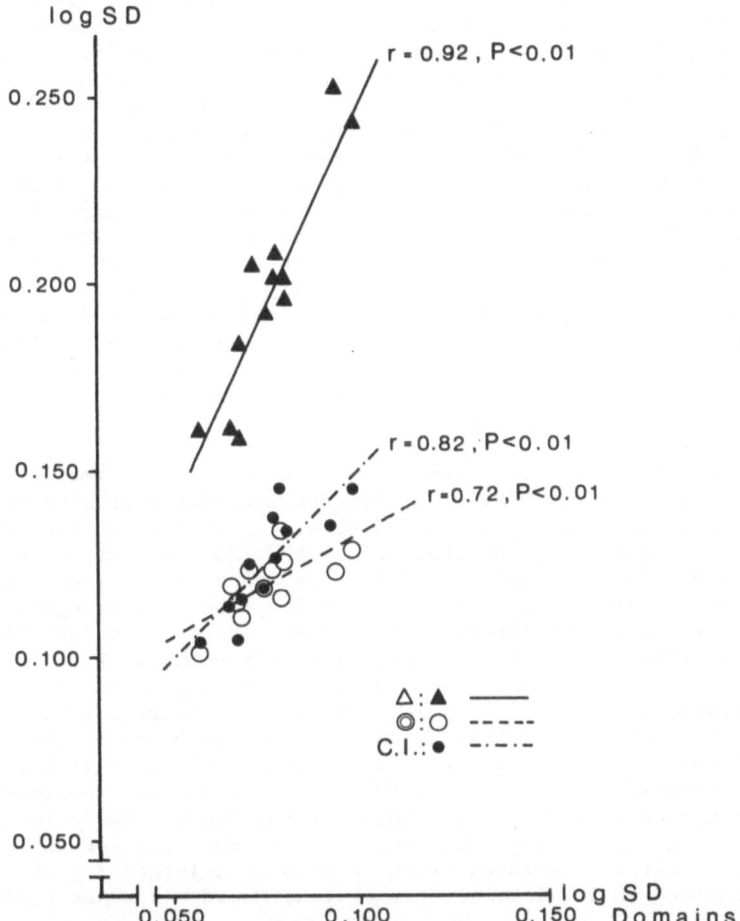

Figure 2.. Regression of log SD, the index of the heterogeneity in capillary spacing, as obtained by the triangular method (△), the method of concentric circles (◎) and the closest-individual method (C.I.), on the log SD derived by the method of capillary domains.

It is difficult to explain all these differences as the situation is complex. The finding that the largest values for the average radius of the tissue cylinder were given by the triangular method in this experiment is rather exceptional. In other experiments no such difference or even smaller values were found (Rakusan and Turek, 1984; Rakusan et al., 1984). It should be pointed out that the triangular method is very subjective, since the triangular network can be drawn in various ways. On the other hand, the log SD found with the triangular method has always been the largest one and this seems to be method-dependent. The method of concentric circles and the closest-individual method both estimate the frequency distribution of distances of tissue points to the nearest capillary. In this particular experiment the former estimated the distances to the edge of the lumen of the nearest capillary, while the latter estimated the distances to the centre of the capillary. Thus, the method of concentric circles may also reflect the variability of the capillary radius, for in the fitting only its average value is used. This method gave only five classes of frequency distribution since only four concentric circles were used in a set. On the other hand, the closest-individual method provides a continuous frequency distribution of the distances between tissue points and the nearest capillaries and the variability of capillary radii is not involved. It is, however, difficult to say whether and to what degree these methodological diversities could have been responsible for the observed differences in the mean value of the tissue cylinder. It should be pointed out that even though the differences were statistically significant they were small, usually less than 1 μm.

Rm obtained by the domain method was also smaller than that derived from the concentric circles but not different from that of the closest-individual method. The value of log SD derived from the domain method was smaller than that of both the concentric circles and closest-individual methods. This has also been found in previous experiments (Turek and Rakusan, 1985). We must recognise that the methods are not unrelated: all tissue points closest to the same capillary must lie within the same capillary domain. Thus, a capillary domain may be also defined as a set of all tissue points 'belonging' to one particular capillary. The fitting procedure for the concentric circles and the closest-individual method is based on the assumption that tissue consists of a series of circular cylinders with a centrally located capillary. In reality this is not the case. Because of this, the assumption of the circular shape of the domains leads in the domain method to a small log SD, whereas in the closest-individual and concentric circles methods this results in a larger log SD. This has been discussed in detail elsewhere (Turek, Hoofd and Rakusan, 1986).

It is reassuring that all the methods used here responded to and depicted satisfactorily the induced changes in the heterogeneity of capillary spacing and thus could all be used for the evaluation of experimental material. There is no theoretical reason why one method should be preferred. However, from the practical point of view, the computerized methods are much faster. Furthermore, the domain method is faster and less cumbersome than the closest-individual method. The choice of which method to use depends on the purpose of the study. When the distribution ,of distances of tissue to the nearest capillary is the final goal, then the closest-individual methods will be the best alternative. If, primarily, the average intercapillary distance and its heterogeneity are needed, the domain method would be the most appropriate. If the mean radius of tissue cylinder and its log SD are to be used as the input data for a model of oxygen transport to cardiac tissue, both computerized methods will deliver these data. However, the absolute values of log SD will be different, for, as mentioned above, they are method-dependent.

ACKNOWLEDGEMENTS

Supported in part by NATO grant number RG. 86/0073 and by the Ontario Heart Foundation.

REFERENCES

Hoofd, L., Turek, Z., Kubat, K., Ringnalda, B.E.M. and Kazda, S. (1985). Variability of intercapillary spacing estimated on histological sections of rat heart. In: Oxygen Transport to Tissue-VII. Eds Kreuzer, F., Cain, S.M., Turek, Z. and Goldstick, T.K., Plenum Press, New York and London, (Adv. Exp. Med. Biol. 191, 239-247).

Kayar, S.R., Archer, P.G., Lechner, A.J. and Banchero, N. (1982). The closest-individual method in the analysis of the distribution of capillaries. Microvasc. Res. 24, 326-341.

Loats, J.T., Sillau, A.H. and Banchero, N. (1978). How to quantify skeletal muscle capillarity. In: Oxygen Transport to Tissue-III. Eds Silver, I.A., Erecinska, M. and Bicher, H.I., Plenum Press, New York and London, (Adv. Exp. Med. Biol. 94, 41-48).

Rakusan, K., Hrdina, P.W., Turek, Z., Lakatta, E.G., Spurgeon, R.A. and Wolford, G.D. (1984). Cell size and capillary supply of the hypertensive rat heart: a quantitative study. Basic Res. Cardiol. 79, 389-395.

Rakusan, K., Korecky, B., Sarkar, K. and Turek, Z. (1986). Merits and pitfalls in morphological assessment of cardiac growth. Fed. Proc. 45, 532-536.

Rakusan, K. and Turek, Z. (1984). A comparison of the methods for assessment of myocardial capillarity. In: Oxygen Transport to Tissue-V. Eds Lubbers, D.W., Acker, H., Leniger-Follert, E. and Goldstick, T.K., Plenum Press, New York and London, (Adv. Exp. Med. Biol. 169, 411-418).

Renkin, E.M., Gray, S.D., Dodd, L.R. and Lia, B.D. (1981). Heterogeneity of capillary distribution and capillary circulation in mammalian skeletal muscles. In: Underwater Physiology, Vol. VII. Eds Bachrach, A.J. and Matzen, M.M., Undersea Medical Society, Bethesda, MD, pp. 465-474.

Turek, Z., Hoofd, L. and Rakusan, K. (1986). Myocardial capillaries and tissue oxygenation. Can. J. Cardiol. 2, 98-103.

Turek, Z. and Rakusan, K. (1981). Lognormal distribution of intercapillary distance in normal and hypertrophic rat hearts as estimated by the method of concentric circles: its effect on tissue oxygenation. Pflugers Arch. 391, 17-21.

Turek, Z. and Rakusan, K. (1985). Heterogeneity of myocardial capillary spacing: comparison of two methods of estimation. Physiologist, 28, 308

Van Haelst, A.C.T.A., Hoofd, L. and Turek, Z. (1985). Log normal distribution of capillary domains in the rat myocardium. J. Physiol. (Lond.) 366, 114P.

IMAGING OF RED BLOOD CELL AND PLASMA DISPERSION IN THE BRAIN CORTEX

A. EKE

Department of Neurology, The University of Alabama at Birmingham, Birmingham, AL 35294, U.S.A. and Experimental Research Department, Semmelweis Medical University, Budapest, Hungary

INTRODUCTION

Most of the methods for measuring regional or local blood flow cannot specify the respective fractions of the measured flow attributable to red blood cell (RBC) flow and plasma flow within the measured volume. It is obviously important to know this because of the decisively different microhaemodynamic behaviour of these two components in the microcirculation and their specialized roles in gas and solute exchange in tissue metabolism. Beyond obtaining a separate measure of RBC and plasma flow for a given volume of tissue, a separate assessment of the complexity of RBC and plasma transit routes within that volume seems equally important. This parameter reveals the dissimilarity between seemingly identical conditions when total volume flows are equal but are realized via vascular transit routes of different geometrical complexity, which could indicate the presence in one, and the absence in the other, of certain conditions such as highly focal oedema, capillary obstruction, vasomotion, etc. It has been repeatedly demonstrated, that the television densitometric method of Eke (1982, 1983, 1984) can yield separate images of RBC and plasma flow at high topographical and temporal resolution by monitoring the transit of RBC and plasma indicator pulses through an array of tissue areas and by analysing these indicator-dilution arrays for arrays of RBC and plasma mean transit time (t) and flow. This method has been recently supplemented by a statistical procedure adopted from Knopp and Bassingthwaighte (1969) to analyse the arrays of RBC and plasma indicator-dilution curves for the scatter of the individual transit times over the mean transit time, t (Eke, 1986). This analysis yields a parameter assessing the spatial complexity of indicator transit pathways confined by the monitored tissue volume and is analogous to the approach reported by Tomita et al. (1983). Their concept is based on some a priori assumptions of the distribution of parenchymal transit times (gamma distribution) and the calculation of the 'gamma index' as the spatial parameter of the gamma density function applied in the procedure to obtain the transport function of the parenchymal region. The major differences between their approach and mine, beyond the mathematics, are attributable to the facts that their indicator-dilution technique does not allow for imaging, requires invasion of the tissue and derives the gamma index for a cerebrocortical tissue volume of around 1 - 2 cm^3 in contrast to the reflectometric measuring volume of 0.01 mm^3.

METHODS

High-resolution two-dimensional images of intraparenchymal RBC and plasma indicator-dilutions were acquired by a computerized television densitometer (Eke, 1983) in a raster of 625 elements and images of volume, mean transit time, volume flow and haematocrit were generated by the in situ television densitometric method of Eke (1984). In essence, this method detects the cerebrocortical transits of separate plasma and RBC indicator (isotonic mannitol solution with and without extremely fine carbon particles dispersed in it) pulses through a square array of adjacent cylindrical volumes of tissue. The region of interest is exposed through a closed cranial window technique to white, cold light epi-illumination; detection is at 589 nm. The spatial resolution of imaging was set to 625 elements/10 mm^2. An image element represented a volume averaged parameter for a tissue volume of 0.01 mm^3 located superficially in the brain cortex. The 25 by 25 parameter arrays were finally zoomed into a 50 by 50 raster by inter-polation and visualized in grey scale on a video monitor. The measure-ments can be repeated at a frequency of 2 measurements/minute and a total of 100 measurements can easily be acquired in each experiment.

Relative dispersion of transit times (Knopp and Bassingthwaighte, 1969) for plasma and RBC was calculated from indicator dilution curves detected in a raster array (relative local dispersion, RLD) and averaged over the parenchymal elements of the array (relative regional dispersion, RRD). Numerically, it was calculated as the ratio of the standard deviation of indicator transit times over the respective mean transit time. The relative dispersion of transit times is interpreted as a parameter describing the spatial complexity of the transit pathways through which the individual indicator particles traverse the volume. At one extreme, when all transit pathways are of different length and diameter, its value is 1; in contrast, in a perfectly homogeneous system with parallel pathways of equal length and diameter, its value is 0. The regionally and locally detected indicator dilution allows further distinctions to be made: RRD reflects the combined spatial complexity of the macro- and microcirculatory pathways distrib-uting plasma and RBCs among and within the microareas confined by the region, while RLD quantifies the distribution complexity of plasma and RBCs within the microcirculatory network of a given microarea in the region. Their difference (RID=RRD-RLD) can be interpreted as an index of the dispersion of transit times via the macrocirculatory pathways of the region connecting the afferent pial system to the local microcirc-ulatory networks, hence, the term relative intraregional dispersion (RID) is introduced.

RESULTS AND DISCUSSION

As Figures 1 and 2 demonstrate, red blood cell and plasma microflows have been found to be distributed highly heterogeneously in cerebrocortical areas as small as 10 mm^2 in the cat anaesthetized by chloralose in a dose of 40 mg/kg. The size of adjacent areas with similar flow values seems to correspond well with the expected size of a microcirculatory control unit (column) in the brain cortex (300 to 800 μm in diameter). It is worth noting that the distribution of RBC flow seems more heterogeneous than that of plasma flow. Figure 3 shows the local haematocrit image for the same area. The local haematocrit was found to be markedly heterogeneous, too, in parenchymal areas seen on both sides of a larger pial vein running across the upper right corner of the image. The range of the local parenchymal haematocrit was below the value of the large arterial haematocrit, which was measured in blood withdrawn from the common carotid artery and found to be 35%.

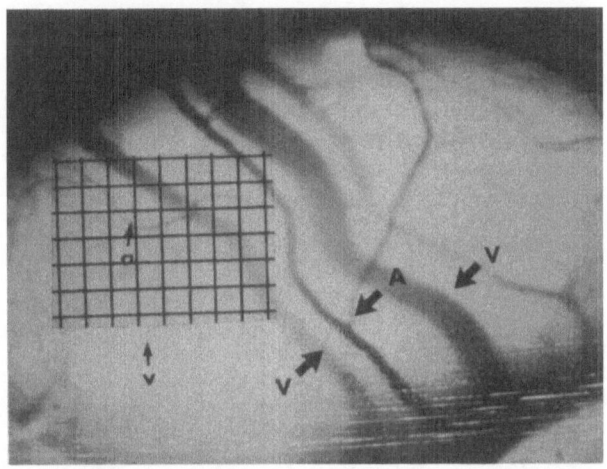

Figure 1. Reflectance image of the cerebrocortical area of an anaesthetized cat under epi-illumination recorded on video tape at 589 ± 5 nm wavelength by a Newicon television camera. Vessels of the pial networks are labelled 'A' and 'a' (arteries), 'V' and 'v' (veins). The area scanned by video densitometry for an array of apparent optical density of the tissue is marked by the crosshatched overlay. Intensity coded parameter images shown in Figs. 2, 3 and 4 correspond to this area of 2.5 x 3 mm.

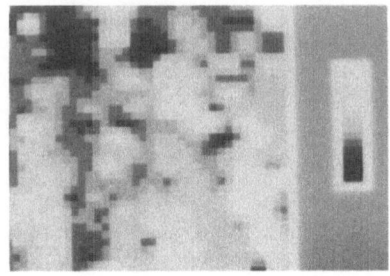
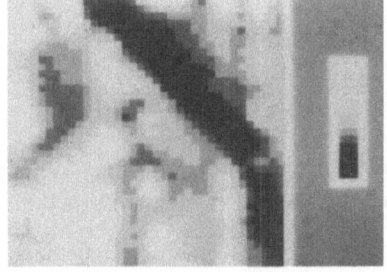

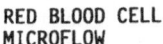

RED BLOOD CELL
MICROFLOW

PLASMA
MICROFLOW

Figure 2. Intensity coded images of red blood cell and plasma local volume flows (microflows) obtained for a cerebrocortical area of 2.5 x 3 mm shown in Fig.1 in an anaesthetized cat (range of pixel data: black = 10, white = 100 arbitrary units).

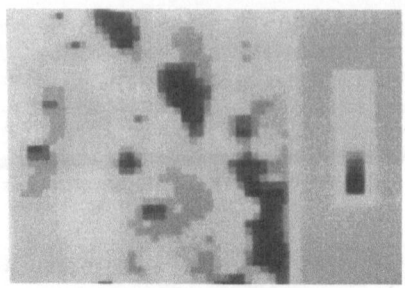

Figure 3. Intensity coded image of local haematocrit for the cerebrocortical area shown in Fig.1. (range of pixel data: black = 0, white = 50%).

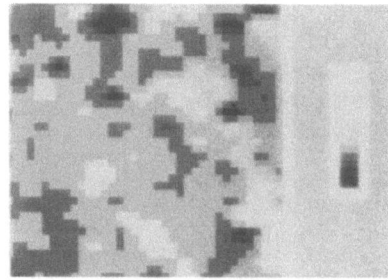

RELATIVE DISPERSION OF
RED BLOOD CELL
LOCAL TRANSIT TIMES

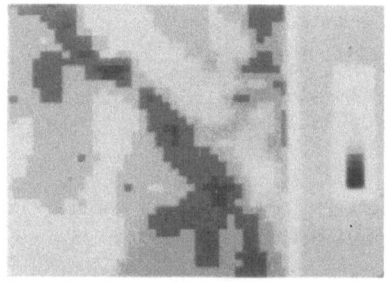

RELATIVE DISPERSION OF
PLASMA
LOCAL TRANSIT TIMES

Figure 4. Intensity coded images of relative local dispersion (RLD) of red blood cell and plasma mean transit times for the area shown in Fig.1. (range of pixel data: black = 0, white = 1.0).

Images of RLD of RBC and plasma transit times were dissimilar and showed greater topographical variation for RBCs than plasma, as can be seen in Figure 4. Plasma dispersion seemed to show more apparent topology with respect to supplying and draining vessels than did RBC dispersion.

Responses of RLD and RRD of red blood cell transit times have been investigated in rhesus monkeys during reactive hyperaemia following a one minute occlusion of the ipsilateral middle cerebral artery (MCA) supplying the monitored cerebrocortical region. This model is especially suitable for the study of the dynamics of the relative dispersion of transit times, for dramatic changes in vascular geometry can be expected during the rapid transient from a no-flow condition to reactive hyperaemia following the release of the MCA-ligature. The results of a representative experiment are shown in Figure 5. Data for RLD are presented as a regional frequency distribution. The RLD-histogram showed a well-defined peak in the control at 0.35. This well-defined control RLD-pattern turned rapidly into a markedly heterogeneous pattern during the first two minutes of reactive hyperaemia but gradually recovered as flows and mean transit times approached the control level. The pattern of RLD seemed to recover sooner than that of mean transit times (see Fig.5).

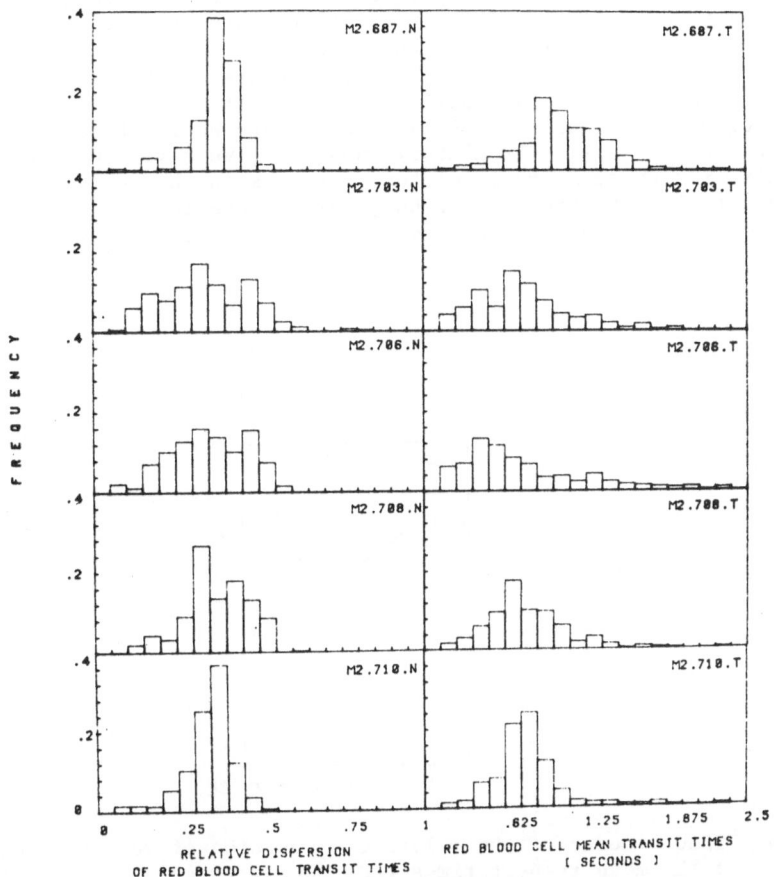

Figure 5. Histogram displays of relative dispersion of red blood cell local transit times (left colum) and that of local mean transit times (right column) generated from indicator-dilution images recorded at 30 second intervals prior to and following a 1 minute temporary occlusion of the ipsilateral middle cerebral (MCA) artery in an anaesthetized squirrel monkey. Top row = control histograms; subsequent rows = serial measurements during the reactive hyperaemia following the release of the MCA-clip. The histograms have unit areas.

Figure 6 demonstrates the responses of the relative dispersion of the regional transit times (RRD, upper trace), the regional mean of the local transit times (RLD, middle trace) and their difference (RID, lower trace). The regional mean of RLD showed a tendency to decrease during reactive hyperaemia, but with great variations in the individual values. A dissociation between RRD and RLD, as represented by RID, was observed during the course of hyperaemia. RID gradually increased by 50% during the first 90 seconds of reactive hyperaemia and then, by 120 seconds, it fell below the control. Its increase between 30 and 90 seconds correlates well with the concurrent peak of the hyperaemia (see the mean transit time histograms) and the phase of markedly increased heterogeneity of RLD. This can be interpreted as the manifestation of the different behaviour of the macro- and microcirculation in adapting spatial geometry to the rapidly changing requirements of tissue perfusion: the macrocirculatory pathways showed an increased geometrical complexity, that may have resulted from readjustments of the vascular dimensions of the existing pathways and also from the opening and closing of some other pathways. The microcirculatory system of the region, however, seems to react by decreasing its geometrical complexity by capillary recruitment, that is by opening up yet un- or under-perfused channels, thereby decreasing heterogeneity of microflow in the microcirculatory unit. As we have seen, this holds true only for those microcirculatory networks in the region, whose RLD parameter falls into the mean of RLD, while there is greater scatter in the behaviour of RLD of the individual microcirculatory networks.

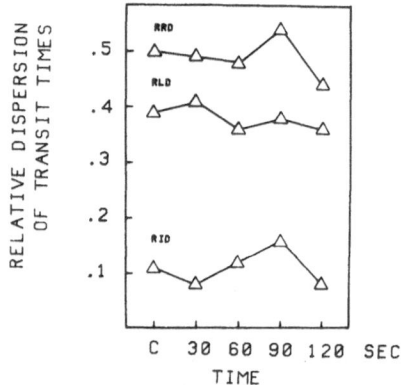

Figure 6. Time course of the relative dispersion of regional red blood cell mean transit times (RRD, upper trace), regional mean of the relative dispersion of red blood cell local transit times (RLD, middle trace) and their difference, the relative intraregional dispersion of red blood cell transit times (RID lower trace) prior to and following a 1 minute temporary MCA-occlusion (referred to in Fig.5) in an anaesthetized squirrel monkey. (C = control. Time scale from the release of the MCA-clip).

On a preliminary basis, the following conclusions can be made: (1) the geometrical complexity of local red blood cell and plasma transit pathways is a heterogeneous parameter in the brain cortex. Its value can vary according to the topographical orientation of the micro-region to its supplying and draining vessels and also as a result of vasoaction (i.e. such as during hyperaemia); (2) the regional frequency distribution of the geometrical complexity of local red blood cell transit pathways seems to show a well defined control pattern, that recovers following disturbances (such as during hyperaemia). This may be an indication of it being a controlled parameter, at least in regions as large as a few square millimetres.

The multiparameter microflow imaging method (Eke, 1984) applied in this study has demonstrated that it can provide not only images and distributions of relevant local microcirculatory parameters, but by means of a simple statistical analysis of the regional and local indicator transits, can also give an insight into the spatial complexity of otherwise inaccessible vascular pathways. It is to be hoped that further data may contribute to a better understanding of the dynamic behaviour of the different vascular segments in adjusting the perfusion of cerebrocortical tissue by red blood cells and plasma to the needs of the actual functional condition.

ACKNOWLEDGEMENTS

The author wishes to thank Ms Lorette Mitchem and Dr Richard Martin for their expertise and contribution to the monkey experiments.

REFERENCES

Eke, A. (1982). Reflectometric mapping of microregional blood flow and blood volume in the brain cortex. J. Cereb. Blood Flow Metabol. $\underline{2}$, 41-53.

Eke, A. (1983). Heterogeneity of cerebrocortical microflow in epileptic seizure. In: Cerebral Blood Flow, Metabolism and Epilepsy. Eds Baldy-Moulinier, M., Ingvar, D.H. and Meldrum, B.S., John Libbey Eurotext, London and Paris

Eke, A. (1984). Repetitive mapping of tissue hematocrit over microareas of the brain cortex. Int. J. Microcirc. Clin. Exp. $\underline{3}$, 548

Eke, A. (1986). Dispersion of red blood cells and plasma in the macro- and microcirculation of the cerebrocortical tissue. Int. J. Microcirc. Clin. Exp. $\underline{5}$, 277

Knopp, T. and Bassingthwaighte, J.B. (1969). Effect of flow on transpulmonary circulatory transport functions. J. Appl. Physiol. $\underline{27}$, 36-43.

Tomita, M., Gotoh, F., Amano, T., Tanahashi, N., Kobari, M., Shinohara, T. and Mihara, B. (1983). Transfer function through regional cerebral cortex evaluated by a photoelectric method. Am. J. Physiol. $\underline{245}$, H385-H398.

DETERMINATION OF CEREBRAL CORTICAL CAPILLARY BLOOD VOLUME FROM MEAN TRANSIT TIME ANALYSIS

J.C. LaMANNA and R.P. SHOCKLEY

Department of Neurology, University Hospitals and Departments of Neurology, and Physiology and Biophysics, Case Western University School of Medicine, Cleveland, Ohio 44106, U.S.A.

INTRODUCTION

A full understanding of the delivery of oxygen to a tissue cannot be attained by study of the blood flow to that tissue alone. An important component of the delivery system is the particular spatial arrangement of the capillaries supplying the tissue. This is especially critical in capillary networks that function through dynamic control of local tissue perfusion by regulating the number of open and closed capillaries. This form of vascular regulation has been generally known as 'capillary recruitment' and this concept implies that there are dynamic changes in tissue blood volume in response to changes in tissue metabolic demand. Capillary recruitment has long been suspected of being an integral part of cerebrovascular function, but until recently, experimental evidence for significant participation has been lacking (Weis et al., 1982; Pardridge and Fierer, 1985).

A contributing factor in the increased awareness of the role of capillary recruitment in brain physiology has been the development of minimally invasive optical methods for estimating regional cerebral blood flow and volume by the analysis of the capillary mean transit time of a haemodilution bolus passing through the cerebral cortex (Eke, Hutiray and Kovach, 1979; Eke, 1982; Tomita et al., 1978, 1983). These methods all make use of the central volume principal (Meier and Zierler, 1954) which is based on the Stewart-Hamilton equation, MTT = BV/BF, which states that the mean transit time (MTT) through the capillary network is equal to the blood volume (BV) divided by the blood flow (BF). This principle was developed by Stewart for determination of cardiac output (Stewart, 1897), and has received a great amount of experimental and theoretical support for its use in description of tissue capillary characteristics (Malindzak et al., 1972; Greenberg et al., 1978; Ramakrishnan, Leonard and Dell, 1984).

While much progress has been made in the theoretical and mathematical basis of the methods, there has been as yet no independent confirmation for their applicability to brain studies. In this paper, we report the results of experiments in which MTT and blood volume were determined by reflectance photometry, and blood flow was independently determined by indicator fractionation of $[^{14}C]$-butanol (Sage et al., 1981; LaManna and Harik, 1985).

METHODS

Animal Preparation

Adult, male Wistar rats (250 g) were anaesthetized with chloral hydrate (400 mg/kg, I.P.). Cannulae were placed in: a) a femoral artery for blood pressure monitoring and arterial blood gas and pH sampling; b) a femoral vein for drug administration; and c) in the right atrium via the external jugular for bolus administration of indicator substances. The arterial cannulae were PE 50, the atrial cannula was .silastic. Anaesthesia was maintained throughout the experiment by intravenous supplements of chloral hydrate, as needed. Rectal temperature was monitored by thermocouple and maintained at 37°C by a feedback-controlled hot water pad. Rats were tracheostomized, paralysed with d-tubocurare, placed on a ventilator, and respired with 30% oxygen in nitrogen. The ventilation rate and volume were adjusted on the basis of the blood gases.

The rat was mounted in a three point fixation headholder. The dorsal surface of the skull was exposed through a midline scalp incision. A 5 x 7 mm area of bone over the parietal cerebral cortex was removed leaving the dura intact. A 3 mm length of large diameter polyethylene tubing was attached to the bone surrounding the opening by acrylic adhesive. Vacuum grease and bone wax were used to ensure a tight seal between the tubing and the bone. The tubing thus formed a reservoir over the dura which was then filled with mineral oil in order to minimize the effects of atmospheric exposure on the blood-tissue gas equilibrium.

Mean Transit Time Determination

Measurements of capillary mean transit time were made from analysis of indicator dilution curves generated from bolus injections of 0.2 ml Dextran solution (10% Dextran, MW 40,000, in normal saline). The passage of these haemodilution transients through the parietal cerebral cortical capillary bed was monitored by reflectance spectrophotometry. The cortical surface was illuminated through the craniectomy with light from a 250 W tungsten-halide lamp by a fibre-optic light guide, after passing through a 590 nm interference filter. Light reflected from the cortex was collected by a microscope through a 5x long-working-distance objective. A photomultiplier tube (PMT; EMI 9798A) was positioned over a 10x objective on the microscope body. The gain of the photomultiplier tube was controlled by a feedback circuit on the high voltage source, so that the anode current was kept constant. Thus, the control voltage of the feedback circuit varied with the log of light intensity (due to the characteristics of the high-voltage/PMT gain relationship), and, thus, was directly proportional to the dilution concentration. This signal was displayed on a strip chart recorder (see Fig. 1).

The mean transit time of the haemodilution bolus was calculated by the following method, adapted from Eke et al., 1979. The concentration-time curve displayed on the strip chart recorder was entered into a Tektronix 4052 desktop graphic system by manual tracing on a digitizing tablet. This curve was baseline corrected, and the frequency distribution curve, H(t), calculated by dividing the cumulative time integral of the dilution curve by the area under the curve between the time of appearance and time of disappearance. The cumulative time integral was calculated by iterative integration of the curve, sequentially displaced by 0.1 sec (bolus injection interval) until a constant value was obtained (Eke et al., 1979). Mean transit time was then calculated by integrating the 1-H(t) function. The calculated MTT from a minimum of 5 transients were averaged at each experimental point.

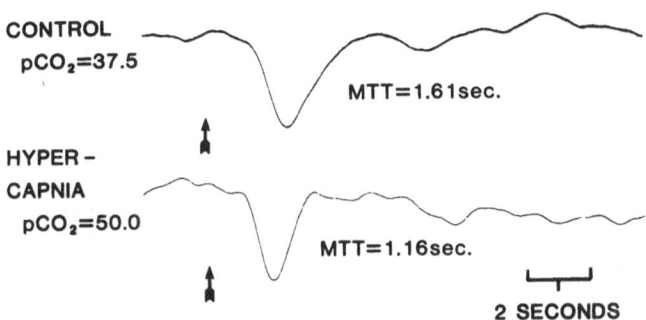

Figure 1. Reflectance record during control and hypercapnic conditions. Arrows indicate injection of dextran bolus.

After some experiments, all the cortical blood was washed out of the rat brain by perfusion with oxygenated saline through the atrial cannula, following clamping of the descending aorta. The absolute value of the output of the high voltage power supply was determined before and after complete perfusion of the cerebral cortex in order to estimate percentage changes in tissue blood volume.

Cerebral Blood Flow

Cerebral blood flow was determined in rats after optical monitoring of haemodilution transients by the [^{14}C]-butanol indicator fractionation method (Sage et al., 1981). At the end of the experiment, a 150 microlitre bolus containing 10 μCi of [^{14}C]-butanol in buffered Ringer's solution was injected into the right atrium while arterial blood from the femoral cannula was withdrawn into a syringe mounted in a constant rate, calibrated withdrawal pump running at 1.6 ml/min. Ten sec after administration of the bolus, the rat was decapitated and the withdrawal pump simultaneously stopped. The withdrawn arterial blood was quantitatively expressed into weighed vials and 100 μl aliquots measured for radioactive content. The brain was rapidly removed and bilateral samples of parietal cortex were weighed, solubilized and counted in a beta scintillation counter. Blood flow was calculated from:

$$CBF = \frac{Fsyr \times dpm[^{14}C]br}{dpm[^{14}C]syr \times Wt} \times 100 \times k$$

Where Fsyr was the calibrated withdrawal rate of the syringe (1.6 ml/min), dpm[^{14}C]br was the dpm in the brain sample, dpm[^{14}C]syr was the total counts of the withdrawn blood sample, Wt was the weight of the brain sample (g), and k was a factor used to account for the incomplete extraction of butanol and was 1.1 for normocapnia and 1.25 for hypercapnia (Sage et al., 1981).

RESULTS AND DISCUSSION

To test our hypothesis that mean transit time analysis could provide quantitative descriptions of cerebral capillary vascular characteristics, we used a simple strategy involving two groups of rats. In the first group of rats, reflectance optical monitoring was used to determine mean transit time and relative blood volume under normocapnic and hypercapnic conditions. From these, we predicted the change in blood flow between the two conditions. In the second group of rats, under similar conditions, we determined cerebral blood flow by a quantitative technique and compared the measured blood flow with the predicted blood flow.

In the first group of 6 rats, capillary mean transit time was determined from a minimum of 5 haemodilution boluses at normocapnia, and the absolute value of the baseline PMT gain control voltage noted. The rat was then made hypercapnic and after a 15 min stabilization period, a second set of haemodilution boluses administered. The baseline PMT gain control voltage for hypercapnia was noted. At the end of the experiment, all the blood from the optical field was washed out of the tissue by perfusion with oxygenated saline and the PMT gain control voltage once more recorded. When the blood was washed from the cortex, the PMT gain control voltage decreased in order to keep the PMT anode current constant because more light was reflected back from the cortex since there was then no light absorption by haemoglobin. The value of this voltage drop was considered to be proportional to the blood volume in the optical field under normal conditions. When carbon dioxide was added to the ventilation gas mixture, the control voltage increased to compensate for the decreased reflected light due to increased absorption by increased blood in the optical field. The voltage difference between control and hypercapnia, divided by the voltage difference between control and saline-perfused tissue, expresses the percentage change in blood volume due to hypercapnia. The mean increase in arterial carbon dioxide tension in this group of rats was 31.2 torr. The relative blood volume increase with hypercapnia was 46 $\pm$ 5%. Similarly, the percentage change in MTT was calculated between control and hypercapnia. If these percentage changes are substituted into the Stewart-Hamilton equation,

$$CBFr = BVr/MTTr$$

(where r signifies the ratio between hypercapnia and normocapnia for the variable), the percent change in CBF can be estimated. The grouped results of these calculations for 6 rats are shown in Table 1. The mean change in $PaCO_2$ was 31.2 mmHg. The MTT ratio was 0.77, meaning that the MTT determined during hypercapnia was faster than in normocapnia. The BV ratio for the same conditions was 1.46, indicating an increased blood volume with hypercapnia. The ratio of these ratios predicts the expected ratio in cerebral blood flow with hypercapnia compared with control. This was calculated to be 92%, or a near doubling of blood flow for just over 30 mmHg increase in carbon dioxide. This calculated increase of 2.95% per torr increase in $PaCO_2$ was well within the expected range from the literature (Lacombe, Meric and Seylaz, 1980).

Table 1. Calculated results from six rats

	$\triangle PaCO_2$	MTTr	BVr	CBFr
5% CO_2	31.2	0.77 $\pm$ 0.04	1.46 $\pm$ 0.05	1.92 $\pm$ 0.09

Values are $\pm$ S.E.M. $\triangle CBF$ = 2.95%/torr CO_2

The blood flow increase with CO_2 predicted by the optical methods for estimating MTT and BV was checked by determination of blood flow by [^{14}C]-butanol in another group of rats, in which MTT was also determined for comparison. These results are shown in Table 2. In these studies the blood flow increased 74% for a mean difference of 26 mmHg $PaCO_2$, or 2.85% per mmHg $PaCO_2$, within 4% of the predicted value. In these latter experiments, a quantitative estimate of cerebral blood volume was made by the Stewart-Hamilton relationship where BV = MTT/CBF. The result is shown as 3.3 ml/100 g in the control rats and 4.6 ml/100 g in the hypercapnic rats.

Table 2.

	$PaCO_2$ (mmHg)	PaO_2 (mmHg)	MTT (sec)	CBF (ml/100g/min)	BV* (ml/100g)	MABP (mmHg)
Control (n = 7)	35.6 ± 1.1	125.3 ± 9.4	1.41 ± 0.07	142.7 ± 14.5	3.3	102 ± 6
Hypercapnic/ normoxic	61.6 ± 6.3	109.9 ± 9.1	1.10 ± 0.04	248.8 ± 17.8	4.6	97 ± 5
% Control			(-22)	(+74)	(+39)	

* Calculated from the Stewart-Hamilton equation.

$\triangle CBF$ = 2.85%/torr CO_2

The results from these experiments demonstrate that the optically-derived values for capillary MTT and cerebral BV are experimentally consistent with the values predicted through an independently determined cerebral BF. The experimental validation of the haemodilution method means that this technique can be used profitably in the study of cerebral capillary recruitment.

REFERENCES

Eke, A. (1982). Reflectometric mapping of microregional blood flow and blood volume in the brain cortex. J. Cereb. Blood Flow Metabol. 2, 41-53.

Eke, A., Hutiray, G. and Kovach, A.G.B. (1979). Induced hemodilution detected by reflectometry for measuring microregional blood flow and blood volume in cat brain cortex. Am. J. Physiol. 236, H759-768.

Greenberg, J.H., Alavi, A., Reivich, M., Kuhl, D. and Uzzell, B. (1978). Local cerebral blood volume response to carbon dioxide in man. Circ. Res. 43, 324-331.

Lacombe, P., Meric, P. and Seylaz, J. (1980). Validity of cerebral blood flow measurements obtained with quantitative tracer techniques. Brain Res. Reviews, 2, 105-169.

LaManna, J.C. and Harik, S.I. (1985). Regional comparisons of brain glucose influx. Brain Res. 326, 299-305.

Malindzak, G.S., Green, H.D., Rapela, C.E. and Gobbee, R.A. (1972). Computer simulation of equations for indicator-concentration curves in parabolic flow model. Ann. Biomed. Eng. 1, 44-55.

Meier, P. and Zierler, K.L. (1954). On the theory of the indicator-dilution method for measurement of blood flow and volume. J. Appl. Physiol. 6, 731-744.

Pardridge W.M. and Fierer, G. (1985). Blood-brain barrier transport of butanol and water relative to N-isopropyl-p-iodoamphetamine as the internal reference. J. Cereb. Blood Flow Metabol. 5, 275-281.

Ramakrishnan, R., Leonard, E.F. and Dell, R.B. (1984). A proof of the occupancy principle and the mean-transit-time theorem for compartmental models. Math. Biosci. 68, 121-136.

Sage, J.I., Van Uitert, R.L. and Duffy, T.E. (1981). Simultaneous measurement of cerebral blood flow and unidirectional movement of substances across the blood-brain barrier: theory, method, and application to leucine. J. Neurochem. 36, 1731-1738.

Stewart, G.N. (1897). Researches on the circulation time and on the influences which affect it. J. Physiol. 22, 159-183.

Tomita, M., Gotoh, F., Amano, T., Tanahashi, N., Kobari, M., Shinohara, T. and Mihara, B. (1983). Transfer function through regional cerebral cortex evaluated by a photoelectric method. Am. J. Physiol. 245, H385-H398.

Tomita, M., Gotoh, F., Sato, T., Amano, T., Tanahashi, N., Tanaka, K. and Yamamoto, M. (1978). Photoelectric method for estimating hemodynamic changes in regional cerebral tissue. Am. J. Physiol. 235, H56-H63.

Weiss, H.R., Buchweitz, E., Murtha, T.J. and Auletta, M. (1982). Quantitative regional determination of morphometric indices of the total and perfused capillary network in the rat brain. Circ. Res. 51, 494-503.

THE MICROPHOTOMETRIC DETERMINATION OF THE VARIABILITY OF OXYGEN

SATURATION OF ERYTHROCYTES LYING WITHIN ROULEAUX

D.W. LUBBERS, H. GRISAR and T.E.J. GAYESKI[*]

Max-Planck-Institut fur Systemphysiologie, 4600 Dortmund, F.R.G. and [*]University of Rochester, Medical Center, Rochester, NY, U.S.A.

SUMMARY

It was found that by using nonlinear multicomponent analysis with the spectra of oxygenated and deoxygenated haemoglobin as basic spectra the mean O_2 saturation of 2 to 3 red cells lying in a rouleau could be determined with an accuracy of 0.5 - 1.0%. At the same PO_2 and PCO_2 distinct differences in O_2 saturation were found in different red cells; for 50% of the saturation values the differences were in the range $\pm$ 1.5%, for 30% of the values, the differences were in the range of $\pm$ 3% and 15% lay in the range of $\pm$ 4.5%. Only 5% of saturation differences were larger than 4.6%. These differences are so large that they have to be considered in calculations of the PO_2 from O_2 saturation measurements.

INTRODUCTION

In 1967 Waldeck published a paper about the P_{50} of single red cells and reported that there were considerable individual variations. He found a mean P_{50} of 25.4 Torr (approx. 3.3 kPa) with a sigma of $\pm$ 4.6 Torr (n = 100). This would correspond to a variation in the O_2 saturation of about $\pm$ 12.5%. The P_{50} values showed an almost normal distribution. All the values covered a P_{50} range between ca. 15 - 35 Torr. Since for O_2 transport calculations the blood PO_2 is often derived from local O_2 saturation measurements, it is of interest to know whether such variations really exist and must be taken into account. In his investigations Waldeck used a microscopic measuring field of 10.2 µm in diameter. Since the mean diameter of a red cell is 7 µm, the red cell covers only about half the measuring field. With such an inhomogeneous distribution in the measuring field there is no longer a linear relationship betweeen extinction and concentration (Hutten, 1969; Lubbers and Wodick, 1969, 1972; Wodick and Lubbers, 1973). Since this was not considered in previous work, we have reinvestigated the problem of the variability of the O_2 saturation of erythrocytes and applied a new evaluation method, nonlinear multi-component analysis (NLMCA; Lubbers and Hoffmann, 1981; Hoffmann et al., 1984).

Because of its form and inhomogeneous haemoglobin distribution, the red cell is a complicated optical system. We have been able to show that the properties of such a complicated biological system can be characterized by a transformation function. This transformation function can be defined in terms of the light path distribution of the skin (Lubbers and Wodick, 1969). Figure 1 shows as an example the transformation function of blood-perfused skin in situ.

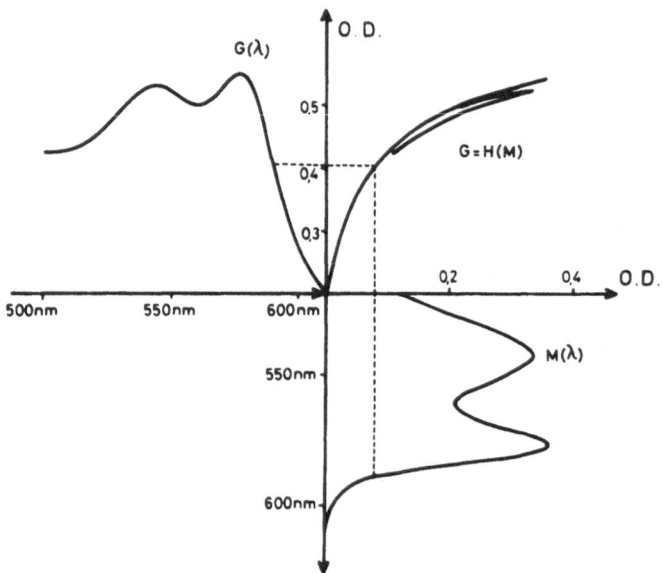

Figure 1. Transformation function of the outer layer of the skin. $G(\lambda)$, measured skin reflection spectra with fully oxygenated blood; $M(\lambda)$, transmission spectra of an oxygenated haemoglobin solution; $H(M)$, transformation curve. The dotted lines connect extinctions of $G(\lambda)$ and $M(\lambda)$ at the same wavelength (Wodick and Lubbers, 1973).

The reflection spectra (left side) of the fully oxygenated blood within the skin - i.e. the extinctions - are compared wavelength by wavelength with the transmission spectra (right side) of a fully oxygenated haemoglobin solution. It demonstrates that there is a bent transformation curve with branches. This bending and branching is caused by inhomogeneous distribution of haemoglobin and by light scattering. Only in small extinction ranges can the transformation be approximated by a straight line; such an approximation is not possible for larger extinction changes.

For evaluation we have applied nonlinear multicomponent analysis, which uses a 3-parametric hyperbola for transformation as developed by Hoffmann et al. (1984). Partly oxygenated haemoglobin is considered as a two component system. Spectra of different degrees of oxygenation can be produced by mixing the two basic spectra, i.e. the spectra of oxygenated (a_1) and deoxygenated (a_2) haemoglobin.

$$c_1 a_1 + c_2 a_2 = \Sigma_i (c_i a_i)$$

c = concentration

The wavelength-dependent scattering is taken into account by multiplying the scattering coefficient, s_1 by the wavelength, λ.

$$a(\lambda) = \Sigma_i(a_ic_i) + s_1\lambda \qquad (1)$$

The measured spectrum, $y(\lambda)$, is a transformation, H, of Equation (1)

$$y(\lambda) = H(a(\lambda)) = H(\Sigma_i(a_ic_i) + s_1\lambda) \qquad (2)$$

The spectra of Equation (1) are obtained by the inverse transformation, H^{-1}. In the NLMCA the transformation is approximated by the following 3-parametric hyperbola.

$$H^{-1}(y(\lambda)) = (b_1y(\lambda) + b_2)/(b_3y(\lambda) + 1) \qquad (3)$$

Since our experiments showed that it was difficult to immobilize red cells over longer times without damage, we worked with rouleaux of red cells using the spectra of oxygenated and deoxygenated haemoglobin as basic spectra. The experiments demonstrated that, indeed, in the wavelength range of 501 – 610 nm the transformation could be well approximated by the 3-parametric hyperbola (Equation 3). The O_2 saturation was determined by calculating the minimum of the least squares error:

$$\min_{\lambda}(\Sigma_\lambda[y(\lambda) - (\Sigma_i(a_ic_i) + s_i\lambda)]^2) \qquad (4)$$

For the photometric measurements we used a Zeiss microscope photometer (UMSP1) with a grid monochromator (600 groves/mm). The spectral half band width was 2.5 nm. The scanning time for the wavelength range from 510 to 610 nm was 25 sec (500 measuring points). Usually, 10 spectra were measured consecutively at the same site, for which a time of about 4.5 min was necessary. The microscope had an Epiplan 40/0.6 (long working distance) objective. A measuring field of 8 μm was used with the corresponding diaphragm aperture of the condenser (15 μm). The measuring chamber could be filled with different gas mixtures. It was covered by two glass microscope slides. One of these was prepared by coating it with a thin layer of varnish, in which small grooves were scratched. A fresh droplet of blood was obtained from a small incision in the fingertip. The blood was mixed with about the same amount of plasma and a small quantity of this mixture put on the prepared glass slide and covered by a gas-permeable membrane; the chamber was then closed. After about 5 min, rouleaux had formed, most of which were lying stationary in the grooves surrounded by plasma. The temperature was continuously monitored.

The microphotometer was connected with a computer on line; after each measurement, the results were presented on a graphics screen. The program was written in Fortran (Hoffmann et al., 1984).

RESULTS

To obtain information about the accuracy of our measurements the measured and the recalculated spectra were compared. Figure 2 shows that when spectra with a small error ($3.9 \cdot 10^{-3}$ OD) were selected an excellent fit between both spectra was obtained. The deviation between the measured and the recalculated spectra, enlarged 25 times, is shown in the insert. The straight lines are the mean error over a wavelength range of 10 nm. It is obvious that there is no wavelength-dependent deviation and that the remaining differences are within the

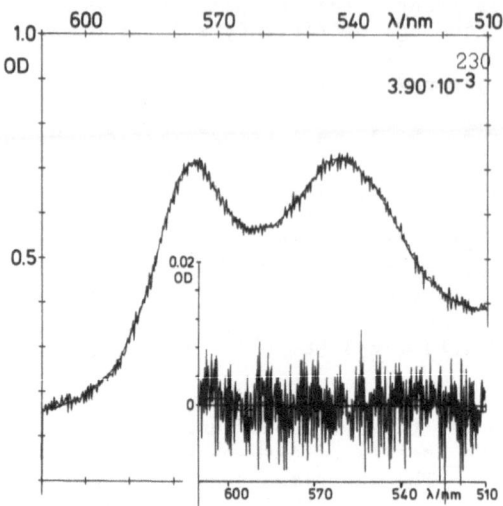

Figure 2. Original and recalculated spectra of a red cell rouleau. The measuring field of 8μm covers 2-3 cells. The recalculated spectrum is depicted as a solid line. There is an excellent fit over the whole wavelength range. The minimum of the least squares error amounts to $3.90 \cdot 10^{-3}$ OD and extends from 64.0% to 64.9% O_2 saturation. The error between both spectra, enlarged 25 times, is shown in the insert. The mean error of the deviation within 10 nm is given as a straight line. It can be seen that only a statistical error remains.

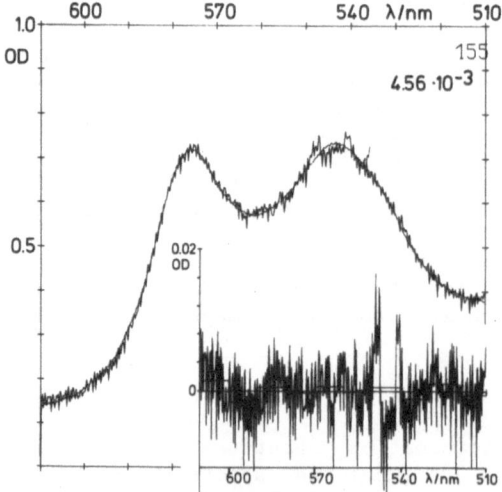

Figure 3. Original and recalculated spectra of a red cell rouleau. This example has a larger error. There are deviations in the 542 nm maximum. The minimum of the least squares error is $4.56 \cdot 10^{-3}$. It extends from 60.0% to 62.5% oxygen saturation. Distinct deviations are visible in the insert.

38

statistical error. The minimum range, in which the mean error had the same value, extended from 64.0% to 64.9% saturation. Figure 3 shows a single spectrum with a larger error ($4.56 \cdot 10^{-3}$ OD). Such larger errors are mainly caused by the instrument, but sometimes also by small movements of the red cells. But in addition in this case there was a clear minimum, which was broader and extended from 60.06 to 62.5% saturation. By averaging over 10 spectra the mean error was more than halved (Fig. 4). This is clearly shown in the insert. The reasons for the small oscillations, which are not always observed, are unknown. In this case the minimum of the mean error ranges from 68.4% to 69.1% saturation. Thus, using the mean of 10 spectra the O_2 saturation of red cells lying in a rouleau can be determined with an absolute accuracy of 0.8 - 1.0%.

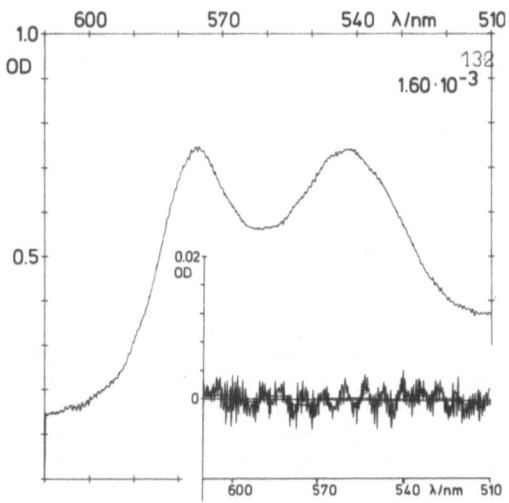

Figure 4. Average of 10 spectra of a red cell rouleau. The minimum of the least squares mean error is reduced to $1.6 \cdot 10^{-3}$ OD; it extends from 68.4% to 69.1% O_2 saturation. Deviations between the averaged spectrum and the recalculated spectrum (insert) are distinctly smaller than in Figs 2 and 3.

Figure 5 shows an experiment lasting 90 minutes, in which the same three different sites (1,2,3) were measured six times (a-f), one after the other. The measuring field covered about 2-3 perpendicularly stacked red cells of a rouleau. The chamber was equilibrated with a gas mixture of 1.94% O_2 and 5% CO_2 at room temperature. The lowest part of the figure shows the O_2 saturations, the middle part, the relative haemoglobin concentrations (relative to the haemoglobin concentration of the basic spectra), and the upper part, the mean least squares errors. As indicated each measurement was repeated ten times consecutively, which took about 4.5 minutes. From the 10 spectra the mean spectrum for the site was calculated and evaluated. The mean saturation and the mean error are marked on the figure by straight lines. It can be seen that in some measuring periods the variation of the calculated O_2 saturation was rather small (e.g. b,1; c,1), but in others it was rather large (e.g. b,3; c,2). Since the temperature changed during the measuring period from 20.6 to 22.3°C the temperature effect had to be corrected. The dotted line shows the expected change of O_2 saturation using a temperature coefficient of 0.0244 (Astrup et al., 1965) and a standard dissociation curve. The deviations from the expected values were large at first, but later approached the expected values. The relative concentration remained almost constant. In spite of the different variations the mean error for 10 measurements remained about the same.

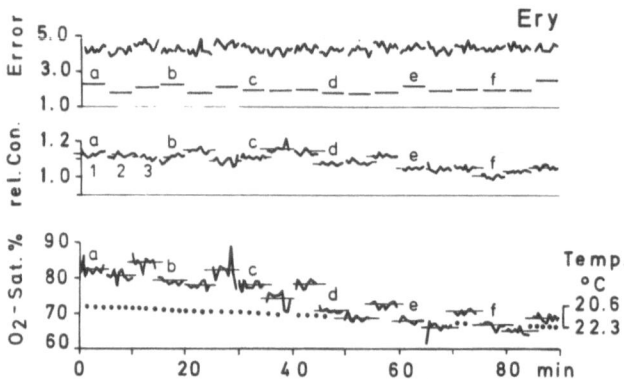

Figure 5. O_2 saturations at three different sites (1,2,3) in rouleaux measured six times (a-f) over 90 min. The measuring starts at a(1,2,3) and is followed by b(1,2,3) and so on up to f. In the lowest part of the figure the calculated O_2 saturations are shown. During each period the O_2 saturation at each site was measured consecutively 10 times; the mean O_2 saturation calculated from the 10 averaged spectra is shown by the straight line. The dotted line corresponds to temperature-corrected O_2 saturation values. In the middle part of the figure the concentrations relative to the concentration of the basic haemoglobin spectra are given. The upper part of the figure shows the least squares error of the single determinations and the mean least squares error of the averaged spectra (straight line).

During most experiments the temperature-corrected O_2 saturation decreased, i.e. the haemoglobin affinity for oxygen increased. It was in order to minimize the influence of this change, that the same three different sites were measured repeatedly, measurements at site 1 being followed by measurements at site 2 and site 3 and then a new measuring cycle started. Thus, the O_2 saturation differences between site 1 and 2, 2 and 3, and 3 and 1 were obtained. Since the O_2 saturation decrease during the experiments was similar at all 3 sites, the mean of 7 measuring cycles was calculated and is depicted in Table 1.

Table 1. Differences in O_2 saturations between three different sites in rouleaux equilibrated with 1.94% O_2 and 5% CO_2 at room temperature (ca. 22°C)

	(82.4 - 65.3)	(68.3 - 54.6)	(71.2 - 57.8)
1-2	-1.7 ± 0.8	-6.3 ± 1.4	-4.7 ± 2.1
	73.0 - 71.3	67.5 - 61.2	64.2 - 59.5
2-3	+4.4	+4.1	-1.6
	71.3 - 75.7	61.2 - 65.3	59.5 - 57.9
3-1	-2.7 ± 0.9	+2.2 ± 1.6	+6.3 ± 1.6
	75.7 - 73.0	65.3 - 67.5	57.9 - 64.2

The values are given in % O_2 saturation. In the uppermost row in brackets the uncorrected O_2 saturation changes at site 1 during 3 experiments are shown, from which the mean O_2 saturations are calculated (Exp. 1, 73.0%; Exp. 2, 67.5%; Exp. 3, 64.2%). 1-2 depicts the difference between site 1 and site 2; similarly 2-3 and 3-1 indicate the differences between those sites. The difference between the pairs of sites was measured 7 times.

There were distinct O_2 saturation differences between the 3 chosen sites; the largest differences were +6.3% and -6.3%.

Because of the changes of O_2 saturation with time, in a larger series of experiments only two consecutively-measured sites were compared. The results of 154 determinations are shown in Table 2. The differences in measured O_2 saturations were arranged in 4 classes: Class 1, ± 0-1.5: all values between -1.5% and +1.5%; Class 2, ± 1.6-3.0: all values between -(3.0% to 1.6%) and +(1.6% to 3.0%); Class 3, ± 3.1-4.5: all values between -(4.5% to 3.1%) and +(3.1% to 4.5%). In Class 4 (4.6-9) all values are larger than -4.5% or +4.5%.

Table 2. O_2 saturation differences between two sites measured consecutively. The O_2 saturations are classified as above

±	0 1.5	1.6 3.0	3.1 4.5	(4.6 9)
n	83	44	21	6
%	53.9	28.4	13.6	4

Class 1 contained 53.9% of all values, i.e. half of all red cells had practically the same O_2 saturation. Of the other half roughly 30% were in Class 2 and 15% in Class 3; in only about 5% were the O_2 saturation differences larger than ± 4.6%; the largest O_2 saturation differences, observed only once, were +8.3% and -8.9%. These saturation differences correspond, at 22°C, to the following PO_2 differences: in Class 1, ΔPO_2 = 0.5 Torr; in Class 2, ΔPO_2 = 1.0 Torr and in Class 3, ΔPO_2 = 1.5 Torr. The saturation change of $\pm$ 8% would correspond to a PO_2 change of 2.7 Torr. The same O_2 saturation differences at 37°C with a P_{50} of 27 Torr would be: in Class 1, ΔPO_2 = 1.2 Torr; in Class 2, ΔPO_2 = 2.3 Torr, and in Class 3, ΔPO_2 = 3.5 Torr. For a saturation change of $\pm$ 8%, ΔPO_2 = 6.4 Torr. Thus, our PO_2 changes are somewhat smaller than the values found by Waldeck (1967), however, one has to take into account that we determined the mean saturation value of 2 to 3 red cells whereby the differences might be reduced. By using red cell rouleaux we could overcome some of the difficulties that occurred previously in our measurements of single red cells and caused large errors (Gayeski et al., 1985).

DISCUSSION

What can be the reason for such variability in O_2 saturation? It is well known that in normal blood, red cells of different ages exist. During ageing the metabolic activity of the red cells decreases, and this is accompanied by a decrease in the concentration of 2,3-DPG, ATP, NADH and NADPH (Fornaini, 1968) and many other changes. The concentration of 2,3-DPG is of particular importance for the relation between PO_2 and O_2 saturation (see Duhm and Gerlach, 1974; Bartels and Baumann, 1977). Edwards and Rigas (1967) found that during in vivo ageing the P_{50} changed by 6.5 Torr (37°C, pH 7.4) and Haidas, Labie and Kaplan, (1971) found a PO_2 difference of 6.1 Torr. Shiga et. al., (1979) reported that in a CO_2-depleted medium (at 37°C, pH 7.36) the P_{50} for younger cells was 25.6 Torr and for older cells 20.4 Torr, the P_{50} for whole blood being ca. 24 Torr. Schmidt (1984) separated young and old red cells by their different densities and found (at 37°C) that the young cells had a P_{50} of 29.2 $\pm$ 3.3 Torr, whereas the old red cells had a P_{50} of 23.8 $\pm$ 2.8 Torr. The difference of 5.4 Torr would correspond to a saturation difference of 13.7%.

REFERENCES

Astrup, P., Engel, K., Severinghaus, J.W. and Munson, E. (1965). The influence of temperature and pH on the dissociation curve of oxyhaemoglobin of human blood. Scand. J. Clin. Lab. Invest. 17, 514-523.

Bartels, H. and Baumann, R. (1977). Respiratory function of hemoglobin. In: Respiratory Physiology II. Ed. Widdicombe, J.G., University Park Press, Baltimore, (Int. Rev. Physiol. 14, 107-134).

Duhm, J. and Gerlach, E. (1974). Metabolism and function of 2,3-diphosphoglycerate in red blood cells. In: The Human Red Cell In Vitro. Eds Greenwalt, T.J. and Jamieson, G.A., Grune & Stratton, Inc., New York, pp. 111-156.

Edwards, M.J. and Rigas, D.W. (1967). Electrolyte labile increase of oxygen affinity during in vivo aging of hemoglobin. J. Clin. Invest. 46, 1579-1588.

Fornanni, G. (1968). Biochemical modifications during the life span of the erythrocyte. Ital. J. Biochem. 16, 258-330.

Gayeski, T.E.J., Hoffmann, J., Grisar, H. and Lubbers, D.W. (1985). The calculation of hemoglobin saturation in single erythrocytes. In: Oxygen Transport to Tissue-VII. Eds Kreuzer, F., Cain. S.M., Turek, Z. and Goldstick, T.K., Plenum Press, New York and London, (Adv. Exp. Med. Biol. 191, 899-908).

Haidas, S., Labie, D. and Kaplan, J.C. (1971). 2,3-diphosphoglycerate content and oxygen affinity as a function of red cell age in normal individuals. Blood, 38, 463-467.

Hoffmann, J., Wodick, R., Hannebauer, F. and Lubbers, D.W. (1984). Quantitative analysis of reflection spectra of the surface of the guinea pig brain. In: Oxygen Transport to Tissue-V. Eds Lubbers, D.W., Acker, H., Leniger-Follert, E. and Goldstick, T.K., Plenum Press, New York and London, (Adv. Exp. Med. Biol. 169, 831-839).

Hutten, H. (1969). Der Einfluss einer inhomogenen Verteilung auf die Messung der Extinktion an monoerythrocytaren Schichten und an einzelnen Erythrocyten. Pflugers Arch. 305, 177-189.

Lubbers, D.W. and Hoffmann, J. (1981). Absolute reflection photometry at organ surfaces. In: Cardiovascular Physiology of Heart, Peripheral Circulation and Methodology. Eds Kovach, A.G.B., Monos, E. and Rubanyi, G., Pergamon Press, Oxford and Akademiai Kiado, Budapest, (Adv. Physiol. Sci. 8, 353-361).

Lubbers, D.W. and Wodick, R. (1969). The examination of multi-component systems in biological materials by means of a rapid scanning photometer. Appl. Optics, 8, 1055-1062.

Lubbers, D.W. and Wodick, R. (1972). Schnelle Photometrie komplizierter biochemischer Mehrkomponentensysteme. Z. Anal. Chem. 261, 271-280.

Schmidt, W. (1984). Sauerstoffbindungseigenschaften von unterschiedlich alten Erythrozyten und ihre Bedeutung bei Ausdauertraining. Dissertation, Hannover.

Shiga, T., Maeda, N., Suda, T., Kon, K. and Sekiya, M. (1979). The decreased membrane fluidity of in vivo aged, human erythrocytes. A spin label study. Biochim. Biophys. Acta, 553, 84-95.

Waldeck, F. (1967). Ein mikrophotometrisches Verfahren zur Aufnahme der Sauerstoffbindungskurve von einzelnen Erythrocyten. Pflugers Arch. 295, 1-14.

Wodick, R. and Lubbers, D.W. (1973). Quantitative Analyse von Reflexionsspektren und anderen Spektren mit inhomogenen Lichtwegen an Mehrkomponentensystemen mit Hilfe der Queranalyse, 1. Das Verfahren der Queranalyse bei Mehrkomponentensystemen mit unbekannten, inhomogenen Lichtwegen. Hoppe-Seyler's Z. Physiol. Chem. 354, 903-915.

KINETICS AND TRANSIENT TIMES OF FLUORESCENCE OPTICAL SENSORS (OPTODES) FOR BLOOD GAS ANALYSIS (O_2, CO_2, pH)

N. OPITZ and D.W. LUBBERS

Max-Planck-Institut fur Systemphysiologie, Rheinlanddamm 201, 4600 Dortmund 1, F.R.G.

Oxygen measurements with optical sensors (O_2 optodes) (Lubbers and Opitz, 1983; Opitz and Lubbers, 1984) are based on fluorescence quenching of certain indicator molecules by molecular oxygen in a diffusion-controlled collisional process (Vaughan and Weber, 1970; Knopp and Longmuir, 1972). The functional dependence follows Stern-Volmer's equation (Stern and Volmer, 1919) : S (PO_2) = S_o/ (1 + $K \cdot PO_2$), where K is overall quenching constant, PO_2, oxygen partial pressure, and S and S_o, relative fluorescence intensity in the presence and absence of oxygen, respectively. Since fluorescence optical sensors incorporate membrane-protected indicator layers, an exponential time course of the PO_2 within these layers can be assumed (Jost, 1960), if a rectangular PO_2 step, ΔPO_2 = PO_2'' - PO_2', is induced in front of the sensor membrane. Insertion of this time course into the hyperbolic calibration curve brings about asymmetrical kinetics of reversible reactions with different transient times, e.g. to 90% of final value (t_{90}):

$$t_{90,02}^{\downarrow} = k^{-1} \cdot \log (10 - (9 \cdot K \cdot \Delta PO_2)/(1 + K \cdot PO_2''))$$

$$t_{90,02}^{\uparrow} = k^{-1} \cdot \log (10 + (9 \cdot K \cdot \Delta PO_2)/(1 + K \cdot PO_2'))$$

$$(1)$$

where $k^{-1} \sim D_{02}/1^2$, D_{02}, oxygen diffusion coefficient and, 1, thickness of the sensor membrane.

Figure 1 demonstrates the asymmetry of the time course of the fluorescence signal, S, during an O_2 step from 0 to 20% and vice versa. The sensor layer was a thin (approx. 20 μm) silicone rubber membrane treated with the O_2-sensitive indicator pyrenebutyric acid, and covered by a 12 μm teflon membrane. Besides these asymmetrical kinetics, dependence of the transient times on the O_2 step height (Fig. 2) as well as on initial and final PO_2 values are found, even for identical step directions and, finally, also dependence on the overall quenching constants of different optical O_2 sensors.

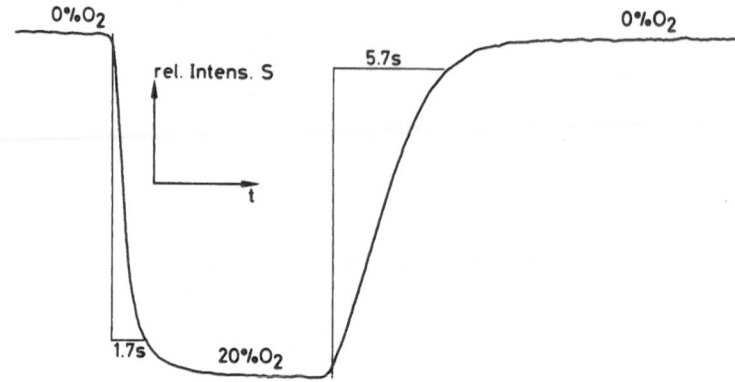

Figure 1. Asymmetrical kinetics of the fluorescence optical O_2 sensor and corresponding transient times (to 90% of final value) for an O_2 step of 20%.

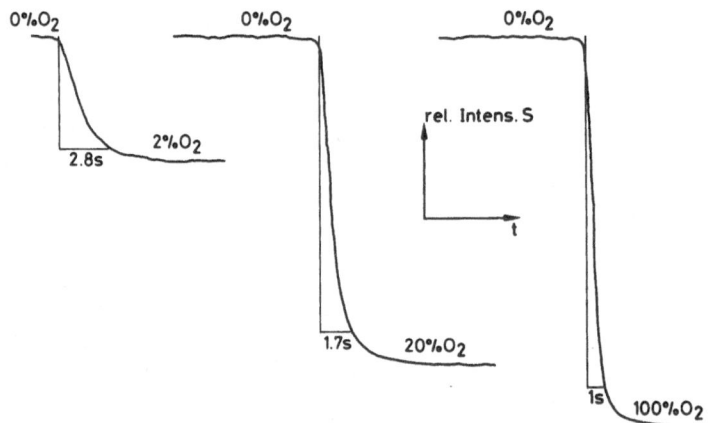

Figure 2. Transient times of the O_2 optode showing dependence on the O_2 step height even in the case of the same direction of step (0 -> 2; 0 -> 20; 0 -> 100).

Thus, the theoretical predictions (Equation 1) agree well with the experimental results. Figure 3 shows a comprehensive representation of the experimental and theoretical data for transient times of both directions according to Equation (1) as a function of the percentage of the O_2 step height.

The asymmetrical kinetics, which result in apparent transient times, are caused by the hyperbolic sensor characteristic of the O_2 optode, which implies an O_2-dependent sensitivity (in contrast to O_2 electrodes). Consequently, a reciprocal plot of the relative fluorescence intensities versus time yields symmetrical kinetics with identical (true) transient times (Fig. 4), since this mode of recording brings about a linearization of the hyperbolic sensor characteristic.

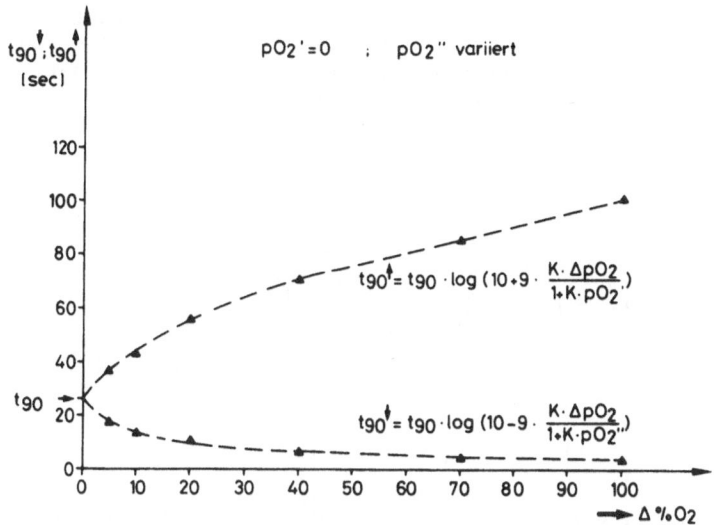

Figure 3. Theoretical course of transient times according to Equation (1) (lines) and experimental results (points). $PO_2' = 0$ Torr; PO_2'' varied.

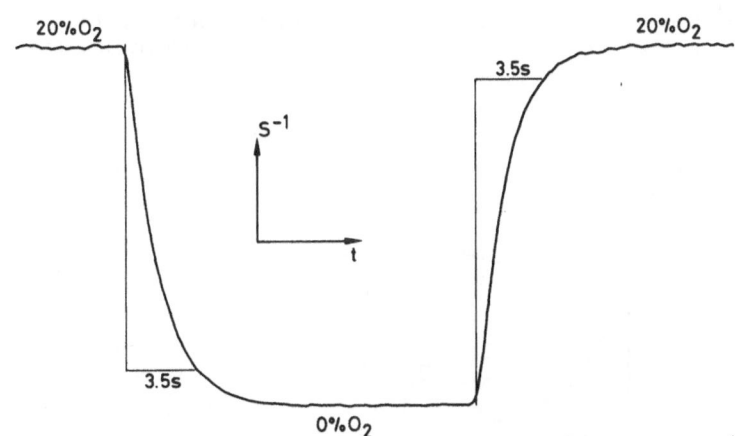

Figure 4. Symmetrical kinetics of the O_2 optode with identical (true) transient times shown by plotting the reciprocal of S vs. t.

The true transient time (to 90% of final value) $t_{90,O_2} = k^{-1}$ ~$D_{O_2}/1^2$ follows from Equation (1) for $K \cdot \Delta PO_2 \to 0$, or, approximately, from $t_{90,O_2} \approx (t_{90,O_2}^\downarrow + t_{90,O_2}^\uparrow)/2$. For $K \cdot \Delta PO_2 \neq 0$ the apparent transient times ($t_{90,O_2}^\downarrow$, $t_{90,O_2}^\uparrow$) differ more from the true value t_{90,O_2}, the bigger the O_2 step height and the overall quenching constant and the smaller the PO_2' value.

47

Since the CO_2 calibration curve is hyperbolic these results are also valid in the case of CO_2 measurements made with pH-sensitive fluorescence indicators dissolved in viscous bicarbonate solutions, and covered by ion-impermeable teflon membranes in order to improve the specificity of measurement.

Nevertheless, because of the different reaction velocities (v^H, R) of the reversible reaction: $H_2CO_3 \rightleftharpoons H^+ + HCO_3^-$, transient times for the two directions differ significantly, even if reciprocal fluorescence signals are monitored (Fig. 5).

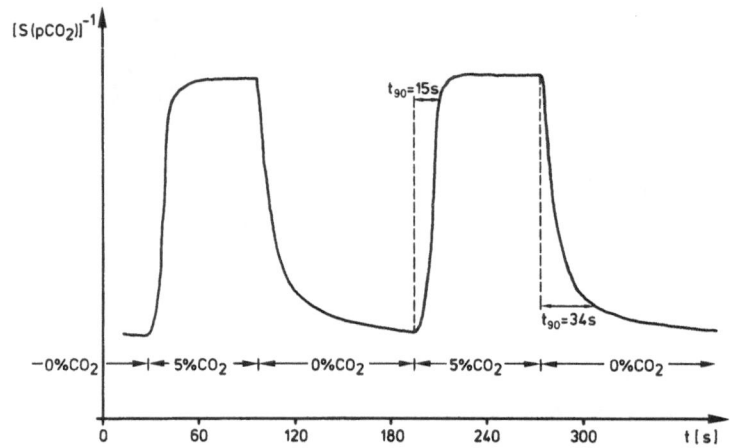

Figure 5. Asymmetric kinetics of the fluorescence optical CO_2 sensor despite recording of reciprocal fluorescence signals (for explanation see text).

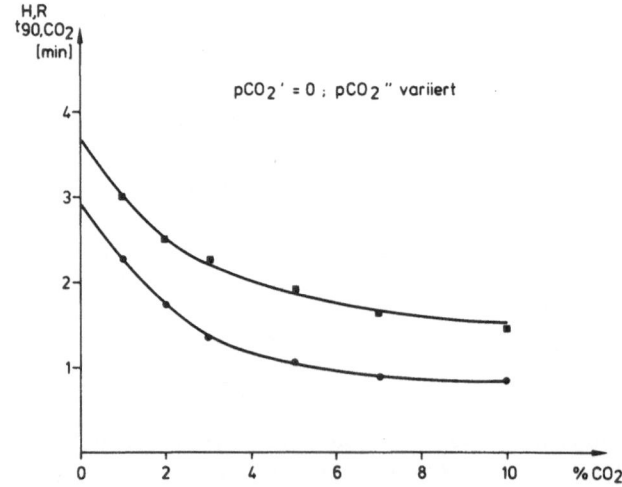

Figure 6. CO_2 dependence of transient times of the CO_2 optode, in spite of a linearized calibration curve, due to the buffering capacity of the pH indicator ($PCO_2' = 0$ Torr; PCO_2'' varied). $c_I = 10^{-2}$ $mol \cdot l^{-1}$.

Moreover, even when observing identical directions of reciprocal signal plots, transient times are still dependent on the CO_2 step height due to the buffering capacity, part at least of which is attributable to the pH indicator itself (Fig. 6). A theoretical analysis of the influence of the buffering capacity of the indicator on the true transient times of CO_2 optodes yields a result, which approximates well to the measured CO_2 course in Figure 5:

$$t_{90,CO2}^{H,R} \approx k^{-1} \cdot \left(1 + \frac{f \cdot c_I \cdot 10^{pK_I}}{(1 + K^* \cdot PCO_2')(1 + K^* \cdot PCO_2'')} \right) + C (v^{H,R}) \tag{2}$$

where c_I = concentration of pH indicator, pK_I = pK value of the pH indicator, K^* = steepness of the linearized CO_2 calibration curve, f = constant, $C (v^{H,R})$ = additional terms due to different reaction velocities (v^H, v^R).

By analogy with CO_2 optodes the following approximation can be derived for pH optodes, provided that the pH calibration curve is linearized:

$$t_{90,pH} \approx k^{-1} \cdot \left(1 + \frac{f \cdot c_I \cdot 10^{pK_I}}{(1 + 10^{pK_I - pH'})(1 + 10^{pK_I - pH''})} \right) \tag{3}$$

These results show (Equations 2 and 3) that in view of the short transient times of fluorescence optical CO_2 and pH measurements relatively low indicator concentrations have to be applied. Thus, the signal-to-noise ratio as well as the resolving power have to be enhanced by other steps (Opitz and Lubbers, 1983), e.g. via the light intensity or the measuring area.

In conclusion it was found that transient times of reversible reactions of PO_2 optodes are identical when linearized with respect to the sensor characteristic, whereas, in contrast to optical O_2 measurements, transient times of pH and CO_2 measurements with optodes still remain dependent on the parameter to be measured, even when corrected for sensor characteristics and the buffering capacity of the pH indicator.

REFERENCES

Jost, W. (1960). Diffusion in Solids, Liquids, Gases. Academic Press, New York-San Francisco-London.

Knopp, J.A. and Longmuir, I.S. (1972). Intracellular measurement of oxygen by quenching of fluorescence of pyrenebutyric acid. Biochim. Biophys. Acta, 279, 393-397.

Lubbers, D.W. and Opitz, N. (1983). Blood gas analysis with fluorescence dyes as an example of their usefulness as quantitative chemical sensors. In: Proc. Int. Meet. Chemical Sensors, Kodansha Ltd, Tokyo; Elsevier, Amsterdam-Oxford-New York, pp.609-619.

Opitz, N. and Lubbers, D.W. (1983). Compact CO_2 gas analyser with favourable signal-to-noise ratio and resolution using special fluorencence sensors (optodes) illuminated by blue LED's. In: Oxygen Transport to Tissue-VI. Eds Bruley, D., Bicher, H.I. and Reneau, D., Plenum Press, New York and London, (Adv. Exp. Med. Biol. 180, 757-762).

Opitz, N. und Lubbers, D.W. (1984). Optische Messverfahren zur Messung von Sauerstoffpartialdrucken in Gasen und biologischen Flussigkeiten. Chem. Ing. Tech., 56, 248-249.

Stern, H. und Volmer, M. (1919). Uber die Abklingzeit der Fluoreszenz. Z. Phys. 20, 183.

Vaughan, W.M. and Weber, G. (1970). Oxygen quenching of pyrenebutyric acid fluorescence in water. A dynamic probe of the microenvironment. Biochemistry, 9, 464.

A METHOD FOR ESTIMATING CONTACT TIME OF RED CELLS IN LUNG CAPILLARIES FROM O_2 AND CO_2 CONCENTRATIONS IN REBREATHING AIR IN MAN

M. MOCHIZUKI, I. SHIBUYA, K. UCHIDA and T. KAGAWA

Department of Physiology, Yamagata University School of Medicine, 990-23 Yamagata, Japan

INTRODUCTION

In a previous paper (Mochizuki et al., 1986) we described the relation between alveolar- and venous-Pco_2 and the contact time (tc). However, at that time, Pco_2-dependency of the arterio-venous difference in O_2 content ($(a-v)Co_2$), the contact-time-dependency of the Haldane effect, and linearity of the relation between the experimental gas exchange ratio and Pco_2 ($R-Pco_2$ line) in rebreathing air were not taken into account. Recently, we have precisely analysed the above correlations from the numerical solutions of the simultaneous O_2 and CO_2 diffusions in the red blood cell (RBC). Based upon the results we have derived a corrected contact time equation. When the time constant of the reaction rate of the extracellular dehydration reaction was less than 0.2 sec, good agreement was observed between the contact time obtained from the pulmonary diffusing capacity for CO (Uchida, Shibuya and Mochizuki, 1986) and that from the present method.

DERIVATION OF THE CONTACT TIME EQUATION

From the numerical solution of the overall diffusions (Mochizuki and Kagawa, 1986), the cellular reaction rate factor ($Fc(CO_2)$, 1/sec·Torr) of the oxygenated RBC was calculated by dividing the CO_2 diffusion quantity (ΔCco_2) by the contact time (tc) and the time average of the Pco_2 gradient across the RBC boundary ($\overline{\Delta Pco_2}$). Since the dispersion of $Fc(CO_2)$ was fairly small, we attempted to derive the tc from the following equation:

$$tc = \Delta Cco_2/(\overline{\Delta Pco_2} \cdot Fc(CO_2)). \tag{1}$$

The gas exchange ratio (R) obtained from the changes in both O_2 and CO_2 concentration in rebreathing air is usually linearly related to the Pco_2 thereof (Mochizuki et al., 1984). Let the gas exchange ratios at the true venous Pco_2 ($tr P\bar{v}co_2$) and any alveolar Pco_2 ($PAco_2$) during the rebreathing process be HEC (Haldane effect coefficient) and RS, respectively. Denoting the arterio-venous O_2 difference at the $PAco_2$ by $(a-v)Co_2$, the veno-arterial CO_2 difference caused by the Pco_2 difference between venous blood and alveolar air, $(v-a)Cco_2(P)$, (where (P) means 'due to the Pco_2 difference') is given by:

$$(v-a)Cco_2(P) = (RS - HEC) \cdot (a-v)Co_2 \qquad (2)$$

Due to the Bohr effect, $(a-v)Co_2$ is influenced by the $PAco_2$. The effect usually decreases as the $PAco_2$ approaches the $tr\bar{P}vco_2$ during the rebreathing, since the intracellular pH (pH_c) decreases with increasing $PAco_2$. Thus, the $(a-v)Co_2$ at the $PAco_2$, where the pH_c in arterial blood equals that in venous blood, was taken to be the standard $(a-v)Co_2$ $((a-v)Co_2^*)$.

Let the mid point between $tr\bar{P}vco_2$ and oxygenated venous Pco_2 $(ox\bar{P}vco_2)$ be Pm_2, the slope of the CO_2 dissociation curve at Pm_2 be $\alpha'(Pm_2)$ (vol%/Torr) and the slope of the R-Pco_2 line obtained from gas analysis of rebreathing air be θ. Mochizuki et al. (1984) estimated the $(a-v)Co_2$ by dividing $\alpha'(Pm_2)$ by θ. Since the $(a-v)Co_2$ thus obtained was not, however, accurate enough, we derived a correcting factor, $F(H)$, from the numerical solution of O_2 and CO_2 diffusions in the RBC and attempted to estimate the $(a-v)(Co_2^*$ as follows:

$$(a-v)Co_2^* = F(H) \cdot \alpha'(Pm_2)/\theta. \qquad (3)$$

$F(H)$ was given by two linear functions of the $(a-v)Co_2^*$ as:

$$F(H) = 1.005 - 0.67 \cdot 10^{-3} \{(a-v)Co_2^* -4.9\} \qquad (4a)$$

when $(a-v)Co_2^* \leq 6.8$ vol%,

and $\quad F(H) = 1 - 0.96 \cdot 10^{-2} \{(a-v)Co_2^* - 7.2\} \qquad (4b)$

when $(a-v)Co_2^* > 6.8$ vol%.

The Pco_2-dependency of $(a-v)Co_2$ was further described by a ratio of $(a-v)Co_2/(a-v)Co_2^*$ $(F(avCo_2))$ as given by:

$$F(avCo_2) = (a-v)Co_2/(a-v)Co_2^*$$

$$= (\frac{2.25}{(a-v)Co_2^*} - 0.09) \frac{tr\bar{P}vco_2 - PAco_2 + 1.2}{100} + 0.99. \qquad (5)$$

Thus, from Equations (2), (3) and (5), the $(v-a)Cco_2(P)$ is given by:

$$(v-a)Cco_2(P) = (RS - HEC) \cdot F(avCo_2) \cdot F(H) \cdot \alpha'(Pm_2)/\theta \qquad (6)$$

Further, from the R-Pco_2 line the $(RS - HEC)$ in the above equation can be expressed as:

$$RS - HEC = \theta \cdot (tr\bar{P}vco_2 - PAco_2) \qquad (7)$$

Substituting $(RS - HEC)$ of Equation (7) in Equation (6), $(v-a)Cco_2(P)$ is further simplified as follows:

$$(v-a)Cco_2(P) = F(avCo_2) \cdot F(H) \cdot \alpha'(Pm_2) \cdot (tr\bar{P}vco_2 - PAco_2) \qquad (8)$$

The CO_2 diffusion quantity of Equation (1), ΔCco_2, does not include the change in the amount of CO_2 dissolved in the plasma.

Therefore, the latter quantity must be subtracted from $(v-a)C_{CO_2}(P)$ to obtain the ΔC_{CO_2}. The change in dissolved CO_2 is calculated from $\alpha(p)'\cdot(tr\bar{P}_{VCO_2} - PA_{CO_2})$, where $\alpha(p)'$ is the CO_2 solubility in the plasma compartment in terms of vol%/Torr. Thus, the ΔC_{CO_2} of Equation (1) is given from Equation (8) by:

$$\{F(avC_{O_2})\cdot F(H)\cdot \alpha'(Pm_2) - \alpha(p)'\}\cdot(tr\bar{P}_{VCO_2}-PA_{CO_2}) \tag{9}$$

The experimental $R-P_{CO_2}$ line has been assumed to be linear against P_{CO_2}, whereas the theoretical one computed from the numerical solution deviates somewhat upwards with decreasing PA_{CO_2} for a constant contact time. Thus, to distinguish the theoretical $(tr\bar{P}_{VCO_2} - PA_{CO_2})$ from the experimental one, the notation for the theoretical P_{CO_2} difference has an asterisk i.e. $(tr\bar{P}_{VCO_2} - PA_{CO_2})^*$. We attempted to eliminate both the differences by using the following ratio:

$$F(\Delta P) = (tr\bar{P}_{VCO_2}-PA_{CO_2})/(tr\bar{P}_{VCO_2}-PA_{CO_2})^*. \tag{10a}$$

The $F(\Delta P)$ mainly depends on the P_{CO_2} difference between venous blood and alveolar air. Now, let the ratio of $(tr\bar{P}_{VCO_2} - PA_{CO_2})^*$ to P_{CO_2} be L, that is:

$$L = (tr\bar{P}_{VCO_2} - PA_{CO_2})^*/\overline{\Delta P_{CO_2}}. \tag{10b}$$

Then, from Equations (10a) and (10b), the L-ratio is expressed as:

$$L = (tr\bar{P}_{VCO_2} - PA_{CO_2})/(F(\Delta P)\cdot\overline{\Delta P_{CO_2}}). \tag{10c}$$

By substituting $(tr\bar{P}_{VCO_2} - PA_{CO_2})$ of Equation (10c) into Equation (9), then ΔC_{CO_2} can be written as:

$$L\cdot F(\Delta P)\{F(avC_{O_2})\cdot F(H)\cdot \alpha'(Pm_2) - \alpha(p)'\}\cdot\overline{\Delta P_{CO_2}} \tag{11}$$

Using the cellular rate factor, $Fc(CO_2)$ of Equation (1), the contact time, tc is given from Equation (11) by:

$$F(\Delta P)\{F(avC_{O_2})\cdot F(H)\cdot \alpha'(Pm_2) - \alpha(p)'\}/(Fc(CO_2)/L) \tag{12}$$

Denoting the mid point between $tr\bar{P}_{VCO_2}$ and PA_{CO_2} by Pm_1, $Fc(CO_2)/L$ is approximated by a hyperbolic function of tc, within a tc range of 0.3 to 1.0 sec:

$$Fc(CO_2)/L = C(Pm_1)/\{tc + D(Pm_1)\}, \tag{13}$$

where $C(Pm_1)$ and $D(Pm_1)$ are quadratic functions of Pm_1, though they depend upon the time constant of the direct extracellular HCO_3^- dehydration reaction. By setting Pm_1 into the equations of $C(Pm_1)$ and $D(Pm_1)$, these values are easily obtained (see Equations 14b, c, d and e of the following paper by Shibuya, Uchida and Mochizuki (1987). Thus, from Equations (12) and (13), the tc equation is simplified to:

$$tc = D(Pm_1)\cdot E/(C(Pm_1) - E), \tag{14a}$$

where

$$E = F(\Delta P)\{F(avC_{O_2})\cdot F(H)\cdot \alpha'(Pm_2) - \alpha(p)'\}. \tag{14b}$$

When trP$\bar{\text{v}}$co$_2$ and oxP$\bar{\text{v}}$co$_2$ are obtained in a rebreathing experiment, Pm$_2$ and α'(Pm$_2$) are readily calculated. Because the PAco$_2$ during rebreathing is arbitrarily decided according to the initial conditions used in the computation of Fc(CO$_2$) and L, the tc can be computed with ease as described below.

EVALUATION OF Fc(CO$_2$)/L AND F(ΔP)

For evaluating the Fc(CO$_2$) down the Pco$_2$ gradient across the RBC membrane, we computed the outward CO$_2$ diffusion by changing the extracellular Pco$_2$ from trP$\bar{\text{v}}$co$_2$ to PAco$_2$. The boundary and initial conditions were determined by referring to our rebreathing experiment at rest in normoxia (Shibuya et al., 1987). The difference between trP$\bar{\text{v}}$co$_2$ and PAco$_2$ was provisionally set at the four levels shown in Figure 2C, where the difference was linearly related to the trP$\bar{\text{v}}$co$_2$. The Fc(CO$_2$) depended on the trP$\bar{\text{v}}$co$_2$, PAco$_2$ and the time constant of the extracellular HCO$_3^-$ dehydration reaction as well as on the contact time. It decreased exponentially as the contact time was prolonged and linearly as the trP$\bar{\text{v}}$co$_2$ was increased. The L-ratio, on the other hand, increased linearly with the increase in contact time. The slope of L against tc also increased with increasing trP$\bar{\text{v}}$co$_2$. Thus, the tc-dependency of Fc(CO$_2$)/L is relatively higher than that of Fc(CO$_2$) alone, as shown in Figure 1, where the time constant of the extracellular dehydration reaction was taken to be 0.1 sec. The parameter values of Equation (13), C(Pm$_1$) and D(Pm$_1$) were given by hyperbolic functions of Pm$_1$ as illustrated in Figures 2A and B, respectively.

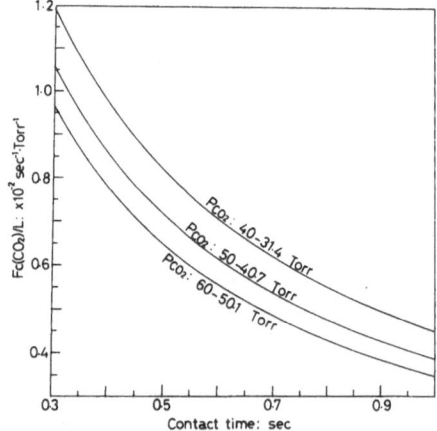

Figure 1. Relation between Fc(CO$_2$)/L and tc. The Pco$_2$ values on each curve show the Pco$_2$ boundary conditions used to compute ΔCco$_2$ and $\overline{\Delta\text{Pco}_2}$.

The dependency of Fc(CO$_2$) on the reaction rate of extracellular HCO$_3^-$ dehydration seemed to be much more complicated than that on Pm$_1$. When the time constant of the dehydration is longer than 0.4 sec, Fc(CO$_2$) increases with the tc. However, when the time constant is shorter than 0.4 sec, Fc(CO$_2$) decreases as the tc increases: the shorter the time constant, the higher the rate of decrease of Fc(CO$_2$) versus the contact time. In contrast to the Fc(CO$_2$), the L-ratio was almost independent of the reaction rate of extracellular dehydration. The dependency of Fc(CO$_2$)/L on the time constant was less than that on the contact time. $Fc(CO_2)/L$ could be expressed in all cases by a hyperbolic equation similar to Equation (13). The parameters ($C(Pm_1)$ and $D(Pm_1)$ computed when $trP\bar{v}co_2 = 50$ Torr and $PAco_2 = 40.7$ Torr are depicted against the time constant in Figure 3.

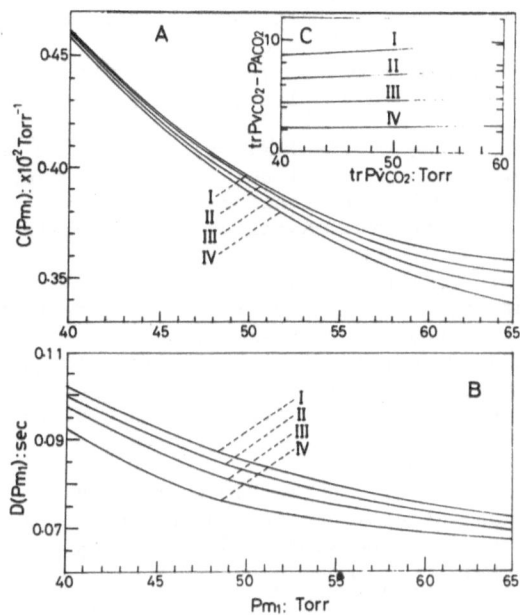

Figure 2. The Pm_1-dependency of the parameters $C(Pm_1)$ and $D(Pm_1)$ of Equation (13). $Fc(CO_2)/L$ values were computed, using the conditions for ($trP\bar{v}co_2 - PAco_2$) of I, II, III, IV depicted in inset C.

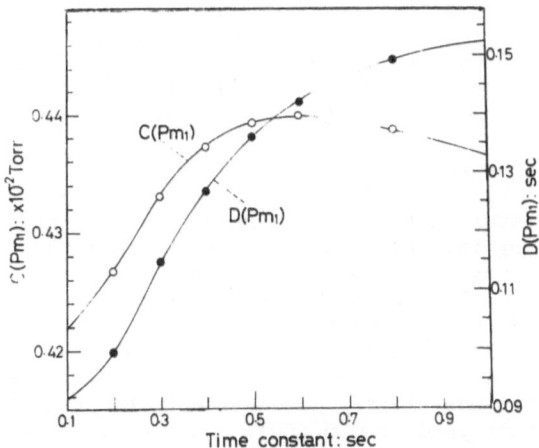

Figure 3. The relation between $C(Pm_1)$ and $D(Pm_1)$ computed from $Fc(CO_2)$ by using the relation of Equation (13) and the time constant of the extracellular HCO_3^- dehydration reaction.

Since the theoretical R-Pco$_2$ line deviates upwards as the PAco$_2$ is decreased, F(ΔP) of Equation (10a) should be greater than unity. Moreover, when diffusion across the capillary wall is taken into account, (trP$\bar{v}$co$_2$ - PAco$_2$) should exceed (trP$\bar{v}$co$_2$ - PAco$_2$)*. For the above two reasons it is conceivable that F(ΔP) exceeds unity and that it approaches unity as PAco$_2$ approaches trP$\bar{v}$co$_2$. In addition, the contact time should not be affected by the change in PAco$_2$ during the course of rebreathing. Taking the above limitations into account, F(ΔP) was determined by checking whether it satisfied the following criteria: (1) F(ΔP) $\geq$ 1, (2) F(ΔP) approaches unity as the PAco$_2$ approaches trP$\bar{v}$co$_2$, and (3) F(ΔP) gives a constant tc value independent of the PAco$_2$. Finally, we compared the tc obtained from Pm$_1$ and Pm$_2$ with that obtained from the pulmonary diffusing capacity for CO (D$_L$(CO)), and confirmed the reliability of the F(ΔP) value as well as that of the values for other parameters.

Uchida et al. (1986) measured the D$_L$(CO) (ml/min·Torr) and the cardiac output ($\dot{Q}$;ml/min), by means of a rebreathing technique (Mochizuki et al., 1984). Furthermore, they estimated the tc by dividing the D$_L$(CO) by a product of $\dot{Q}$, the fractional haematocrit and the rate factor of the CO replacement reaction for O$_2$ from oxygenated red blood cells (Fc(CO); 1/sec·Torr). Thus, using the same R-Pco$_2$ line we attempted to obtain the tc value from Pm$_1$ and Pm$_2$.

First, the trP$\bar{v}$co$_2$ was obtained from the R-Pco$_2$ line using the HEC value which was determined from O$_2$ uptake ($\dot{V}$o$_2$) by referring to the experimental relation between HEC, (a-v)Co$_2^*$ and $\dot{V}$o$_2$ (Shibuya et al., 1987) as given by:

$$HEC = 0.294 - 0.57 \cdot 10^{-2} \cdot (a-v)Co_2^* \qquad (15)$$

and

$$(a-v)Co_2^* = 9.5 \cdot \dot{V}o_2^{0.35}. \qquad (16)$$

From the measured trP$\bar{v}$co$_2$, we calculated four different PAco$_2$ values according to the relation shown in Figure 2C, and the midpoint between trP$\bar{v}$co$_2$ and PAco$_2$ (Pm$_1$). Then C(Pm$_1$) and D(Pm$_1$) were calculated from the respective equations as illustrated in Figures 2A and B. Next, by extrapolating the R-Pco$_2$ line to the abscissa, oxP$\bar{v}$co$_2$ was evaluated to calculate Pm$_2$. From Pm$_2$, the slope of the CO$_2$ dissociation curve, α'(Pm$_2$) was obtained, by inserting Pm$_2$ into the following equation (Tazawa et al., 1983):

$$\alpha'(Pm_2) = 3.962 \cdot Pm_2^{-0.5857}, \text{ (vol\%/Torr)}. \qquad (17)$$

After the determination of C(Pm$_1$), D(Pm$_1$) and α'(Pm$_2$), the tc was computed by inserting them together with F(H) of Equations (4a) or (4b) and F(avCo$_2$) of Equation (5) into Equations (14a) and (14b). The F(ΔP) was determined by a trial-and-error method so that the criteria mentioned before were satisfied. When the time constant of the extracellular HCO$_3^-$ dehydration reaction was taken to be 0.1 sec, and F(ΔP) was given by the following linear equation of (trP$\bar{v}$co$_2$ - PAco$_2$), the tc was fairly independent of the PAco$_2$ and agreed with the tc value obtained from D$_L$(CO):

$$F(\Delta P) = 1 + 0.422 \cdot 10^{-2} \cdot (trP\bar{v}co_2 - PAco_2). \qquad (18)$$

Figure 4 shows the tc at rest and during exercise plotted against (trP$\bar{v}$co$_2$ - PAco$_2$). The plotted points represent the means and the bars, the S.D. The shaded area delineates the S.D. around the mean tc obtained from the $D_L(CO)$ in individual subjects (Uchida et al., 1986). Both the tc values coincided well with each other, regardless of the relation between $trP\bar{v}co_2$ and $PAco_2$, or the difference between $trP\bar{v}co_2$ and $PAco_2$. In Figure 5, the tc values obtained from $C(Pm_1)$ and $D(Pm_1)$ are plotted against the tc obtained simultaneously from the $D_L(CO)$ and $\dot{Q}$ (Uchida et al., 1986). The correlation coefficient was 0.91 and the regression line coincided fairly well with the identity line. In a sitting position at rest tc was about 0.7 sec, and during exercise of about 1.3 l(STPD)/min in $\dot{V}o_2$, it was around 0.4 sec. The difference in tc between the two methods has an S.D. of about 0.05 sec, suggesting the reliability of analysing the CO_2 diffusion process out of the RBC.

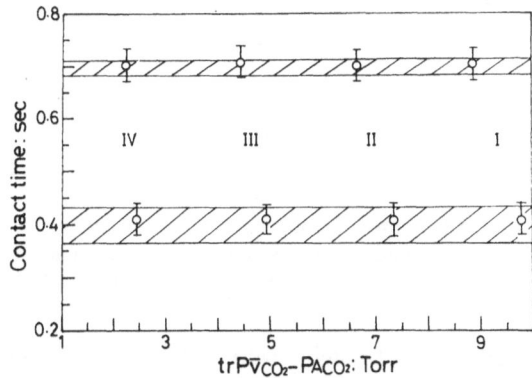

Figure 4.   The contact time calculated at 4 different $PAco_2$ levels during the course of rebreathing. $C(Pm_2)$ and $D(Pm_1)$ were evaluated from the experimental Pm_1 by using the relation illustrated in Fig. 2.

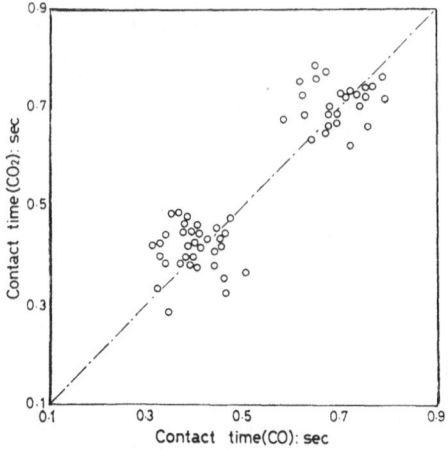

Figure 5.   Relation between tcs obtained from $D_L(CO)$ and $\dot{Q}$ and those from Equations (14a) and (14b) in the same rebreathing experiment.

Figure 6 shows the mean and S.D. of the tc computed by changing the time constant of the extracellular HCO_3^- dehydration reaction from 0.1 sec to 0.6 sec. Good agreement between the tcs obtained from Pm_1 and Pm_2 and $D_L(CO)$ was observed, when the time constant was shorter than 0.2 sec.

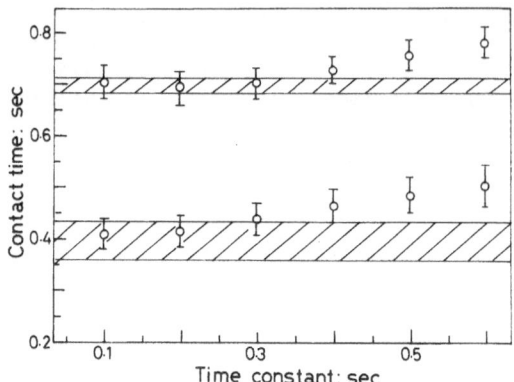

Figure 6.   Dependency of the tc upon the time constant of the extra-cellular HCO_3^- dehydration reaction. The shaded area represents the mean $\pm$ S.D. of the tc obtained from the $D_L(CO)$ in the same subject and at the same time. The upper and lower plots were obtained at rest and during exercise, respectively.

The theoretical evidence that the time constant of carbonic anhydrase in the capillary wall is less than 0.2 sec, as shown in Figure 6 is one of the most important findings in this study. In our previous study (Mochizuki et al., 1986), we assumed that the time constant of extracellular HCO_3^- dehydration was 1.5 sec according to the experimental data of Klocke (1978). At the time, the requirement that tc should be independent of the $PAco_2$ during the rebreathing, was not taken into account. To meet the above requirement, the correcting factor, $F(\Delta P)$, was determined as given by Equation (18). In addition, to satisfy the requirement that the tc must coincide with that estimated from the $D_L(CO)$, a time constant for extracellular dehydration of less than 0.2 sec was proposed. Bidani, Mathew and Crandall (1983) measured the reaction rate of extracellular HCO_3^- dehydration in isolated rat lungs, and obtained a rate similar to ours, which suggests the validity of our theoretical data.

REFERENCES

Bidani, A., Mathew, S.J. and Crandall, E.D. (1983). Pulmonary vascular carbonic anhydrase activity. J. Appl. Physiol: Respir. Envir. Exercise Physiol. 55, 75–83.

Klocke, R.A. (1978). Catalysis of CO_2 reactions by lung carbonic anhydrase. J. Appl. Physiol. 44, 882–888.

Mochizuki, M. and Kagawa, T. (1986). Numerical solution of partial differential equations describing the simultaneous O_2 and CO_2 diffusions in the red blood cell. Jpn. J. Physiol. 36, 43–63.

Mochizuki, M., Kagawa, T., Uchida, K. and Shibuya, I. (1986). Relation between the contact time and venous and alveolar P_{CO_2} at rest. In: Oxygen Transport to Tissue-VIII. Ed. Longmuir, I.S., Plenum Press, New York and London, (Adv. Exp. Med. Biol. <u>200</u>, 27-34).

Mochizuki, M. Tamura, M., Shimasaki, T., Niizeki, K. and Shimouchi, A. (1984). A new indirect method for measuring arteriovenous O_2 content difference and cardiac output from O_2 and CO_2 concentrations by rebreathing air. Jpn. J. Physiol. <u>34</u>, 295-306.

Shibuya, I., Uchida, K. and Mochizuki, M. (1987). Quantitative relations between gas exchange parameters including contact time at rest and during exercise. This volume.

Tazawa, H., Mochizuki, M., Tamura, M., and Kagawa, T. (1983). Quantitative analyses of the CO_2 dissociation curve of oxygenated blood and the Haldane effect in human blood. Jpn. J. Physiol. <u>33</u>, 601-618.

Uchida, K., Shibuya, I. and Mochizuki, M. (1986). Simultaneous measurement of cardiac output and pulmonary diffusing capacity for CO by a rebreathing method. Jpn. J. Physiol. <u>36</u>, 657-670.

QUANTITATIVE RELATIONS BETWEEN GAS EXCHANGE PARAMETERS INCLUDING

CONTACT TIME AT REST AND DURING TREADMILL EXERCISE

I. SHIBUYA, K. UCHIDA and M. MOCHIZUKI

Department of Physiology, Yamagata University, School of
Medicine, 990-23, Yamagata, Japan

INTRODUCTION

Studies on quantitative relations between gas exchange parameters
at rest and during exercise have provided many reports on the relation
between oxygen uptake ($\dot{V}o_2$), arterio-venous O_2 difference
($(a-v)Co_2$) and cardiac output ($\dot{Q}$) (Reeves et al., 1961; Bevegard,
Holmgren and Jonsson, 1963; Astrand et al., 1964; Hermansen, Ekblom and
Saltin, 1970). However, there are only a few reports on the relations
between contact time (tc), pulmonary capillary blood volume (Vc) and
the other related parameters. Johnson et al. (1960) measured tc, $\dot{Q}$ and
Vc at rest and during exercise by using a gas mixture containing carbon
monoxide, helium and acetylene. However, their method was complicated
because of difficulties in the quantitative measurement of these gases.

Recently, Mochizuki et al. (1984) developed a reliable method for
determining the $(a-v)Co_2$, $\dot{V}o_2$, and $\dot{Q}$ from the O_2 and CO_2
concentrations in rebreathing air. Furthermore, using the solutions of
differential equations for O_2 and CO_2 diffusion into and out of red
cells (Mochizuki and Kagawa, 1986), they (Mochizuki et al., 1987)
developed a method for estimating tc from the true ($trP\bar{v}co_2$) and
oxygenated ($oxP\bar{v}co_2$) venous Pco_2 and alveolar Pco_2 ($PAco_2$).
These two methods enabled us to measure gas exchange parameters
including tc and Vc by an air rebreathing technique, which is simple,
safe and non-invasive. In the present study, using these methods we
attempted to clarify the quantitative relations between the parameters
at rest and during exercise in normal subjects.

METHODS

1) <u>Experimental procedure</u>
Measurements at rest were made on seven normal male subjects, aged
from 21 to 48 years, in a standing position at least two hours after a
meal. After 10 minutes bed rest, the subject breathed air for 3 min.
Then, he inspired 1.5 l of air from a rubber bag, and rebreathed at a
rate of 30/min and with a tidal volume of more than 1.0 l. The $\dot{V}o_2$
during rebreathing was measured by the O_2 injection method (Mochizuki
et al., 1984). From a syringe connected to the rebreathing bag, 90 ml

of pure O_2 were injected into the bag at the 10th expiratory period. The O_2 and CO_2 concentrations in the respiratory gas were measured continuously with a glow-discharge gas analyser (San-ei, 1HO2-4; Nitta and Mochizuki, 1969). The gas sample was introduced into the analyser from a point in the rebreathing circuit close to the mouth piece. Figure 1 shows actual records of O_2 and CO_2 concentrations in rebreathing air at rest.

After the measurement at rest, exercise was performed on a treadmill (Tatebe, DR-5A-T) at 5 different levels of exercise, Ex-1 to Ex-5. They were all submaximal. First, the subject walked on a treadmill for 5 min and then started the rebreathing experiment. The amount of rebreathing air was increased at each step-increase of exercise by about 200 ml. During exercise, the $\dot{V}o_2$ was measured by injecting 200 or 300 ml of pure O_2 into the rebreathing bag prior to the 9th, 8th or 7th expiratory phase. The series of experiments were repeated 5 to 7 times on different days.

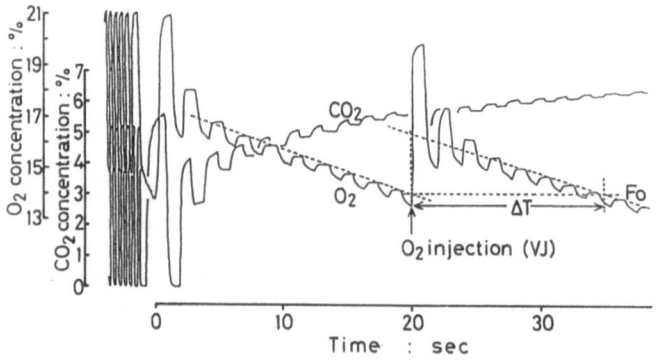

Figure 1. Actual records of O_2 and CO_2 concentrations in rebreathing air at rest.

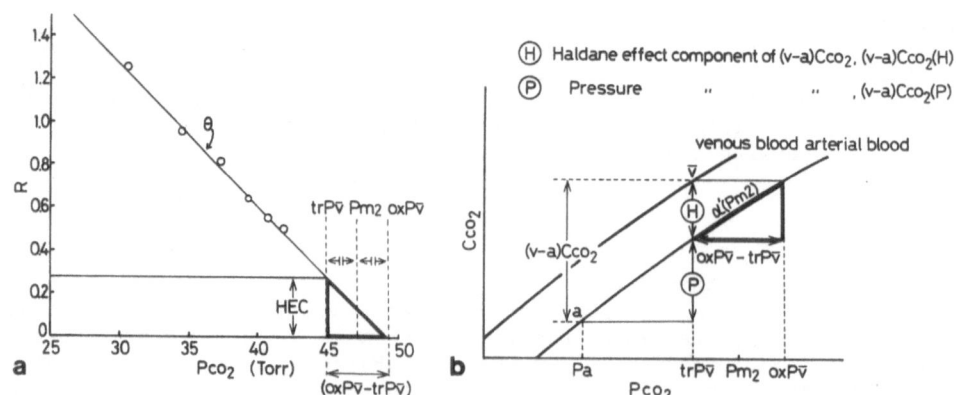

Figure 2. (A) An example of R-Pco_2 relation obtained during rebreathing at rest. (B) Schematic illustration of CO_2 dissociation curves for venous and arterial blood.

2) Computation of the parameter values

(i) Arterio-venous O_2 difference, $(a-v)C_{O_2}$

From the O_2 and CO_2 concentrations (C_{O_2} and C_{CO_2}) in rebreathing air, the gas exchange ratio (R) was calculated according to Mochizuki et al. (1984):

$$R(i) = \frac{\Delta G(i) - \Delta F(i) \cdot G(i+1) - \Delta G(i) \cdot F(i+1)}{\Delta F(i) - \Delta F(i) \cdot G(i+1) - \Delta G(i) \cdot F(i+1)} \tag{1}$$

where $R(i)$ is the R at the i'th inspiration and $F(i+1)$ and $G(i+1)$ represent, respectively, O_2 and CO_2 fractions at the (i+1)'th inspiration. Furthermore, $\Delta F(i)$ and $\Delta G(i)$ are the difference in O_2 and CO_2 fractions, respectively, between the (i-1)'th and (i+1)'th inspirations. Figure 2A shows the relationships between the R and P_{CO_2} calculated from the data in Figure 1. The R value linearly decreases with an increase in P_{CO_2}. The slope of the R-P_{CO_2} line (θ) was calculated by the method of least squares. The $oxP\bar{v}_{CO_2}$ and trP_{CO_2} can be determined as the P_{CO_2} at R = 0 and the Haldane effect coefficient (HEC), respectively. From the R-P_{CO_2} relation in Figure 2A, HEC is expressed as:

$$HEC = (oxP\bar{v}_{CO_2} - trP\bar{v}_{CO_2}) \cdot \theta. \tag{2}$$

Alternatively HEC is defined as:

$$HEC = (v-a)C_{CO_2}(H) \ / \ (a-v)C_{O_2}, \tag{3}$$

where $(v-a)C_{CO_2}(H)$ is the Haldane effect component of the veno-arterial CO_2 difference. Figure 2B shows the CO_2 dissociation curves for venous and arterial blood. Let the slope of the curve for arterial blood at the mid point of $trP\bar{v}_{CO_2}$ and $oxP\bar{v}_{CO_2}$ (Pm_2) be $\alpha'(Pm_2)$, then $(v-a)C_{CO_2}(H)$ can be expressed as:

$$(v-a)C_{CO_2}(H) = (oxP\bar{v}_{CO_2} - trP\bar{v}_{CO_2}) \cdot \alpha'(Pm_2), \tag{4}$$

From Equations (3) and (4), the HEC is rewritten as follows:

$$HEC = (oxP\bar{v}_{CO_2} - trP\bar{v}_{CO_2}) \cdot \alpha'(Pm_2 \ / \ (a-v)C_{O_2}. \tag{5}$$

From Equations (2) and (5), $(a-v)C_{O_2}$ is given by:

$$(a-v)C_{O_2} = \alpha'(Pm_2)/\theta, \qquad (vol\%) \tag{6}$$

where $\alpha'(Pm_2)$ (Tazawa et al., 1983) is given by:

$$\alpha'(Pm_2) = 3.962 \cdot Pm_2^{-0.5857}, \quad (vol\%/Torr). \tag{7}$$

The $(a-v)C_{O_2}$ is influenced not only by the arterio-venous P_{O_2} difference but also by the $P_{A_{CO_2}}$, due to the Bohr-effect. The $(a-v)C_{O_2}$ usually decreases as the $P_{A_{CO_2}}$ approaches the $trP\bar{v}_{CO_2}$ with elapse of rebreathing time. Kagawa and Mochizuki (1986) designated the $(a-v)C_{O_2}$ at the $P_{A_{CO_2}}$, where the intracellular pH is equal to that in venous blood, as the standard $(a-v)C_{O_2}$, and denoted this by $(a-v)C\overset{*}{_{O_2}}$. Further, they calculated the $(a-v)C\overset{*}{_{O_2}}$ by the same method as Mochizuki et al. (1984):

$$(a-v)C\overset{*}{_{O_2}} = F(H) \cdot \alpha'(Pm_2)/\theta, \qquad (vol\%), \tag{8}$$

where F(H) is a factor for correcting the ratio, $\alpha'(Pm_2)/\theta$, as follows:

$$F(H) = 1.005 - 0.67 \cdot 10^{-3} \cdot \{(a-v)Co_2^* - 4.9\}, \tag{9a}$$

when $(a-v)Co_2^* \leq 6.8$ vol%; and

$$F(H) = 1 - 0.96 \cdot 10^{-2} \cdot \{(a-v)Co_2^* - 7.2\}, \tag{9b}$$

when $(a-v)Co_2^* > 6.8$ vol%.

To determine the F(H) from Equations (9a) and (9b), we assumed a tentative value for $(a-v)Co_2^*$ from $\dot{V}o_2$ using the standardized relation between the $(a-v)Co_2^*$ and $\dot{V}o_2$ (Fig. 3) as given by:

$$(a-v)Co_2^* = 9.5 \cdot \dot{V}o_2^{0.35}, \quad (vol\%) \tag{10}$$

Hitherto, the HEC in vivo has been considered to remain constant at a specific value, 0.32 (Kim, Rahn and Farhi, 1966). However, Kagawa and Mochizuki (1986) showed that it usually ranged between 0.2 and 0.27 and decreased with an increase in $(a-v)Co_2^*$ as given by:

$$HEC = 0.294 - 0.57 \cdot 10^{-2} \cdot (a-v)Co_2^* \tag{11}$$

In Equation (11) the value for $(a-v)Co_2^*$ was calculated according to Equation (10) and to derive HEC the $tr\bar{P}vco_2$ was estimated from the R-Pco_2 line.

(ii) O_2 uptake, and $\dot{V}o_2$

$\dot{V}o_2$ was calculated from the volume of O_2 injected (VJ, ml) and the time interval (ΔT, sec, in Fig. 1) during which the O_2 concentration, increased by the O_2 injection, returns to the level just before the injection (Shibuya, Niizeki and Uchida, 1986):

$$\dot{V}o_2 = \frac{VJ}{\Delta T} \cdot \frac{1 - FO}{1 - (1 - \bar{R}) \cdot FO} \cdot 60, \quad (ml/min), \tag{12}$$

where FO and $\bar{R}$ are the O_2 concentration just before the O_2 injection, and the mean gas exchange ratio during ΔT, respectively.

(iii) Contact time, tc

Mochizuki et al. (1987) computed the relation between the amount of CO_2 diffusing out of the RBC and the Pco_2 gradient across the RBC boundary from the numerical solution of simultaneous O_2 and CO_2 diffusions. By dividing the quantity of CO_2 that diffused by a product of the tc and the time average of the Pco_2 gradient during the contact time ($\overline{\Delta Pco_2}$), they derived the reaction (or diffusion) rate factor ($Fc(CO_2)$). Then they attempted to estimate the tc inversely by dividing the diffusion quantity by the product of the $Fc(CO_2)$ and ΔPco_2. In the practical computation, they introduced a contact time equation, which was solved by using three experimental values of $tr\bar{P}vco_2$, $ox\bar{P}vco_2$, and $PAco_2$. First, they computed the ratio (L) of ($tr\bar{P}vco_2 - PAco_2$) to $\overline{\Delta Pco_2}$ from the numerical solution. Then the quotient of $Fc(CO_2)/L$ was computed by referring to the standardized relation between $PAco_2$ and $tr\bar{P}vco_2$, as given by:

$$PAco_2 = 0.936 \cdot tr\bar{P}vco_2 - 6.03, \quad (Torr) \tag{13}$$

In a tc range of 0.3 to 1.0 sec, the $Fc(CO_2)/L$ was approximated by a hyperbolic equation as:

$$Fc(CO_2)/L = C(Pm_1)/\{tc + D(Pm_1)\}, \tag{14a}$$

where the Pm_1 is the mid point between the $trP\bar{v}co_2$ and $PAco_2$. $C(Pm_1)$ (vol%/Torr) and $D(Pm_1)$ (sec) were calculated from the following equations:

For $Pm_1 \leq 50$ Torr:

$$C(Pm_1) = 1.0483 - 0.2115 \cdot 10^{-1} \cdot Pm_1 + 0.162 \cdot 10^{-3} \cdot Pm_1{}^2, \tag{14b}$$

$$D(Pm_1) = 0.2487 - 0.5247 \cdot 10^{-2} \cdot Pm_1 + 0.395 \cdot 10^{-4} \cdot Pm_1{}^2, \tag{14c}$$

and for $Pm_1 > 50$ Torr:

$$C(Pm_1) = 1.0775 - 0.2222 \cdot 10^{-1} \cdot Pm_1 + 0.172 \cdot 10^{-3} \cdot Pm_1{}^2, \tag{14d}$$

$$D(Pm_1) = 0.231 - 0.452 \cdot 10^{-2} \cdot Pm_1 + 0.32 \cdot 10^{-4} \cdot Pm_1{}^2. \tag{14e}$$

Using the above $Fc(CO_2)/L$, Mochizuki et al. (1987) ultimately derived the contact time equation:

$$tc = D(Pm_1) \cdot E / \{C(Pm_1) - E\}, \tag{15a}$$

where

$$E = \{F(avCo_2) \cdot F(H) \cdot \alpha'(Pm_2) - \alpha(p)'\} \cdot F(\Delta P) \tag{15b}$$

$F(avCo_2)$ and $F(\Delta P)$ are factors for correcting the Pco_2-dependency of $(a-v)Co_2$ and the linearity of the experimental $R-Pco_2$ line, respectively. $\alpha(p)'$ is the CO_2 solubility in the plasma fraction, being 0.0377 vol%/Torr. From Kagawa and Mochizuki (1986), $F(avCo_2)$ equals:

$$(\frac{0.0225}{(a-v)Co_2^*} - 0.9 \cdot 10^{-3}) \cdot (trP\bar{v}co_2 - PAco_2 + 1.2) + 0.99 \tag{16}$$

The $F(\Delta P)$ is a linear equation of $(trP\bar{v}co_2 - PAco_2)$ (Mochizuki et al., 1987) as given by:

$$F(\Delta P) = 1 + 0.442 \cdot 10^{-2} \cdot (trP\bar{v}co_2 - PAco_2) \tag{17}$$

Thus, from the measured data for O_2 and CO_2 concentrations in rebreathing air, we computed Pm_1 and Pm_2. By inserting them together with $F(H)$, $F(avCo_2)$ and $F(\Delta P)$ into Equations (15a) and (15b), tc was evaluated.

(iv)Cardiac output, $\dot{Q}$, and capillary blood volume, Vc

Cardiac output, $\dot{Q}$ (l/min), was obtained by dividing $\dot{V}o_2$ (l/min) by $(a-v)Co_2^*$ (vol%)/100; and capillary blood volume, Vc (ml), was calculated by multiplying $\dot{Q} \cdot 1000/60$ (ml/sec) by tc (sec).

All these computations were made by means of a computer, DEC-VAX-11/750, in which the measured O_2 and CO_2 concentrations in rebreathing air were put directly through a digitizer to obtain the R-Pco$_2$ line.

RESULTS

The mean and S.D. of resting $\dot{V}o_2$ and (a-v)Co$_2^*$ were 0.32 $\pm$ 0.05 1/min and 6.44 $\pm$ 0.69 vol%, respectively. At the maximal exercise tested, Ex-5, the $\dot{V}o_2$ was in a range of 1.5 to 2 1/min and the (a-v)Co$_2^*$ was in a range of 10 to 13 vol%. Figure 3 shows the relationship between (a-v)Co$_2^*$ and $\dot{V}o_2$, the regression line being given by Equation (10). The correlation coefficient between the logarithms of (a-v)Co$_2^*$ and $\dot{V}o_2$ was 0.904. The $\dot{Q}$ obtained by dividing the $\dot{V}o_2$ by the (a-v)Co$_2^*$ has the relationship shown in Figure 4, the regression line being given by:

$$\dot{Q} = 10.52 \cdot \dot{V}o_2{}^{0.65} \quad (1/min) \tag{18}$$

The correlation coefficient between log $\dot{Q}$ and log $\dot{V}o_2$ was 0.973.

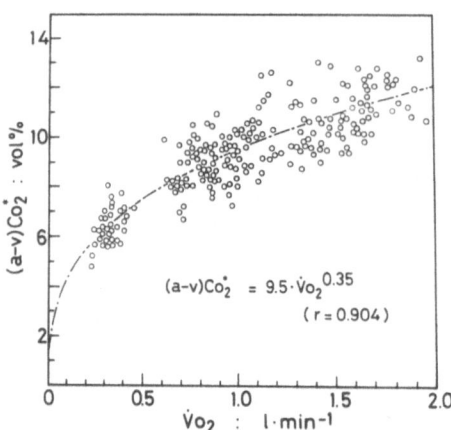

Figure 3. The relation between (a-v)Co$_2^*$ and $\dot{V}o_2$ in a standing position at rest and during treadmill exercise.

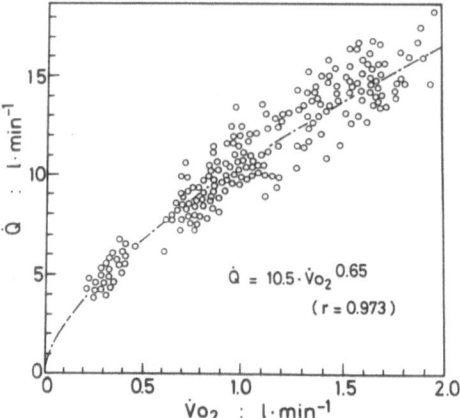

Figure 4. The relation between $\dot{Q}$ and $\dot{V}o_2$ in a standing position at rest and during treadmill exercise.

The relationship between the tc and (a-v)Co$_2^*$ at rest and during exercise, is shown in Figure 5, where the regression line is given by the following hyperbolic equation:

$$tc = 4.86 \cdot (a-v)Co_2^*{}^{-1.025}, \quad (sec) \tag{19}$$

The mean and S.D. of tc in a standing position at rest was 0.75 $\pm$ 0.05 sec, which was shortened to 0.3 to 0.4 sec during exercise at the level of Ex-5. The correlation coefficient between the logarithms of tc and

$(a-v)Co_2^*$ was 0.911. The Vc calculated from the $\dot{Q}$ and tc ranged from 55 to 65 ml in a standing position at rest, and increased to 100 to 120 ml during level Ex-5 exercise. Figure 6 shows the relationship between the Vc and $\dot{Q}$. The regression line is given by:

$$Vc = 31.2 \cdot \dot{Q}^{0.432}, \text{ (ml)} \qquad\qquad (20)$$

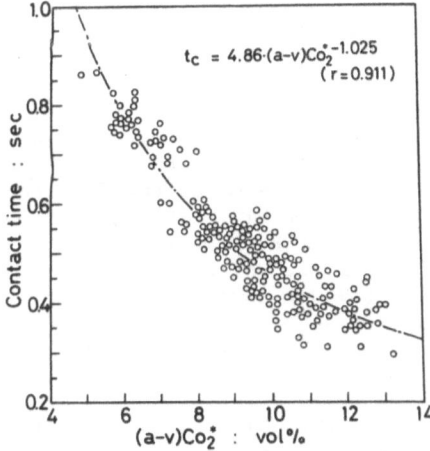

Figure 5. The relation between contact time (tc) and $(a-v)Co_2^*$ in a standing position at rest and during treadmill exercise.

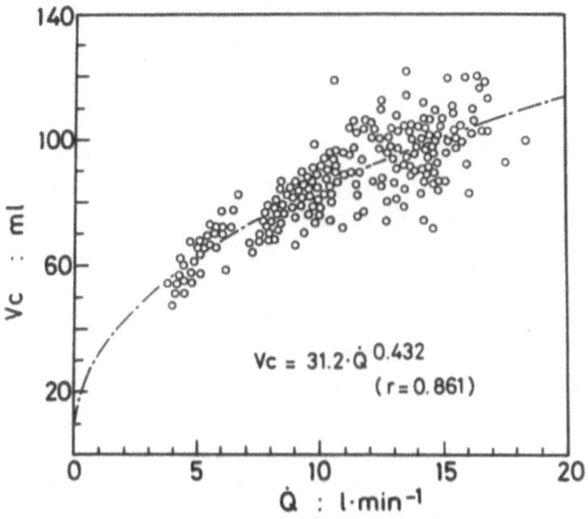

Figure 6. The relation between Vc and $\dot{Q}$ in a standing position at rest and during treadmill exercise.

Table 1. Summarized data on relationships between $\dot{Q}$ and $\dot{V}o_2$ at rest and during exercise.

Author	Exercise Method	n	Exponential fitting $\dot{Q} = A \cdot \dot{V}o_2{}^B$			Linear fitting $\dot{Q} = C+D \cdot \dot{V}o_2$		
			A	B	r	C	D	r
This study	TM RB	226	10.52	0.650	0.973	3.20	6.98	0.946
Reeves et al.(1961)	TM FK	47	9.78	0.668	0.978	3.21	6.21	0.976
Damato et al.(1966)	TM FK	84	10.05	0.646	0.976	3.43	6.15	0.965
Hermansen et al.(1970)	TM DL	39	9.25	0.700	0.975	4.05	5.12	0.972
Astrand et al.(1964)	BE DL	126	10.37	0.591	0.972	4.81	5.05	0.964

n: Data number; TM: Treadmill walking; BE: Bicycle ergometer; FK: Direct Fick; RB: Rebreathing; DL: Dye dilution; r: correlation coefficient.

DISCUSSION

Table 1 summarizes the relationships between $\dot{Q}$ and $\dot{V}O_2$ obtained in this study and by other authors using the direct Fick or dye dilution method. Good agreement is seen between their and our data, suggesting reliability of our method. The $\dot{Q}$ at rest and during exercise has until now been expressed by a linear equation against the $\dot{V}O_2$ (Bevegard et al., 1963; Astrand et al., 1964; Hermansen et al., 1970). However, the correlation coefficient in the exponential fitting was higher than that in the linear fitting in all the 5 cases shown in Table 1. In fact, the $\dot{Q}$ at rest deviated significantly from the linear relation as shown in Figure 4. Therefore, as far as the relation between $\dot{Q}$ and $\dot{V}O_2$ in a physiological range is concerned, the exponential expression seems to be advantageous.

The tc value at rest was in a range of 0.7 to 0.8 sec, which was shortened to 0.3 to 0.4 sec during exercise at the level of Ex-5. The values coincided well with those obtained from the measurement of the pulmonary diffusing capacity for CO (Johnson et al., 1960; Uchida et al., 1986), suggesting adequacy of our method for estimating tc. The Vc value at rest was in the range of 55 to 65 ml, which was similar to the value reported by Johnson et al., (1960). During exercise, it increased about 2-fold, indicating that recruitment and/or dilatation of the pulmonary vascular bed occurred.

In conclusion, the present study not only clarifies quantitative relations between gas exchange parameters but also shows that the parameters which have been difficult to estimate can be obtained by a simple, safe and non-invasive rebreathing method.

ACKNOWLEDGEMENTS

We are grateful to Drs Doi and Shimouchi and Messrs Niizeki, Iwabuchi and Sendo for helpful cooperation in this study.

REFERENCES

Astrand, P.-O., Cuddy, T.E., Saltin, B. and Stenberg, J. (1964). Cardiac output during submaximal and maximal work. J. Appl. Physiol. 19, 268-274.

Bevegard, S., Holmgren, A. and Jonsson, B. (1963). Circulatory studies in well trained athletes at rest and during heavy exercise, with special reference to stroke volume and the influence of body position. Acta Physiol. Scand. 57, 26-50.

Damato, A.N., Galante, J.G. and Smith, W.M. (1966). Hemodynamic response to treadmill exercise in normal subjects. J. Appl. Physiol. 21, 959-966.

Hermansen, L., Ekblom, B. and Saltin, B. (1970). Cardiac output during submaximal and maximal treadmill and bicycle exercise. J.Appl. Physiol. 29, 82-86.

Johnson, R.L. Jr., Spicer, W.S., Bishop, J.M. and Forster, R.E. (1960). Pulmonary capillary blood volume, flow and diffusing capacity during exercise. J. Appl. Physiol. 15, 893-902.

Kagawa, T. and Mochizuki, M. (1986). Theoretical analyses for arterial-venous O_2 content difference and Haldane effect during rebreathing. In: Overall Gas Exchange through Red Blood Cells. Ed. Mochizuki, M., pp. 407-426.

Kim, T.S., Rahn, H. and Farhi, L.E. (1966). Estimation of true venous and arterial Pco_2 by gas analysis of a single breath. J. Appl. Physiol. 21, 1338-1344.

Mochizuki, M. and Kagawa, T. (1986). Numerical solution of partial differential equations describing the simultaneous O_2 and CO_2 diffusions in the red blood cell. Jpn. J. Physiol. 36, 43-63.

Mochizuki, M., Shibuya, I., Uchida, K. and Kagawa, T. (1987). A method for estimating contact time of red blood cells through lung capillary from O_2 and CO_2 concentrations in rebreathing air in man. This volume.

Mochizuki, M., Tamura, M., Shimasaki, T., Niizeki, K. and Shimouchi, A. (1984). A new indirect method for measuring arteriovenous O_2 content difference and cardiac output from O_2 and CO_2 concentrations by rebreathing air. Jpn. J. Physiol. 34, 295-306.

Nitta, K. and Mochizuki, M. (1969). A continuous method for measuring O_2 and CO_2 in expired air. Jpn. J. Physiol. 19, 41-54.

Reeves, J.T., Grover, R.F., Blount, S.G. Jr. and Filley, G.F. (1961). Cardiac output response to standing and treadmill walking. J. Appl. Physiol. 16, 283-288.

Shibuya, I., Niizeki, K. and Uchida, K. (1986). Change in O_2 uptake in hyperoxia during rebreathing in man. Jpn. J. Physiol. 36, (in press).

Tazawa, H., Mochizuki, M., Tamura, M. and Kagawa, T. (1983). Quantitative analysis of the CO_2 dissociation curve of oxygenated blood and the Haldane effect in human blood. Jpn. J. Physiol. 33, 601-618.

Uchida, K., Shibuya, I. and Mochizuki, M. (1986). Simultaneous measurement of cardiac output and pulmonary diffusing capacity for CO by a rebreathing method. Jpn. J. Physiol. 36, 657-670.

A VERSATILE AND SENSITIVE METHOD FOR MEASURING OXYGEN

D.F. WILSON, J.M. VANDERKOOI, T.J. GREEN, G. MANIARA,
S.P. DEFEO and D. C. BLOOMGARDEN

Department of Biochemistry and Biophysics, Medical School,
University of Pennsylvania, Philadelphia, PA 19104, U.S.A.

SUMMARY

Oxygen dependence of the lifetime of the excited triplet state of phosphorescent molecules can be used to measure the oxygen concentration in aqueous media. These measurements are insensitive to much of the optical interference that limits the usefulness of measurements based on the oxygen dependent quenching of luminescence intensity. The measurements also extend to significantly lower oxygen concentrations than are normally attainable using oxygen electrodes. The phosphorescence lifetimes can be accurately measured from a few microseconds to seconds, permitting a wide dynamic range of oxygen concentration measurements. With currently available probes, for example, it is possible to make continuous measurement of oxygen concentrations from 10^{-4} M to 10^{-8} M in a single experiment.

INTRODUCTION

Our understanding of the oxygen dependence of tissue metabolism requires accurate measurements of oxygen concentrations both in vitro and in vivo. These measurements are complicated in that the concentrations of critical importance are rather low, from 0.1 uM to 20 uM. Oxygen electrodes, the most widely used and successful method to date (see Silver, 1984 for review) are of limited value in this range of oxygen concentrations due to instabilities in the electrode surfaces and in the oxygen diffusion barriers. An optical method based on the oxygen dependent quenching of the fluorescence of pyrene butyric acid has also been used to measure oxygen (Vaughan and Weber, 1970; Knopp and Longmuir, 1972; Opitz and Lubbers, 1984) but with limited success, and a method based on the luminescence intensity of Ruthenium(II) tris(bipyridyl) ion has very recently been proposed (Sasso, Quina and Bechara, 1986). Luminescence intensity measurements in general are affected by many experimental parameters other than oxygen concentration and the pyrene butyric acid fluorescence is relatively insensitive to low oxygen concentrations. Physiological oxygen indicators such as myoglobin have proven useful in some tissues (see for examples Tamura et al., 1978; Figulla, Hoffmann and Lubbers, 1984;

Wittenberg and Wittenberg, 1985) but are limited in the range of oxygen sensitivity and, in the case of myoglobin, are not present in all cells.

At the 1985 ISOTT meeting we introduced a new optical method for measuring oxygen utilizing oxygen-dependent quenching of phosphorescence (Vanderkooi and Wilson, 1986). In order to avoid most of the experimental limitations inherent in intensity measurements, phosphorescence lifetime was used as the indicator of oxygen concentration. In the present paper we report successful development of an instrument utilizing this method, some performance characteristics of the instrument and the first measurements in biological samples.

RATIONALE

The lifetime of the triplet state, and thereby the phosphorescence lifetime, is dependent on oxygen concentration and can be used as a measure of the latter. Quenching by oxygen is diffusion limited and thus can be described by the Stern-Volmer relationship as modified for phosphorescence or fluorescence lifetimes:

$$I_0/I = T_0/T = 1 + K_Q T_0 [O_2]$$

where I_0 and T_0 are respectively the phosphorescence intensity and lifetimes when the oxygen concentration $[O_2]$ is zero. I and T are the phosphorescence intensity and lifetime at any oxygen concentration greater than zero. K_Q is related to the physical constants of the system by the expression:

$$K_Q = 4r\pi RDN \times 10^3$$

where R is the radius of interaction between O_2 and phosphor, r is the efficiency of quenching (usually approaching 1), D is the sum of the diffusion constants of phosphor and oxygen and N is Avagadro's number. For most experimental conditions (solvent, ionic strength, ion composition, temperature and lumiphor binding) K_Q is constant. In order to use phosphorescence lifetime to measure oxygen concentration it is necessary to determine the values of T_0 and K_Q by measurement of T at two different oxygen concentrations (such as anoxia and air saturation) and at any other value of T the oxygen concentration can be calculated. The caveat should be added that the Stern-Volmer relationship does not hold for I_0/I if the quencher promotes intersystem crossing but the expression remains true for T_0/T.

APPARATUS DESIGN

The lifetime of a lumiphore is usually measured by either of two methods:
A. _Phase modulation_. In this method the intensity of the exciting light is periodically varied either by modulating the light itself or by modulating an optical attenuator in the path of the exciting light. When the modulation frequency is near that of the lifetime of the lumiphore the emitted light is modulated at the same frequency but is delayed in time (phase shifted) with respect to the exciting light. The frequency and phase shift define the lifetime of the lumiphore.
B. _Pulsed excitation_. In this method the lumiphore is excited by an intense light flash of duration much less than T and the decay of

luminescence following the flash is measured. The lifetime of the phosphor can be directly calculated since in the absence of exciting light the luminescence intensity is proportional to the number of molecules in the triplet state.

Using pulsed excitation we have obtained a versatile instrument which, in addition to measuring oxygen concentration, has wide applicability to the study of other biophysical and chemical problems utilizing phosphorescence.

A schematic diagram of the instrument is shown in Figure 1. Excitation is by a xenon flash lamp of approximately 0.3 J/flash and a half time for decay of the light output of less than 5 usec. The light flash is passed through a monochromator and impinges on the sample. The light emitted at 90° to the exciting light is passed through a filter and onto a photomultiplier mounted approximately 5 cm from the sample. The photomultiplier output is amplified and then digitized by an 8 bit, 20 MHz A/D board. The A/D board digitizes up to 4096 points, retaining the data in its resident memory, and then transfers this data to an AT&T 6300 microcomputer by direct memory access. The computer averages the input from a series of flashes (1-200) and calculates the lifetime of phosphorescence decay from the averaged data.

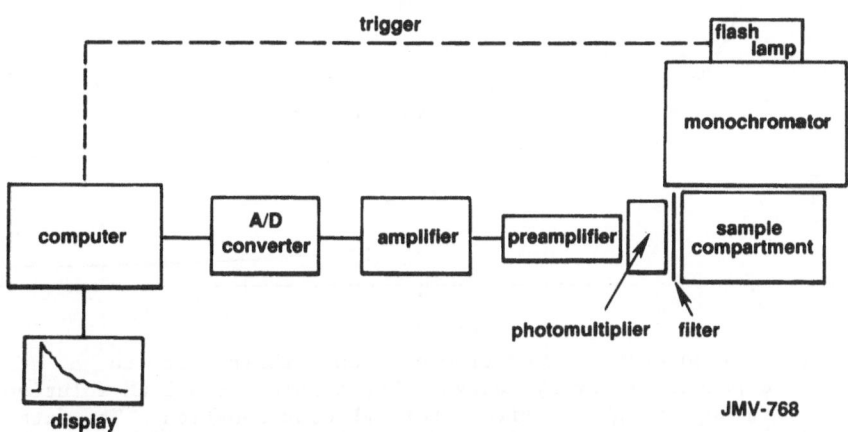

Figure 1. Schematic diagram of the design used for the phosphorescence lifetime/oxygen measuring instrument used in these studies.

A plot of data taken for the phosphorescence of Palladium-coproporphrin in the absence of oxygen is shown in Figure 2. The data are the average value for 20 flashes (16 msec between flashes). A solid line is given for the best fit by a single exponential decay. The correlation coefficient was greater than 0.999. Similar measurements with other probes have indicated that in general the decay curves follow a single exponential. This is useful because where rapid calculations are needed, the assumption of a single exponential decreases the time required for computation of the lifetime.

In our previous paper (Vanderkooi and Wilson, 1986) several classes of molecules were listed which have phosphorescence suitable for oxygen measurement. In this paper we will use Pd-coproporphrin to demonstrate operation of the instrument. When this compound is added to a 2% solution of bovine serum albumin (final concentration of 1 μM) the value of T_0 is 1.2 msec and K_Q is 1 x 10^8 1 mole^{-1}sec^{-1}. At 260 μM oxygen, T is 37 μsec and the oxygen concentration for 50% decrease in T_0 (to 600 μsec) is 8.3 μM. The instrument in its present design measures the lifetime with an accuracy of better than $\pm$ 1% ($\pm$ S.D.) at all oxygen concentrations when averaging 20 flashes. This means one measurement is obtained every second. Approximately 0.3 seconds is required for 20 flashes at 60 Hz and the remaining time is used for computation and graphing. The latter could be significantly shortened by using a faster computer.

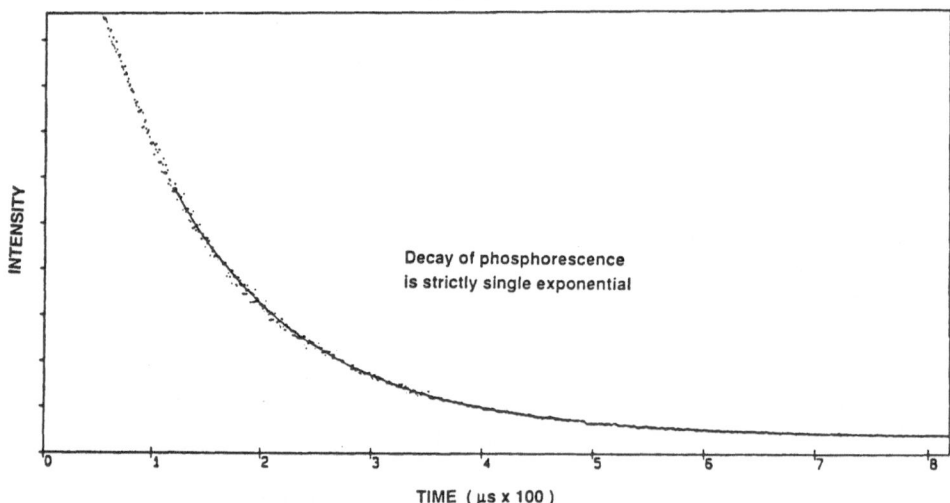

Pd Coproporphrin O. D. 0.981

INTENSITY

Decay of phosphorescence is strictly single exponential

TIME (μs x 100)

Figure 2.   Phosphorescence measurements and their fit to a single exponential decay curve. The sample was a 1 uM solution of Pd-coproporphrin after thorough deoxygenation. The data for 20 flashes were averaged and then a best fit of the data to a single exponential computed for the part from 0 to 80% of maximum. The best fit curve is drawn as a solid line while the data are given as discrete points. The fit to a single exponential was calculated as a least squares fit to the logarithmic form (straight line) and the fit has a correlation coefficient of greater than 0.999.

When 1 μM Pd-coproporphrin is added to a medium, such as a suspension of mitochondria, the phosphorescence lifetime can be used to monitor the oxygen concentration. The result of such an experiment is shown in Figure 3. Bovine serum albumin was added to bind the Pd-copro-porphrin and prevent any possible interactions with the mitochondrial membranes. Since the value of T in air saturated medium at 25° was approximately 35 μsec, it was possible to begin the oxygen measurements at air saturation. The mitochondrial suspension contained oxidizable substrates (8 mM glutamate and 8 mM malate) and inorganic phosphate (8 mM). Addition of 650 μM ADP stimulated respiration and when this was phosphorylated to ATP the respiratory rate returned to that prior to ADP addition. A second addition of ADP resulted in a stimulation of respiration similar to that of the first addition. These measurements demonstrate that the instrument can readily follow oxygen consuming reactions even when relatively rapid (seconds) changes in the reaction rate occur.

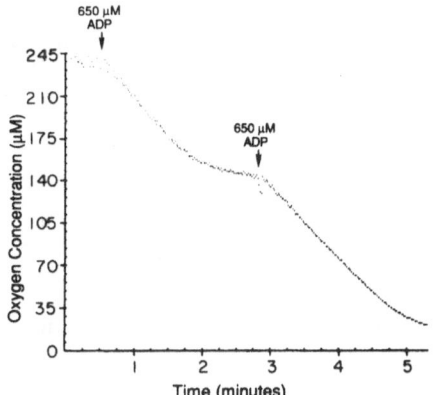

Figure 3. Respiratory control in mitochondria using phosphorescence lifetime to measure oxygen concentration. Rat liver mitochondria were suspended in a mannitol-sucrose medium with 2% bovine serum albumin, glutamate (8 mM), malate (8 mM) and phosphate (10 mM). 1 μM Pd-coproporphrin was added as an oxygen indicator and measurements of the phosphorescence lifetime made each second. The oxygen concentration was then calculated from the Stern-Volmer relationship. At the indicated points 650 μM ADP was added to initiate ATP synthesis.

The value of the method becomes more apparent when low oxygen concentrations are measured. In the above experiments, bovine serum albumin was used to bind the probe; in the absence of bovine serum albumin the probe is sensitive to lower concentrations of oxygen

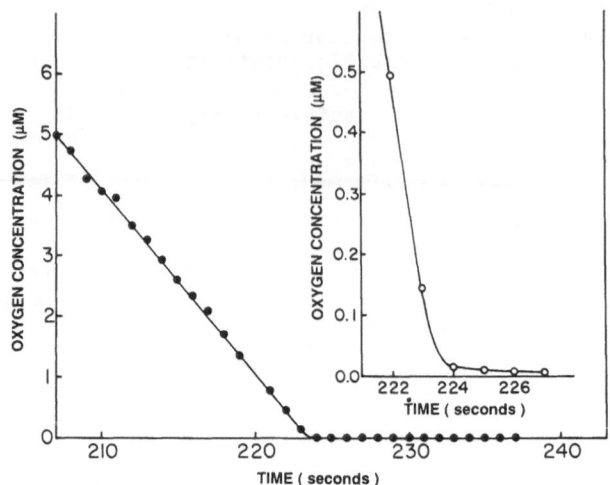

Figure 4. The respiration of mitochondria at low oxygen concentrations. Rat liver mitochondria were suspended as described for Figure 3 but in the absence of bovine serum albumin. ADP (3 mM) was added after the oxygen concentration was below 0.1 mM. The measurements below 0.5 μM O_2 are given in the insert with an expanded scale.

(K_Q is approximately 2×10^9 1 mole^{-1}sec^{-1}). When mitochondria are respiring in a 'low energy' state, such as with high ADP and inorganic phosphate and low ATP, the affinity for oxygen is very high. Figure 4 shows the oxygen uptake for such a suspension of mitochondria in the absence of bovine serum albumin, presenting data obtained for oxygen concentration less than 5 μM. Each measurement is plotted and it may be seen that the respiratory rate remained linear to less than 1 μM. The data for less than 0.5 μM oxygen are shown in the inset and indicate a noise level of less than $\pm$ 0.02 μM. It should be noted that each measurement represents the mean oxygen concentration during the 0.3 seconds used to take data from 20 light flashes. The probe molecules are in solution in the medium and respond to the oxygen in their environment within microseconds. The limiting response time is the time of data collection from a single flash, approximately one lifetime of the probe molecule (usually less than 2 msec).

ACKNOWLEDGEMENT

Supported by grant GM-39363 from the US National Institutes of Health.

REFERENCES

Figulla, H.R., Hoffmann, J. and Lubbers, D.W. (1984). Evaluation of reflection spectra of the isolated heart by multicomponent spectra analysis in comparison to other evaluating methods. In: Oxygen Transport to Tissue-V. Eds Lubbers, D.W., Acker, H., Leniger-Follert, E. and Goldstick, T.K., Plenum Press, New York and London, (Adv. Exp. Med. Biol. 169, 821-830).

Knopp, J.A. and Longmuir, I.S. (1972). Intracellular measurement of oxygen by quenching of fluorescence of pyrene butyric acid. Biochim. Biophys. Acta, 279, 393-397.

Opitz, N. and Lubbers, D.W. (1984). Increased resolution power in PO_2 analysis at lower PO_2 levels via sensitivity enhanced optical PO_2 sensors (PO_2 optodes) using fluorescence dyes. In: Oxygen Transport to Tissue-VI. Eds Bruley, D., Bicher, H.I. and Reneau, D., Plenum Press, New York and London, (Adv. Exp. Med. Biol. 180, 261-267).

Sasso, M.G., Quina, F.H. and Bechara, E.J.H. (1986). Ruthenium(II) tris(bipyridyl) ion as a luminescent probe for oxygen uptake. Analyt. Biochem. 156, 239-243.

Silver, I.A. (1984). Polarographic techniques of oxygen measurements. In: Oxygen: An In-depth Study of its Pathophysiology. Eds Gottlieb, S.F., Longmuir, I.S. and Totter, J.R. Undersea Med. Soc. pub No. 62(ws) 3-1-84, Bethesda, MD. pp. 215-238.

Tamura, M., Oshino, N., Chance, B. and Silver, I.A. (1978). Optical measurements of intracellular oxygen concentration of rat heart in vitro. Arch. Biochem. Biophys. 191, 8-22.

Vanderkooi, J. and Wilson, D.F. (1986). A new method for measuring oxygen concentration in biological systems. In: Oxygen Transport to Tissue-VIII. Ed. Longmuir, I.S., Plenum Press, New York and London, (Adv. Exp. Med. Biol. 200, 189-193).

Vaughan, W.M. and Weber, G. (1970). Oxygen quenching of pyrenebutyric acid fluorescence in water. A dynamic probe of the microenvironment. Biochemistry, 9, 464-473.

Wittenberg, B.A. and Wittenberg, J.B. (1985). Oxygen pressure gradients in isolated cardiac myocytes. J. Biol. Chem. 260, 6548-6554.

CRITICAL OXYGEN DELIVERY LEVELS DURING SHOCK FOLLOWING NORMOXIC AND

HYPEROXIC HAEMODILUTION WITH FLUOROCARBONS OR DEXTRAN

N.S. FAITHFULL[*] and S.M. CAIN

Department of Physiology and Biophysics, University of Alabama at Birmingham, Birmingham, Alabama 35294, U.S.A. [*]Work performed on leave from Department of Anaesthesia, Erasmus University, Rotterdam, The Netherlands. Present address: Department of Anaesthesia, University of Manchester, Hope Hospital, Manchester M6 8HD, U.K.

SUMMARY

Fluosol-DA 20% (FDA), an emulsion of fluorocarbons in a glucose electrolyte solution can deliver physiologically significant amounts of oxygen (O_2) to the tissues and improve ischaemic hypoxia. To investigate its effect on critical oxygen delivery level (QO_2c), four groups of six phenobarbitone anaesthetised air-ventilated splenic clamped mongrel dogs were haemodiluted to a haematocrit (Hct) of 25%; two groups with FDA and two with 6% dextran solution. Oxygen consumption (VO_2) was derived from expired gas measurement and analysis, or by using a spirometer and carbon dioxide absorption. Whole body O_2 flux (QO_2) was calculated from mixed venous and arterial O_2 contents and the Fick-derived cardiac output. QO_2 was progressively decreased by haemorrhaging in steps of 1.5 - 2.5 ml per kg. Hct was kept at 25% using packed cells. VO_2/QO_2 pairs were calculated at each step and QO_2c was determined for each animal by least squares fitting of data with 2 straight lines. Analyses of variance were performed.

QO_2c was significantly less in the FDA and O_2 (F+O) group than either the dextran and O_2 (D+O) or dextran and air (D+A) groups. Analysis of O_2 extraction at QO_2c, which effectively normalized results for differences in resting VO_2, had significantly better extraction in the FDA and air (F+A) than the D+A group. When fluorocarbon- and plasma-dissolved oxygen was subtracted, the O_2 extraction in the F+A group was significantly better than in the D+A and F+O groups. The results imply that normoxic FDA haemodilution in animals respiring air can improve O_2 delivery and that hyperoxia interferes with this process.

INTRODUCTION

When whole body oxygen delivery (QO_2 = cardiac output x arterial O_2 concentration) is reduced, whole body oxygen consumption (VO_2) usually remains constant until the critical oxygen delivery level (QO_2c) is reached. The value obtained for QO_2c may be similar when different methods are employed to reach the point though mixed venous oxygen tension (PvO_2) may be very different (Cain, 1977). QO_2c can be approached by progressive haemodilution, hypoxic hypoxia or progressive haemorrhage (Cain, 1977; Schwarz, Frantz and Shoemaker, 1981; Becker et al., 1985).

Perfluorochemicals (PFCs) have a high solubility for respiratory gases and are available in emulsified form in the plasma substitute Fluosol-DA 20% (FDA), which is currently undergoing clinical trials in various countries. Because oxygen solubility of FDA is only 0.75 ml per 100 ml per 100 mmHg (Zander and Makowski, 1982; Grote et al., 1985) high inspired concentrations of oxygen (FIO_2s) are usually administered concurrently.

PFC emulsions have low viscosity, particularly at low shear rates. This, combined with the very small particle size in the emulsion, may explain their ability to improve tissue oxygenation in the presence of ischaemic hypoxia during myocardial infarction (Rude et al., 1982; Biro, 1983; Faithfull, Erdmann and Fennema, 1986), cerebral infarction (Sutherland, Farrar and Peerless, 1984) and during perfusion of extremities following prolonged traumatic amputation (Smith et al., 1985). In addition, one report (Erdmann, 1982) indicates that FDA may improve oxygen diffusion coefficients in the tissue and this may also contribute to better oxygenation. If these effects are significant then haemodilution with FDA should lower QO_2c by improving oxygenation during progressive haemorrhagic shock as diffusion pathways are increased in the presence of sludging and decreased capillary perfusion. In these experiments, we compare QO_2c in FDA haemodiluted animals under normoxia and hyperoxia with similar groups haemodiluted with dextran solutions.

MATERIALS AND METHODS

Four groups of six mongrel dogs (12.3 to 23.1 kg body weight) were anaesthetized with 30 mg kg^{-1} sodium pentobarbitone. Subsequent administration of about 30 mg h^{-1} was given as needed. The animals were intubated and ventilated with a Harvard apparatus respiratory pump to maintain end-expiratory carbon dioxide concentration at betweeen 4.5 and 5%. After surgery was complete, the animals were paralysed with suxamethonium chloride (30 mg I.V.) and maintained with 0.1 mg min^{-1} I.V.

When the dogs were ventilated with room air, mixed expired gas was passed through a Harvard dry test gas meter and samples were analysed for oxygen and carbon dioxide content using an Applied Electrochemistry Inc. and a Beckman LB-2 medical gas analyser. The outputs of the meter and the gas analysers were interfaced with a Cyborg Corp. ISAAC analog/digital converter and processed by an Apple IIe microprocessor. Raw gas exchange data were displayed on a minute to minute basis and later corrected to standard temperature and pressure. When the animals were ventilated with O_2, oxygen uptake was measured in a closed circuit respirometer that could be accessed by stopcocks.

The right femoral artery was cannulated and a catheter was placed in the pulmonary artery via the right external jugular vein. The pressures were measured and displayed on a paper chart recorder. The heart rate was measured with a cardiotachometer. The spleen was gently exposed via a midline laparotomy. After minimum handling to prevent capsule contraction and erythrocyte release into the circulation, the hilum was ligated with multiple ligatures and the incision closed with haemostatic clips.

In view of the frequent occurrence of hypotensive reactions to an initial dose of a fluorocarbon emulsion, all dogs were given a 2 ml intravenous dose of either FC-43 (an emulsion containing perfluorotributylamine) or Fluosol-DA 20% (containing a mixture of perfluorodecalin and perfluorotripropylamine). If a reaction occurred the protocol was not begun till full recovery had taken place.

The animals were heparinized (1000 U kg^{-1}) and the experimental protocol began with a 20 minute stabilization period. Samples of femoral and pulmonary arterial blood were taken at 15 minutes. In all samples blood PO_2, PCO_2 and pH were measured at 37°C on a Radiometer BMS3 Mark 2 Blood Micro System or a Radiometer ABL 30 Acid/Base Laboratory. Haemoglobin concentration and oxygen content were measured using an Instrumentation Laboratory Co-oximeter 282, the output of which was corrected by reference to a calibrated Lexington Instruments Corp. Lex O_2 Con-TL.

Haemodilution was performed with either 6% dextran (m.w. 70,000) in Tyrode solution or Fluosol-DA 20% by arterial bleeding and simultaneous intravenous administration of diluent. The volume required to reach the target haematocrit (hct) of 25% was calculated based on multiple previous haemodilution experiments performed in this laboratory.

After stabilization, blood samples were taken at 45 minutes and haemorrhage was started at 50 minutes. The initial bleed was 5 ml per kg body weight and a calculated volume of red cells was administered to maintain the haematocrit near 25%. The above steps were repeated every 20 minutes. The approach to QO_2c could be detected by the minute to minute oxygen consumption, obtained from the Apple IIe, and bled volumes were reduced accordingly. The experiments were continued till the demise of the animals.

The critical level of oxygen delivery (QO_2c) for each animal was obtained by fitting two regression lines to the VO_2/QO_2 data with a least squares method that minimized variance from the lines (Mellits, 1968). The intersection point of the plateau and slope line was taken as QO_2c. Statistical significance was assessed using analysis of variance and was accepted at the 5% level.

RESULTS

The animals in the different groups were comparable with respect to body weights and starting hcts. The post-haemodilution hcts were not significantly different, the values being 23.0 $\pm$ 2.8 (S.E.M.) for the dextran and air (D+A) group, 23.1 $\pm$ 1.2 for the dextran and oxygen (D+O) group, 26.1 $\pm$ 1.5 for the FDA and air (F+A) group and 24.6 $\pm$ 1.8 for the animals treated with FDA and oxygen (F+O) group. The addition of packed red cells resulted in the values being well maintained and the values at the end of the experiments were 23.0 $\pm$ 2.1, 26.5 $\pm$ 4, 25.1 $\pm$ 1.1 and 23.8 $\pm$ 0.8 respectively.

Blood gas and acid-base status of the animals in the different groups at the start of haemorrhage was comparable ; arterial oxygen tensions in the O_2 ventilated animals were of course elevated with respect to the air-ventilated groups. There were no significant differences in cardiovascular parameters at this time.

Calculated QO_2c values for each group are presented in Table 1. The lowest values were seen in the F+O group and these were significantly lower than the D+A or D+O groups. The corresponding values for the F+A group were significantly lower than in the D+O group. However, there were significant differences in the baseline values for VO_2 (Table 1); these were estimated as the mean of all pre- and post-dilutional VO_2 values. Haemodilution had no significant effect on VO_2 in any group.

Table 1.   Values   (mean $\pm$ S.E.M.)   for   critical   oxygen   delivery (QO_2c),   baseline   oxygen   consumption   (VO_2),   oxygen extraction   coefficient   at QO_2c (OEC) and oxygen extraction from erythrocytic haemoglobin at QO_2c (HOEC)

	D+A	D+O	F+A	F+O
QO_2c	8.1 ± 1.42	9.3 ± 0.85	7.9 ± 0.47	6.8 ± 0.67
VO_2	5.7 ± 0.36	7.4 ± 0.26	6.6 ± 0.22	6.0 ± 0.19
OEC at QO_2c	0.62 ± 0.05	0.75 ± 0.03	0.79 ± 0.02	0.80 ± 0.04
HOEC at QO_2c	0.61 ± 0.04	0.69 ± 0.05	0.75 ± 0.05	0.68 ± 0.05

D+A = dextran and air; F+A = Fluosol and air; D+O = dextran and oxygen; F+O = Fluosol and oxygen.

QO_2c is partly dependent on the level of resting 'plateau' VO_2 and hence Table 1 also gives values for oxygen extraction coefficients (OECs) at QO_2c - these are independent of VO_2. OECs were significantly higher in the F+O group than in the D+O and D+A group. Interestingly the values were significantly higher in the F+A group than in the D+A group, in spite of the fact that under ambient air conditions FDA carries negligible amounts of oxygen.

The arterial oxygen tension in the F+O group during the experimental protocol was 519.9 mmHg ($\pm$ 11.5 S.E.M.) (68.3 kPa). Under these conditions PFC present in the blood would have transported significant quantities of O_2 and, because of the straight line relationship between PaO_2 and content, the oxygen extraction of oxygen in physical solution would have been over 90%. When dissolved oxygen is subtracted from the results, a figure for oxygen extraction from erythrocyte haemoglobin (HOEC) is obtained. HOEC at QO_2c was significantly greater in the F+A group than in the other groups. The better HOEC in the F+A group was confirmed by examination of the values for mixed venous oxygen tension (PvO_2) at QO_2c. These values are shown in Figure 1.

DISCUSSION

The data presented above suggest that haemodilution with Fluosol-DA 20%, with or without supplemental oxygen, can result in greater oxygen extraction when QO_2 is limited and can hence reduce QO_2c. The explanation for this is probably multifactorial.

Haemodilution with FDA produces marked reduction in in vitro viscosity, particularly at low sheer rates (Naito and Yokoyama, 1978) and, at microcirculatory sheer rates, this was less than when 6% dextran (70,000 MW) was used as diluent (Biro, 1983). Though haemodilution decreased blood viscosity and total peripheral resistance (PR), Fan et al. (1980) demonstrated increases in vascular hindrance, particularly following haemodilution with dextran (Biro, 1982). In this study significant falls in PR were seen in the F+A group following haemodilution but not in the D+A group.

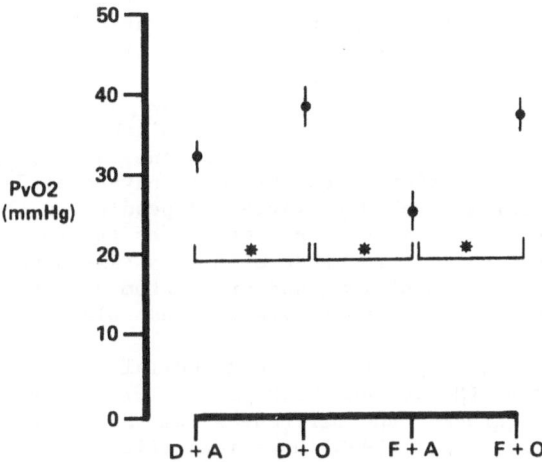

Figure 1. Mixed venous oxygen tensions (PvO_2) at critical oxygen delivery levels during haemorrhagic shock in dogs following haemodilution with 6% dextran or Fluosol-DA 20%.
D+A = dextran and air; F+A = Fluosol and air; D+A = dextran and oxygen; F+O = Fluosol and oxygen. Statistical significance indicated by asterisks (* = $p<0.05$).

Analysis of variance of post-dilutional PRs revealed significantly lower values in the F+A group than in the D+A or D+O groups; the F+O group had significantly lower resistance than the D+O group. Hence dilution with FDA reduced PR more than dextran. Hyperoxia has been shown to decrease capillary perfusion by causing arteriolar vasoconstriction and shunting of blood from nutritive channels (Duling, 1972; Lindbom and Afors, 1984). This may have caused the decrease in VO_2 observed by Chapler, Cain and Stainsby (1984) when anaemic animals are ventilated with 100% O_2. The latter effect was not observed in our experiments. However these hyperoxic microcirculatory effects may have caused lower extraction from haemoglobin-bound oxygen

in the F+0 than in the F+A group.   When total O_2 extraction was
examined in these two groups the effect of hyperoxia was not evident,
due to the substantial amounts of O_2 extracted from PFCs. Blood
diluted with Fluosol-DA 20% had a greater oxygen carrying capacity than
blood at the same haematocrit containing dextran. This extra oxygen
would contribute to oxygen consumption and result in a sparing effect
on haemoglobin-bound oxygen, particularly at high arterial oxygen
tensions.   Indeed, at maximum limitation of QO_2, oxygen extracted
from PFCs accounted for 31.3% of VO_2 in the F+0 group as against 1.7%
in the F+A group. The oxygen solubility of 6% dextran is 0.26 ml per
100 ml per 100 mmHg (Zander, 1978) which differs little from that of
plasma.   Hence in the D+A groups the oxygen delivery characteristics
were almost identical to those of plasma-diluted blood. The occurrence
of higher OECs in the F+0 group than the D+A or D+0 groups is thereby
understandable; but the higher OEC in the F+A group than in the D+A
group needs explanation.

Haemoglobin-bound oxygen reaches the tissues by diffusion through
the red cell, across the red cell membrane, through plasma and through
the capillary endothelial membrane.   Traditional physiology (Krogh,
1919) considers the capillary as a point source of oxygen and regards
distance from the capillary as the primary variable determining tissue
oxygen tesion. This implies a small transcapillary gradient. However,
Hellums (1977) has shown that resistance to mass transfer of oxygen is
about the same in both blood and tissue and recent mathematical
analysis (Clark et al., 1985) has established the presence of a
boundary layer near the red cell membrane acting as a significant
barrier to oxygen transfer. Resistance to oxygen diffusion across the
capillary endothelium probably varies depending on the capillary bed
concerned.   For instance, the endothelium of the rete mirabilis has a
diffusion resistance 500 times that of plasma (Rasio and Goresky, 1979)
whereas the endothelium of lung and of omentum does not appear to have
diffusion resistance substantially greater than plasma (Sinha, 1969).

Fluorocarbons easily traverse endothelial layers and are excreted
through the lungs (Naito and Yokoyama, 1978).   Passage of pure or
emulsified PFCs through the peritoneal membrane occurs (Clark et al.,
1979) and Kolodgie et al. (1985) have identified fluorocarbon particles
of various sizes within vascular endothelial cells, interstitium and
necrotic myocytes in the necrotic zone of myocardial infarction
following treatment with FDA.

If indeed FDA improves oxygen extraction by improving diffusion to
the tissues, higher tissue oxygen tensions should occur after
administration of fluorocarbons. Following administration of 100%
oxygen to dogs that had received either FDA or dextran, dramatic 285%
increases in myocardial oxygen tensions were measured in the FDA group
(Rude, Bush and Tilton, 1984).   In the dextran group the increases
amounted to a mean of 136%. Conversely, higher oxygen extractions
would lower the venous partial pressures of oxygen as we saw in our
experiments.

In conclusion, the present study has demonstrated that
haemodilution with Fluosol-DA 20%, with or without additional oxygen,
can result in significant increases in whole body oxygen extraction as
compared with that in animals similarly diluted with 6% dextran in
Tyrode solution.   This effect was not due to greater O_2 transport in
blood afforded by PFCs but appeared to be more in the nature of
facilitated diffusion. Hyperoxia itself detrimentally affected oxygen
extraction, possibly by its direct microcirculatory effects. When

ambient air is breathed the effects may be caused by a combination of the superior rheological properties of FDA and a postulated diffusion facilitation phenomenon, leading to decreases in critical oxygen delivery levels during haemorrhagic shock.

ACKNOWLEDGEMENTS

This work was supported in part by NIH grant HL 14693. The authors are grateful to Dr C.E. King, W.E. Bradley and D.W. Reynolds for their assistance in these experiments. Alpha Therapeutic Corporation very generously donated the Fluosol-DA used in these studies.

REFERENCES

Becker, L., Schumacker, P., Nelson,D.P., Saltz, S.A., Rowland, J., Long G.R. and Wood, L.D.H. (1985). Influence of FIO_2 on the relationship between oxygen delivery and uptake (VO_2) in the dog. Fed. Proc. 44, 908.

Biro, G.P. (1982). Comparison of acute cardiovascular effects and oxygen-supply following haemodilution with dextran, stroma-free haemoglobin solution and fluorocarbon suspension. Cardiovasc. Res. 16, 194-204.

Biro, G.P. (1983). Fluorocarbon and dextran hemodilution in myocardial ischemia. Canad. J. Surg. 26, 163-168.

Cain S.M. (1977). Oxygen delivery and uptake in dogs during anemic and hypoxic hypoxia. J. Appl. Physiol.: Respirat. Environ. Exercise. Physiol. 42, 228-234.

Chapler, C.K., Cain, S.M. and Stainsby, W.N. (1984). The effects of hyperoxia on oxygen uptake during acute anemia. Canad. J. Physiol. Pharmac. 62, 809-814.

Clark, A., Federspiel, W.J., Clark, P.A.A. and Cokelet, G.R. (1985). Oxygen delivery from red cells. Biophys. J. 47, 171-181.

Clark, L.C., Moore, R.E., Diver, S. and Miller, M. (1979). A new look at the vapour pressure problem in red cell substitutes. In: Proceedings of the IVth International Symposium and Perfluorochemical Blood Substitutes, Excerpta Medica, Amsterdam. pp. 55-67.

Duling, B.R. (1972). Microvascular responses to alterations in oxygen tension. Circ. Res. 31, 481-489.

Erdmann, W. (1982). O_2 diffusion coefficients. In: Oxygen Carrying Colloidal Blood Substitutes. Eds Frey, R., Beisbarth, H., Stossek, K., W. Zuckschwerdt Verlag, Munich, p. 143.

Faithfull, N.S., Erdmann, W. and Fennema, M. (1986). Oxygen tensions in the ischaemic myocardium following haemodilution with fluorocarbons or dextran. Brit. J. Anaesth. 58, 1013-1040.

Fan, F-C., Chen, R.Y.Z., Schuessler, G.B. and Chien, S. (1980). Effects of hematocrit variations on regional hemodynamics and oxygen transport in the dog. Am. J. Physiol. 238, H545-H552.

Grote, J., Steuer, K., Muller, R., Sontgerath, C. and Zimmer, K. (1985). O_2 and CO_2 solubility of fluorocarbon emulsion Fluosol-DA 20% and O_2 and CO_2 dissociation curves of blood-Fluosol-DA 20% mixtures. In: Oxygen Transport to Tissue-VII. Eds Kreuzer, F., Cain, S.M., Turek, Z. and Goldstick, T.K., Plenum Press, New York and London, (Adv. Exp. Med. Biol. 191, 453-461).

Hellums, J.D. (1977). The resistance to oxygen transport in the capillaries relative to that in the surrounding tissue. Microvasc. Res. 13, 1131-136.

Kolodgie, F.D., Dawson, A.K., Forman, M.B. and Virmani, R. (1985). Effect of perfluorochemical (Fluosol-DA) on infarct morphology in dogs. Virchows Arch. B, 50, 119-134.

Krogh, A. (1919). The number and distribution of capillaries in muscles with calculations of the oxygen partial pressure head necessary for supplying the tissue. J. Physiol. 52, 409-415.

Lindbom, L. and Arfors, K-E. (1984). Non-homogeneous blood flow distribution in the rabbit tenuissimus muscle. Acta Physiol. Scand. 122, 225-233.

Mellits, D.E. (1968). Statistical methods. In: Human Growth. Ed. Cheek, D.B., Lea and Febiger, Philadelphia, pp. 19-38.

Naito, R. and Yokoyama, K. (1978). Perfluorochemical Blood Substitutes. Technical Information Series No.5, Green Cross Corporation, Osaka.

Rasio, E.A. and Goresky, C.A. (1979). Capillary limitation of oxygen diffusion in the isolated rete mirabilis of the eel (Anguilla anguillaris). Circ. Res. 44. 495-503.

Rude, R. E., Bush, L.R. and Tilton, G.D. (1984). Effects of fluorocarbons with and without oxygen supplementation on cardiac hemodynamics and energetics. Am.J. Cardiol. 54, 880-883.

Rude, R.E., Glogar, D., Khuri, S.F., Kloner, R.A., Karaffa, S., Muller, J. E., Clark, L.C. and Braunwald, E. (1982). Effect of intravenous fluorocarbons during and without O_2 enhancement on acute myocardial ischemic injury assessed by measurement of intramyocardial gas tensions. Am. Heart. J. 103, 986-995.

Schwartz, S., Frantz, R.A. and Shoemaker, W.C. (1981). Sequential hemodynamic and oxygen transport responses in hypovolemia, anemia and hypoxia. Am. J. Physiol. 241, H864-H871.

Sinha, A.K. (1969). Oxygen Uptake and Release by Red Cells through Capillary Wall and Plasma Layer. Thesis, University of California, San Francisco, U.S.A.

Smith, A.R., van Alphen, W., Faithfull, N.S. and Fennema, M. (1985). Limb preservation in replantation surgery. J. Plast. Reconstr. Surg. 75, 227-237.

Sutherland, G.R., Farrar, J.K. and Peerless, S.J. (1984). The effect of Fluosol-DA on oxygen availability in focal ischemia. Stroke, 15, 829-835.

Zander, R. (1978). Oxygen transport by solutions for blood replacement in comparison with other infusion solutions. Klin. Wschr. 56, 567-573.

Zander, R. and Makowski, H.V. (1982). Life without haemoglobin? In: Oxygen Carrying Colloidal Blood Substitutes. Eds Frey, R., Beisbarth, H., Stossek, K., W. Zuckschwerdt Verlag, Munich, pp. 133-141.

PROPHYLAXIS AND TREATMENT OF MYOCARDIAL ISCHAEMIA BY HAEMODILUTION WITH FLUOROCARBON EMULSIONS

N.S. FAITHFULL[*], M. FENNEMA and W. ERDMANN

Department of Anaesthesia, Erasmus University Rotterdam, The Netherlands.[*]Present address: University Department of Anaesthesia, Hope Hospital, Manchester M6 8HD, U.K.

INTRODUCTION

Fluorocarbons (PFCs), largely inert chemicals with a high solubility for respiratory and other gases, are available in emulsified form in the plasma substitute Fluosol-DA 20% (FDA). This has an oxygen solubility of 0.75 ml per 100 mmHg (Zander and Makowski, 1982; Grote et al., 1985), about two and a half times that in plasma. To take advantage of this difference a high inspired concentration of oxygen is usually administered.

Clinical studies have confirmed the efficacy of FDA in raising whole body oxygen transport (Mitsuno, Ohyanagi and Naito, 1982; Tremper et al., 1982; Nishimura, Hiranuma and Sugi, 1983; Nishimura and Sugi, 1984; Ohyanagi et al., 1984) and haemodilution with FDA has been shown to exert a beneficial effect in various tissues subjected to ischaemic hypoxia. Treatment with FDA in a middle cerebral artery occlusion model markedly reduced cerebral damage (Peerless et al., 1985) and encouraging clinical results have been obtained with cerebral ischaemia of various aetiologies (Oda, Murata and Uchida, 1982). Decrease in the severity of myocardial infarction has been observed, particularly in relation to the volume of myocardium at risk (Glogar et al., 1981; Nunn et al., 1983; Kodolgie et al., 1985).

The present study presents work in which myocardial oxygen tensions were measured in a pig myocardial ischaemia model treated with FDA either before or after induction of ischaemia.

METHODS

Four groups of five juvenile Yorkshire pigs (body weight 23.5-27 kg) were anaesthetized with intraperitoneal thiopentone, 30 mg/kg. The trachea was intubated and the lungs ventilated with 66% nitrous oxide in oxygen and 1.0-1.5% halothane. Muscular relaxation was obtained by using a continuous infusion (0.5 mg/kg·h^{-1}) of pancuronium bromide. Systemic arterial pressure was monitored via the left femoral artery, and pulmonary artery pressure, pulmonary capillary wedge pressure and cardiac output were obtained using a 7-French gauge thermodilution

Swan-Ganz catheter inserted via the left femoral vein. Central venous pressure was monitored via a catheter in the right femoral artery. All the above pressures were transduced and displayed on a paper chart recorder. Arterial and mixed venous blood-gas and acid-base state, haemoglobin concentration and oxyhaemoglobin saturation were regularly estimated using a Radiometer ABL1 Acid Base Laboratory and a Radiometer OSMI cooximeter. In cases where FDA had been administered, readings from the cooximeter were inaccurate and the oxyhaemoglobin saturation was calculated from the computer subroutine described by Kelman (1966).

A thoracotomy was performed and the pericardium opened. To measure myocardial oxygen tension (PmO_2), four steel-protected gold microelectrodes 200 μm in diameter were inserted approximately 3mm into the myocardium of the lower anterior wall of the left ventricle. These were placed so that two were in and two were outside the infarct area.

One group of animals was bled 20 ml/kg body weight and the volume deficit was replaced by FDA. After stabilization of electrodes, the left anterior descending coronary artery (LAD) was occluded at the junction of its lower and middle thirds in all groups. One hour later, two of the groups of animals were bled 20 ml/kg body weight and the volume deficit was replaced in one group by FDA and in the other by 5% dextran solution (mol. wt. 40000). Haemodilution procedures were completed by 2 hours after LAD occlusion. The fourth (control) group received no treatment. The experiments were terminated 3 hours after haemodilution and the electrodes were recalibrated to assess drift. All animals were ventilated with 0.5% halothane in 100% oxygen for at least half an hour prior to occlusion.

Cardiovascular and oxygenation variables were calculated using a Phillips P2000 microcomputer and statistical analysis was performed with an Apple IIe microcomputer. Statistical significance was assessed using paired or unpaired Student's t tests and the null hypothesis was rejected at a p value of less than 0.05

RESULTS

Before LAD occlusion cardiac output was significantly higher and systemic vascular resistance was significantly lower in the group prediluted with FDA than in the control group. Otherwise there were no significant differences in systemic cardiovascular parameters between any of the groups. There were no significant differences in whole body oxygen consumption (VO_2) or flux (QO_2). Following LAD occlusion, no significant changes in cardiovascular or oxygenation parameters were seen in any group, even though macroscopically obvious areas of infarction were created.

Following postinfarction haemodilution with FDA the cardiac output was significantly higher than in the control group. In addition, at 3 hours (1 hour after completion of haemodilution), mean arterial pressure and systemic vascular resistance were significantly lower than in the control group, though not lower than before LAD occlusion.

After haemodilution with 5% dextran very significant decreases in systemic vascular resistance were seen and mean arterial pressure was lower than in the control group at 1 and 3 hours after haemodilution. In contrast to the Fluosol group the cardiac output showed no significant rise. As a result, very significant decreases in QO_2 occurred — no decreases occurred in the postinfarction FDA group which at all times after haemodilution had a significantly higher QO_2 than in the dextran group. Changes in haemodynamic and oxygen transport parameters are described in greater detail elsewhere (Faithfull, Fennema and Erdmann, 1986).

Analysis of PmO_2 in each animal was confined to the electrode indicating the greatest percentage decrease in PmO_2 over the first hour of occlusion. This was taken to be in the 'most hypoxic area' of the infarct. Figure 1 shows percentage changes in PmO_2 in the control (no treatment) and postinfarction FDA and dextran haemodilution groups. Significant decreases in PmO_2 occurred following LAD occlusion. In the control group these continued to decrease slowly till the end of the experiments. Slightly greater falls occurred in the dextran group. In the FDA group, however, the values at the end of the haemodilution procedures (2 hours post occlusion) were no longer significantly lower than preinfarction. PmO_2 continued to improve in this group and, at the end of the experiments, had recovered to 92.9% of preinfarction values. In contrast, values in the dextran and control groups were reduced 33.7% and 27.8% respectively.

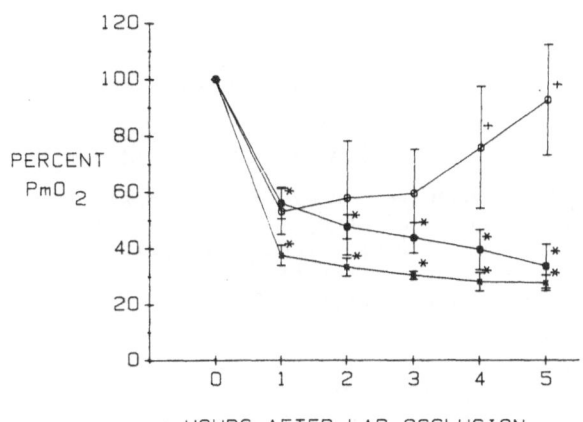

HOURS AFTER LAD OCCLUSION

Figure 1. Percentage changes in myocardial oxygen tension in the most ischaemic area of a fresh myocardial infarct in pigs. Haemodilution with dextran or Fluosol-DA 20% was performed between hours 1 and 2. ■, Dextran group; □, Fluosol group; ✗, Control group. Significant difference from preinfarct value (0 hours) indicated by asterisks (* $p < 0.05$) and difference from the control group by crosses (+ $p < 0.05$).

In Figure 2 percentage changes for the control group and for the preinfarction FDA haemodilution group are presented. It should be noted that though PmO_2 values were significantly reduced after 15 min of occlusion in the control group, they were not significantly reduced till 45 min after occlusion in the FDA group. At the end of the experiments PmO_2 values in the FDA group were significantly higher than in the control group.

DISCUSSION

PFCs carry more oxygen than other available haemodilutants and they have been shown to provide better oxygenation under conditions of haemodilution than hydroxyethyl starch (Okada et al., 1979; Watanabe et al., 1979), dextran solutions (Kohno et al., 1979; Biro, 1982) or stroma-free haemoglobin solutions (Biro, 1982). In addition, PFC emulsions have other important properties that help to oxygenate ischaemically hypoxic tissues.

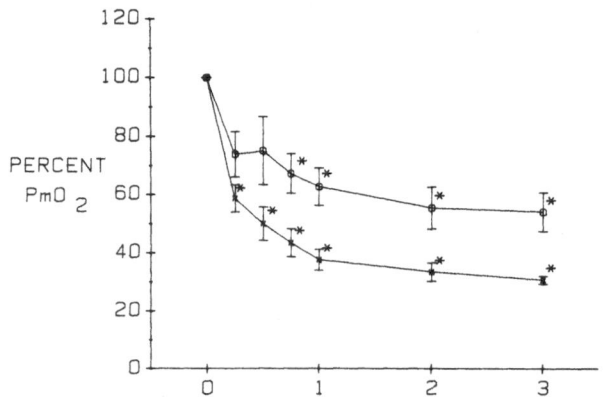

Figure 2. Percentage changes in myocardial oxygen tension in the most ischaemic area of a myocardial infarct in pigs. *X*, Control group; □, group receiving prior haemodilution with Fluosol-DA 20%. Significant decreases in PmO_2 are indicated by asterisks (* p<0.05).

The viscosity of FDA is lower than that of normal blood particularly at the low shear rates occurring in the microcirculation (Naito and Yokoyama, 1978). Moreover under these conditions mixtures of blood and FDA have viscosities (at similar haematocrits) that are much lower than blood/dextran mixtures (Biro, 1983). Haemodilution with FDA thus tends to produce greater microcirculatory flow for the same perfusion pressure than does haemodilution with dextran. Due to the higher oxygen content of blood/PFC mixtures, more oxygen is available for tissue oxygenation at similar haematocrits. Its lower viscosity, by speeding flow through the microcirculation, increases PO_2 at the venous end of capillary channels and consequently improves diffusion gradients in this area. This would decrease the occurence of critical hypoxia in the 'lethal corner' of the tissue cylinder. This effect would be greatly augmented by the rise in the arterial oxygen partial pressure (PaO_2) that occurs when PFCs are administered (Mitsuno et al., 1982; Tremper et al., 1982; Cefalo et al., 1984; Nishimura and Sugi, 1984).

The particle size of fluorocarbons in FDA is extremely small, the mean diameter is claimed to be approximately 0.1 μm (Naito and Yokoyama, 1978). However, following storage the size may increase somewhat (S.K. Sharma and K.C. Lowe, personal communication). During conditions of ischaemic hypoxia these small oxygen-transporting particles can probably bypass erythrocytes that may be impacted in the smallest capillary vessels due to their pH-dependent loss of flexibility (Schmid-Schoenbein, Weiss and Ludwig, 1973). Reoxygenation of capillary beds and dispersal of microaggregates is thus possible with subsequent improvement in microcirculation and tissue oxygenation.

The above mechanisms may be responsible for the improvement in circulation that occurs when PFCs are used in the treatment of myocardial infarction (Rude et al., 1982; Biro, 1983), cerebral infarction (Sutherland, Farrar and Peerless, 1984) and during perfusion of extremities following prolonged traumatic amputation (Smith et al., 1985). It has recently become apparent however that the fluorocarbons may be able to facilitate oxygen diffusion from the red cell into the tissues (Faithfull and Cain, 1987). Small pieces of suggestive evidence for this are scattered through the literature and recent studies have revealed that haemodilution with FDA may reduce the level for critical oxygen delivery in dogs even in the absence of oxygen supplementation (Faithfull, King and Cain, 1986).

In conclusion these experiments have revealed that haemodilution with Fluosol-FDA 20% may help to improve myocardial oxygenation following myocardial infarction. Prophylactic haemodilution can reduce the severity of myocardial ischaemia following infarction. These effects may be caused by a combination of the rheological properties of FDA and a possible oxygen diffusion facilitation effect.

REFERENCES

Biro, G.P. (1982). Comparison of acute cardiovascular effects and oxygen-supply following haemodilution with dextran, stroma-free haemoglobin solution and fluorocarbon suspension. Cardiovasc. Res. 16, 194-204.

Biro G.P. (1983). Fluorocarbon and dextran hemodilution in myocardial ischemia. Canad. J. Surg. 26, 163-168.

Cefalo, R.C., Seeds, J.W., Proctor, H.J. and Baker, V.V. (1984). Maternal and fetal effects of exchange transfusion with a red cell substitute. Am. J. Obstet. Gynecol. 148, 859-867.

Faithfull, N.S. and Cain, S.W. (1987). Critical oxygen delivery during shock following normoxic and hyperoxic haemodilution with fluorocarbons or dextran. This volume.

Faithfull, N.S., Fennema, M. and Erdmann, W. (1986). Haemodilution and myocardial ischaemia - studies with fluorocarbons and dextran in pigs. In: Anaesthesia - Innovation in Management III. Eds Droh, R., Erdmann, W. and Sprintger, R., Springer, Berlin.

Faithfull, N.S., King, C.E. and Cain, S.W. (1986). Critical oxygen delivery during haemorrhagic shock following haemodilution with dextran or fluorocarbons. Brit. J. Anaesth. 58, 817P-818P.

Glogar, D.H., Kloner, R.A., Muller, J., DeBoer, L.W.V. and Braunwald, E. (1981). Fluorocarbons reduce myocardial ischemic damage after coronary occlusion. Science, 211, 1439–1441.

Grote, J., Steuer, K., Muller,R., Sontgerath,C. and Zimmer, K. (1985). O_2 and CO_2 solubility of the fluorocarbon emulsion Fluosol-DA 20% and O_2 and CO_2 dissociation curves of blood-Fluosol-DA 20% mixtures. In: Oxygen Transport to Tissue-VII. Eds Kreuzer, F., Cain,S.M., Turek, Z. and Goldstick, T.K., Plenum Press, New York and London, (Adv. Exp. Med. Biol., 191, 453–461).

Kelman, G.R. (1966). Digital computer subroutine for conversion of oxygen tension into saturation. J.Appl. Physiol., 21, 1375–1376.

Kodolgie, F.D., Dawson, A.K., Forman, M.B., and Virmani, R. (1985). Effect of perfluorochemical (Fluosol-DA) on infarct morphology in dogs. Virchow Arch. B., 50,119–134.

Kohno, S., Baba, T., Miamoto, A. and Niiya, K. (1979). Effect of exchange transfusion with Fluosol-DA 20%, on delivery and consumption of oxygen in rabbits under normal air breathing. In: Proc. IVth Int. Symp. on Perfluorocarbon Blood Substitutes. Ed. Naito, R., Exerpta Medica, Amsterdam, pp. 361–371.

Mitsuno, T., Ohyanagi, H. and Naito, R. (1982). Clinical studies of a perfluorochemical whole blood substitute (Fluosol-DA). Ann. Surg. 195, 60–69.

Naito, R. and Yokoyama, K. (1978). Fluosol-DA. In: Perfluorochemical Blood Susbstitutes. Technical Information Series No.5. The Green Cross Corporation, Osaka.

Nishimura, N., Hiranuma, N. and Sugi, T. (1983). Cardiovascular effects of various colloidal solutions during major surgery. Crit. Care Med. 11, 940–942.

Nishimura, N. and Sugi, T. (1984). Changes of haemodynamics and O_2 transport associated with perfluorochemical blood substitute Fluosol-DA. Crit. Care Med. 12, 36–38.

Nunn, G.R., Dance, G., Peters, J. and Cohn, L.H. (1983). Effect of exchange transfusion on myocardial infarction size in dogs. Am. J.Cardiol. 52, 203–205.

Oda, Y., Murata, T., and Uehida, Y. (1982). Clinical evaluation of artificial blood substitute (Fluosol-DA 20%) in patients with cerebral ischaemia. Neurol. Surg. 10, 637–644.

Ohyanagi, H., Nakaya, S., Okumura,S. and Saitoh, Y. (1984). Surgical use of Fluosol-DA in Jehovah's witness patients. Artificial Organs, 8, 10–18.

Okada, K., Takagi, Y., Kosugi, I. and Kitagaki, T. (1979). Effect of Fluosol-DA tissue oxygenation during normovolemic hemodilution. In: Proc. IVth Int. Symp. on Perfluorochemical Blood Substitutes. Ed. Naito, R., Exerpta Medica, Amsterdam, pp. 391–399.

Peerless, S.J., Nakamura, R., Rodriguez-Salazar, A. and Hunter, I.G. (1985). Modification of cerebral ischemia with Fluosol. Stroke, 16, 38–43.

Rude, R.E., Glogar, D., Khuri, S.F., Kloner, R.A, Karaffa, S., Muller, J.E., Clark, L.C. and Braunwald, E. (1982). Effect of intravenous fluorocarbons during and without O_2 enhancement on acute myocardial ischemic injury assessed by measurement of intramyocardial gas tensions. Am. Heart J. <u>103</u>, 986-995.

Schmid-Schoenbein, H., Weiss, J. and Ludwig, H. (1973). A simple method for measuring red cell deformability in models of the microcirculation. Blut, <u>16</u>, 369-379.

Smith, A.R., van Alphen, W., Faithfull, N.S. and Fennema, M. (1985). Limb preservation in replantation surgery. J. Plast. Reconstr. Surg. <u>75</u>, 227-237.

Sutherland, G.R., Farrar, J.K. and Peerless, S.J. (1984). The effect of Fluosol-DA on oxygen availability in focal cerebral ischemia. Stroke, <u>15</u>, 829-835.

Tremper, K.K., Friedman, A.E., Levine, E.M., Lapin, R. and Camarillo, D. (1982). The preoperative treatment of severely anemic patients with a perfluorochemical oxygen-transport fluid, Fluosol-DA. N. Eng. J. Med. <u>307</u>, 277-283.

Watanabe, M., Hanada, S., Yano, K., Yokoyama, K., Suyama, T. and Naito, R. (1979). Long-term survival of rats severely exchange-transfused with Fluosol-DA. In : Proc. IVth Int. Symp. on Perfluorochemical Blood Substitutes. Ed. Naito, R., Exerpta Medica, Amsterdam, pp. 347-357.

Zander, R. and Makowski, H.V. (1982). Life without hemoglobin? In : Oxygen Carrying Blood Substitutes. Eds Frey, R., Beisbarth, H. and Stossek, K.W., Zuckschwerdt Verlag, Munich, pp. 133-141.

EMULSIFIED PERFLUOROCHEMICALS AS PHYSIOLOGICAL OXYGEN-TRANSPORT FLUIDS:

ASSESSMENT OF A NOVEL FORMULATION

S.K. SHARMA, A.D. BOLLANDS*, S.S. DAVIS and
K.C. LOWE*

Departments of Pharmacy and *Zoology, University of
Nottingham, University Park, Nottingham NG7 2RD, U.K.

INTRODUCTION

Emulsions of perfluorochemicals (PFC) in isotonic electrolyte solutions are known to have properties which make them attractive as physiological oxygen-transport fluids. Such properties include: the ability to dissolve substantial volumes of oxygen and other respiratory gases; and their small particle sizes (< 0.25 μm) which enable them to pass through capillary beds. In addition, due to the strength of the carbon-fluorine bond (ca. 116 kcal/mol), the PFC themselves are generally regarded as being both chemically and biologically inert (Riess and Le Blanc, 1982). Some of the physiological effects of emulsified PFC have been studied in several species and also in experiments using mammalian cells in culture (Lowe and Bollands, 1985; Lowe, 1986).

A commercial emulsion, Fluosol-DA 20% (Green Cross Corp. Japan), which contains perfluorodecalin (FDC) and perfluorotripropylamine (FTPA) emulsified with Pluronic F-68 has already been tested in human trials in Japan, Canada and the U.S.A. (Mitsuno, Ohyanagi and Naito, 1982; Tremper et al., 1982; Waxman et al., 1984; Stefaniszyn, Wynards and Salerno, 1985). However, one inherent problem with Fluosol-DA is the tendency for destabilization on storage, resulting in an increase in particle size (Riess and Le Blanc, 1982). This effect is minimized by storage of the stem emulsion component of Fluosol-DA in a frozen state, as recommended by the manufacturers (Naito and Yokoyama, 1978).

The major cause of droplet growth in emulsions is coalescence, but this can normally be retarded using emulsifying agents which form electrostatic and mechanical barriers at the oil-water interface. However, a more subtle means of instability can occur by a process of molecular diffusion known as Ostwald Ripening and this can occur, even if particles have excellent barriers to coalescence (Davis, Round and Purewal, 1981).

Previous work has demonstrated that while the characteristics of FDC make it suitable as the basis of a PFC emulsion for in vivo use, it is generally unstable in an emulsified form (Naito and Yokoyama, 1978;

Le Blanc, 1982). In this paper, we report the development and preliminary physiological assessment of novel compositions of FDC emulsions stabilized against Ostwald Ripening. A preliminary report of some physico-chemical properties of these formulations has already been published (Sharma, Davis and Lowe, 1986).

MATERIALS AND METHODS

Composition and preparation of emulsions

FDC (Flutec PP5) was supplied by I.S.C.Chemicals Ltd (Avonmouth, Bristol). It was emulsified by sonication (Dawe Automatic 7532A Soniprobe; Dawe Instruments Ltd, London) for 30 min with 4% Pluronic F-68 (I.C.I.Ltd, Macclesfield) in an aqueous phase, to give a final 20% (w/v) preparation containing 1% of one of the following polycyclic, perfluorated, higher boiling point oil (HBPO) additives to enhance stability: perfluoroperhydroacenaphthylene (C-12), perfluoroperhydrofluorene (C-13), perfluoroperhydrophenanthrene (C-14) or perfluoroperhydrofluoranthene (C-16) (I.S.C. Chemicals Ltd.). Some of the physico-chemical characteristics of FDC and the HBPO additives are given in Table 1. Control FDC emulsions contained no oil additives but were otherwise identical to test preparations.

Table 1. Physical properties of FDC and perfluorinated HBPO additives

COMPOUND	MOLECULAR FORMULA	MOLECULAR STRUCTURE	BOILING POINT (°C)	V.P.-25°C (m bar)
Perfluoro-decalin	$C_{10}F_{18}$		142	9.0
Perfluoroper-hydroacenaph-thylene	$C_{12}F_{20}$		173-175	1.8
Perfluoroper-hydrofluorene	$C_{13}F_{22}$		192-193.5	0.5
Perfluoroper-hydrophen-anthrene	$C_{14}F_{24}$		215-216	0.15
Perfluoroper-hydrofluor-anthene	$C_{16}F_{26}$		242-245	0.03

Assessment of emulsion stability

Emulsions were sized on preparation and periodically thereafter. Changes in mean particle diameter and size distribution with storage time was assessed periodically by Photon Correlation Spectroscopy (PCS; Malvern Instruments, Malvern). Emulsions were stored either at 4 or 37°C.

A stability parameter to indicate the ageing process of the FDC emulsions was defined as:

$$\text{Stability parameter} = Dt \: / \: Do \qquad\qquad (1)$$

where Do and Dt correspond to the mean particle size on emulsion preparation and at time 't' days later respectively.

Physiological studies

Female Wistar rats (body weight: 140–160 g, n = 37) were used. They were maintained in the laboratory animal house under controlled conditions (13 h light, 11 h dark; temperature 24 ± 1°C) and fed on a standard food concentrate diet (Rat and Mouse Breeding Diet, Haygates, Birmingham) ad libitum. Prior to experimentation, they were allocated randomly into one of six experimental groups as follows:

Group I (n = 14): saline controls, intravenous injection. Group II (n = 5): saline controls, intraperitoneal injection. Group III (n = 4): FDC emulsion with C-16 oil additive, intravenous injection. Group IV (n = 4): FDC emulsion with C-16 oil additive, intraperitoneal injection. Group V (n= 5): Fluosol-DA (F-DA), intravenous injection. Group VI (n = 5): F-DA, intraperitoneal injection.

All animals were initially anaesthetized with ether. Intraperitoneal (I.P.) injections were then performed in the inguinal region while intravenous (I.V.) injections were made into a tail vein. A 20% FDC emulsion containing 1% of the C-16 oil additive was prepared 12–24 h before injection. The aqueous phase was 0.9% (w/v) NaCl. The emulsion was passed through a Millipore filter (0.22 μm) before use. F-DA (Table 2) was freshly prepared by mixing stem emulsion and annex solutions immediately before injection. Emulsions were prewarmed to ca. 37°C before administration. Control animals were injected with sterile saline solution (0.9% w/v NaCl). The dose of emulsion or saline injected was 10 ml/kg body weight throughout. Twenty-four h later, all animals were given a single I.P. injection of ca. 5×10^8 double-washed sheep red blood cells (SRBC) suspended in 1.0 ml Hank's standard balanced saline solution.

After seven days, all animals were anaesthetized with ether and exsanguinated by cardiac puncture. The weights of liver, spleen, thymus and mesenteric lymph nodes were then measured following careful dissection of individual tissues. Plasma antibody titres were measured by means of a conventional serial-dilution haemagglutination assay using a 0.5% suspension of cells (Pritchard and Eady, 1980).

Statistical analyses

Statistical comparisons of tissue weight changes were carried out according to the methods of Snedecor and Cochran (1980). Means and standard errors (S.E.M.) have been used and statistical significance between mean values was assessed using a conventional Student's t test. A probability of $P < 0.05$ was considered significant.

Emulsion stability data

Changes in stability parameter plotted against storage time for both experimental and control emulsions are shown in Figure 1. This illustrates the way in which the particulate growth of the emulsions is retarded and hence stability improved for those emulsions containing HBPO additives. A trend in emulsion stability correlation to the type of additive is clearly indicated. The least stable emulsions were the controls (i.e. those without oil additives) while the greatest stability was observed in emulsions containing the C-16 oil.

Table 2. Composition of Fluosol-DA 20% (Green Cross Corporation, Japan)

*Perfluorodecalin	14.0
*Perfluorotripropylamine	6.0
+Pluronic F-68	2.7
+Yolk phospholipids	0.4
Glycerol	0.8
NaCl	0.600
KCl	0.034
$MgCl_2$	0.020
$CaCl_2$	0.028
$NaHCO_3$	0.210
Glucose	0.180
++Hydroxyethyl starch	3.0

All values are w/v (%); total osmotic pressure: 410 m-osM; total oncotic pressure: 380-395 mmH_2O; *gas-carrying emulsion; +emulsifying agents; ++ oncotic pressure agent.

In addition to retarding the ageing process of FDC emulsions, the HBPO also helped to maintain a narrow size distribution. This is illustrated in Figure 2 which shows the variation of particle size distribution for emulsions stored at 37°C for up to 44 days. Control emulsions showed a greater increase in particle size distribution on storage compared with those containing HBPO additives; the C-16 oil was found to be the most effective in maintaining a narrow particle size distribution within the FDC emulsion (Fig. 2).

Tissue weight changes

Mean liver weight was increased by 12-15% (P < 0.001) following I.P. or I.V. injection of emulsion (Table 3). In contrast, liver weight was unchanged at 8 days after injection of F-DA, irrespective of route of administration. Spleen weight was also increased to a maximum of 20% (P < 0.01) in animals injected I.P. with the novel FDC emulsion but this was less than increases of up to 44% (P < 0.001) which occurred in F-DA-injected rats. Thymus weight decreased (P < 0.05) following I.P. injection of FDC emulsion but was similar to the mean control value (0.27 ± 0.01% body weight) in all other animals. Mesenteric lymph node weight following administration of FDC emulsion was not significantly different from control (0.08 ± 0.01% body weight) but was increased by 38% in response to I.V. injection of F-DA.

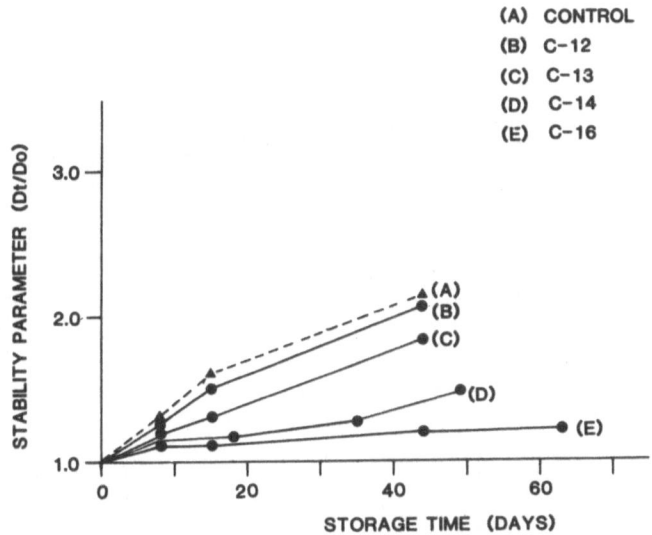

Figure 1. Relative stability of FDC emulsions stabilized with 1% HBPO and stored at 4°C.

Mean plasma antibody titres to SRBC (expressed as $\log_2$ titre) showed progressive increases up to day 5 in all groups of animals (Fig. 3). Titres measured on day 8 following I.V. injection of either FDC emulsion or F-DA were similar to the corresponding mean control value. However, day 8 titres were increased in rats injected I.P. with emulsified FDC (8.9 ± 0.3) or F-DA (7.8 ± 0.4).

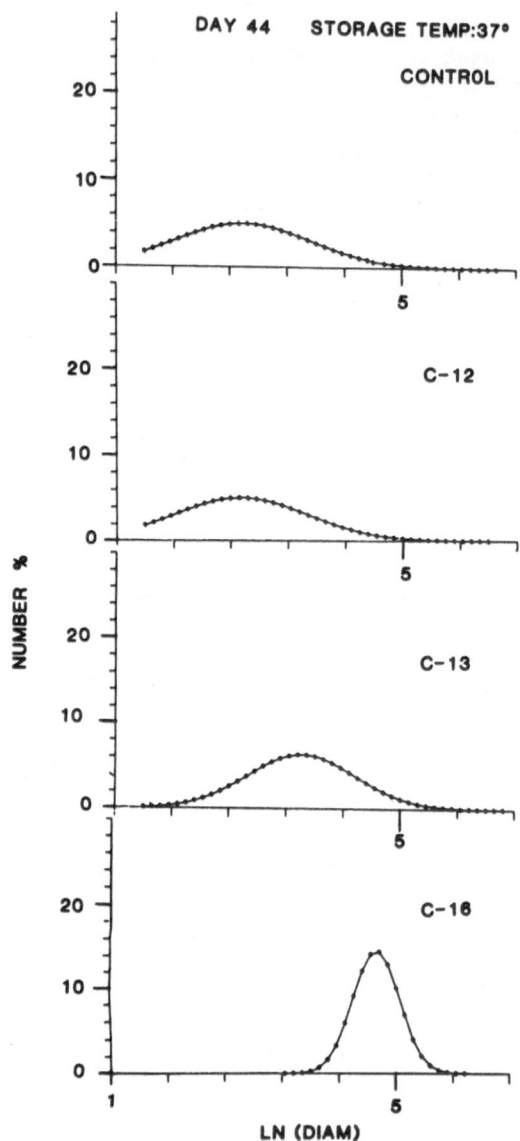

Figure 2. Polydispersity data: log-normal distribution of particle size for novel FDC emulsions stabilized with 1% HBPO and stored at 37°C for up to 44 days.

Table 3. Lymphoid tissue weights in rats after treatment with either Fluosol-DA 20% or FDC emulsion containing C-16 oil

| Treatment | N | Tissue Weights (% body weight) | | | |
		Liver	Spleen	Thymus	Mesenteric lymph node
Controls	19	4.62+0.05	0.25+0.01	0.27+0.01	0.08+0.01
Fluosol-DA					
I.V.	5	4.65+0.12	0.36+0.03^X	0.27+0.02	0.11+0.01*
I.P.	5	4.53+0.04	0.35+0.03^X	0.29+0.01	0.09+0.01
FDC emulsion					
I.V.	4	5.32+0.10^X	0.27+0.01	0.24+0.02	0.06+0.01
I.P.	4	5.18+0.07^X	0.30+0.01*	0.22+0.01!	0.07+0.01

Values are given as Mean + S.E.M.

!$P < 0.05$;   *$P < 0.01$; $^X P < 0.001$

DISCUSSION

The present experiments demonstrate that stability of FDC emulsions can be enhanced by the addition of polycyclic perfluorinated HBPO. The degree of stability achieved in preparations stored either at 4 or 37°C appeared to be directly related to the boiling point and hence, vapour pressure, of the oil additive. A significant stabilization effect was produced by the addition of perfluoroperhydro-fluoranthene ($C_{16}F_{26}$).

Ostwald Ripening is a process leading to an irreversible increase in particle size of the larger particles in an emulsion at the expense of the smaller ones. It depends on the solubility of the disperse phase of the emulsion in the continuous phase (Liftshitz and Slezov, 1959; Higuchi and Misra, 1962) and differs from the normal route of emulsion breakdown in that there is no physical contact between the disperse phase particles. Whilst commercial preparations such as F-DA are stabilized against the normal route of emulsion breakdown, they do not contain additives which protect against Ostwald Ripening. The present results show that Ostwald Ripening is equally important as an ageing process in FDC emulsions.

The present preliminary results show that an FDC emulsion containing a C-16 HBPO additive can be used in vivo in rats with no obvious adverse effects. The increases in tissue weights which followed injection of the novel FDC emulsion were in general agreement with previous findings that PFC particles can accumulate in reticuloendothelial and other tissues (Lowe and Bollands, 1985; Bollands and Lowe, 1986a,b,c). The variation in lymphoid tissue responses to the new emulsion and F-DA probably reflects the

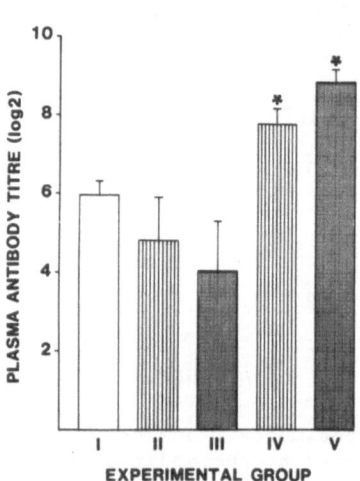

Figure 3. The mean plasma haemagglutination response (expressed as $\log_2$ titre) of rats at seven days after I.P. injection of SRBC. Experimental pretreatments were as follows:

Group I   : Saline-injected controls (n = 19)
Group II   : F-DA I.V. (n = 5)
Group III   : FDC emulsion containing C-16 oil I.V. (n = 4)
Group IV   : F-DA I.P. (n = 5)
Group V   : FDC emulsion containing C-16 oil I.P. (n = 4)

Vertical bars represent S.E.M.; *P< 0.01 compared with mean control value.

differences in their compositions, since the latter contains 6% FTPA (Naito and Yokoyama, 1978; Table 2). The estimated whole body half-life of FTPA in rats is approximately 60 days compared with a corresponding value of about 7 days for FDC (Naito and Yokoyama, 1978). Thus, it is reasonable to conclude that FTPA will make a significant contribution to differences in lymphoid tissue responses to the new emulsion and to F-DA.

Variations in liver and spleen responses to different PFC emulsions have been reported previously: for example, Caiazza and others (Caiazza, Fanizza and Ferrari, 1984) noted that no significant alterations in tissue weights occurred in rats following I.V. injection of emulsified perfluorotributylamine (FC-43; Green Cross Corporation, Japan) in doses comparable to those administered in the present experiments.

In addition to the effects of different PFC preparations, the responses of individual lymphoid tissues also vary according to route and dose of emulsion administered (Bollands and Lowe, 1986c, 1987a). It is likely that the lack of significant hepatomegaly after injection with either emulsified FDC or F-DA reflected the use of relatively low doses of these emulsions; in previous studies, injection of greater doses of either F-DA or FC-43 in rats or mice produced proportionately greater increases in liver weights (Lutz and Metzenauer, 1980; Lutz et al., 1982b; Mason, Withers and Steckel, 1985).

Species variability is also an important determinant of lymphoid tissue responses to PFC emulsions (Lowe and Bollands, 1985). For example, there are marked differences in the extent to which identical doses (per unit body weight) of F-DA can produce increases in liver or spleen weights in rats and mice (Bollands and Lowe, 1986b, 1987a). This has inevitably introduced complications in identifying the effects of different PFC emulsions and their components upon normal physiological variables, especially in relation to immune system function.

Previous work using rats and primates has shown that in addition to producing changes in tissue weights, F-DA can also induce transient impairment of the reticuloendothelial system (RES) clearance function (Lutz, Barthel and Metzenauer, 1982a; Castro et al., 1983; Castro, Nesbitt and Lyles, 1984). The suppression of RES phagocytic function in rats produced by F-DA was intermediate between that induced by soybean oil emulsion and that caused by the potent RES-blocking agent, ethyl palmitate (Castro et al., 1984). Such an inhibition of RES clearance function by F-DA may lead to reduced immunological competence and would account for the decreased resistance to Escherichia coli toxin observed in mice injected with the emulsion (Lutz et al., 1982a).

The increases in plasma antibody titres to SRBC following I.P. injection of the novel emulsion or F-DA were in agreement with previously published observations of immunopotentiating effects produced by emulsified PFC (Bollands and Lowe, 1986b, 1987a). One explanation for these findings is that the PFC or other emulsion constitutents act as an adjuvant when injected into the peritoneal cavity of either rats or NIH mice 24 h prior to SRBC (Bollands and Lowe, 1986b, 1987a). However, inconsistencies have been noted in the humoral immune responses in mice injected with F-DA since Shah and others reported that pretreatment of Balb/c mice with F-DA or FC-43 I.V. led to a decrease in the in vivo production of antibodies to SRBC (Shah, Yamamura and Usuba, 1984). While this difference may reflect strain variations in response, other factors, especially timing and route of presentation of PFC relative to the immune challenge, must also be considered. In this regard, Mitsuno and others observed that in rats injected with > 20 ml/kg F-DA, the emulsion enhanced antibody production when given after immunization whereas it suppressed antibody production when injected before immunization (Mitsuno, Ohyanagi and Yokoyama, 1984). More recently, experiments have demonstrated that changes in lymphoid tissue weights and plasma antibody titres in rats immunized with SRBC vary according to the time of a previous or subsequent injection of F-DA via the same route (Bollands and Lowe, 1987b). Further work is in progress to determine the mechanism(s) by which emulsified PFC can alter immune system function and to identify the active principle(s) involved.

Some concern has been expressed about possible adverse physiological effects of emulsified PFC being attributable, at least in part, to the Pluronic F-68 surfactant found in all commercial preparations (Lowe and Bollands, 1985; see also Gronow et al., 1987). Experiments in rabbits have shown that infusion of F-DA produced haematological disturbances comparable to those seen in animals receiving plasma containing biologically-active complement fractions (Vercellotti et al., 1982). Pluronic was implicated as an active principle in these reactions since it reproduced the complement-activating effects of whole F-DA both in vivo and in vitro (Vercellotti et al., 1982). In subsequent work it was proposed that peroxide derivatives of Pluronic, formed during steam sterilization of the emulsion, could contribute to, but not fully account for, adverse effects of F-DA on arterial endothelial ultrastructure (McCoy et al., 1984). The extent to which Pluronic is responsible for the adverse physiological reactions to commercial PFC emulsions (Lowe and Bollands, 1985) remains to be determined. However, we are aware that if such emulsions are to become widely used in clinical medicine, it may be necessary to use other, biocompatible, emulsifiers and this possibility is currently under investigation.

We conclude that FDC emulsions stabilized against Ostwald Ripening with polycyclic perfluorinated HBPO may have potential value for in vivo use as components of oxygen-transport fluids. Work is in progress to assess physico-chemical and biocompatibility characteristics of other, similar, preparations including those containing different surfactants.

ACKNOWLEDGEMENTS

This work was supported by research grants from I.S.C.Chemicals Ltd, Avonmouth. Novel emulsions described in this paper are covered by U.K. Patent No. 85 04916 and International Patent No. 86 04709. We are grateful to Drs T.Suyama and K. Yokoyama of the Green Cross Corporation, Japan, for their generous gifts of Fluosol-DA. A.D.B. is the recipient of a Medical Research Council Studentship.

REFERENCES
Bollands, A.D. and Lowe, K.C. (1986a). Responses of rat lymphoid tissue to a perfluorocarbon blood substitute. Br.J.Pharmac. 87, 118P.

Bollands, A.D. and Lowe, K.C. (1986b). Effects of a perfluorocarbon emulsion, Fluosol-DA on rat lymphoid tissue and immunological competence. Comp. Biochem. Physiol. 85C, 309-312.

Bollands, A.D. and Lowe, K.C. (1986c). Comparative effects of emulsified perfluorocarbons on lymphoid tissue in rodents. Br.J.Pharmac. 89, 664P.

Bollands, A.D. and Lowe, K.C. (1987a). Lymphoid tissue responses to perfluorocarbon emulsion in mice. Comp. Biochem. Physiol. (in press).

Bollands, A.D. and Lowe, K.C. (1987b). Lymphoid tissue responses to Fluosol-DA in rats: time course effects relative to immunological challenge. Br.J.Pharmac. (in press).

Caiazza, S., Fanizza, M. and Ferrari, M.(1984). Fluosol-43 particle localization pattern in target organs of rats. Virchows Arch. Path. Anat. Physiol. A. 404, 127-137.

Castro, O., Nesbitt, A.E. and Lyles, D. (1984). Effect of a perfluoro-carbon emulsion (Fluosol-DA) on reticuloendothelial system clearance function. Am. J. Hemat. 16, 15-21.

Castro, O., Reindorf, C.A., Socha, W.W. and Rowe, A.W. (1983). Perfluorocarbon enhancement of heterologous red cell survival: a reticuloendothelial block effect? Int. Arch. Allergy Appl. Immunol. 70, 88-91.

Davis, S.S., Round, H.P. and Purewal, T.S. (1981). Ostwald Ripening of emulsion systems. An explanation for the effect of an added third component. J. Coll. Interface Sci. 80, 508-511.

Gronow, G., Kelting, Th., Skrezek, Ch., Plas, J. v.d. and Bakker, J.C. (1987). Oxygen transport to renal tissue: effect of oxygen carriers. This volume.

Higuchi, W.I. and Misra, J. (1962). Physical degradation of emulsions via the molecular diffusion route and the possible prevention thereof. J. Pharm. Sci. 51, 459-466.

Liftschitz, I.M. and Slezov, V.V. (1959). Kinetics of diffusive decomposition of supersaturated solid solutions. Sov. Phys. JETP, 35, 331-339.

Lowe, K.C. (1986). Blood transfusion or blood substitution? Vox Sang. 51, 257-263.

Lowe, K.C. and Bollands, A.D. (1985). Physiological effects of perfluorocarbon blood substitutes. Med. Lab. Sci. 42, 367-375.

Lutz, J. and Metzenauer, P. (1980). Effects of potential blood substitutes (perfluorochemicals) on rat liver and spleen. Pflugers Arch. 387, 175-181.

Lutz, J., Barthel, U. and Metzenauer, P. (1982a). Variation in toxicity of Escherichia coli endotoxin in rats after treatment with perfluorated blood substitutes in mice. Circ. Shock, 9, 99-106.

Lutz, J., Metzenauer, P., Kunz, E. and Heine, W.D. (1982b). Cellular responses after use of perfluorinated blood substitutes. In: Oxygen-carrying Colloidal Blood Substitutes. Eds Frey, R., Beisbarth, H. and Stosseck, K., Zuckschwerdt, Munich, pp. 73-81.

Mason, K.A., Withers, H.R. and Steckel, R.J. (1985). Acute effects of a perfluorochemical oxygen carrier on normal tissues of the mouse. Radiat. Res. 104, 387-394.

McCoy, L.E., Becker, C.A., Goodin, T.H. and Barnhart, M.I. (1984). Endothelial responses to perfluorochemical emulsion. Scan. Electron Micros. 6, 311-319.

Mitsuno, T., Ohyanagi, H. and Naito, R. (1982). Clinical studies of a perfluorochemical blood substitute. Ann. Surg. 195, 60-69.

Mitsuno, T., Ohyanagi, H. and Yokoyama, K. (1984). Development of a perfluorochemical emulsion as a gas carrier. Artif. Org. 8, 25-33.

Naito, R. and Yokoyama, K. (1978). Perfluorochemical Blood Substitutes. Technical Information Series No. 5, Green Cross Corporation, Osaka.

Pritchard, D.I. and Eady, R.P. (1980). Some aspects of immunology in the nude rat. In: Immunodeficient Animals for Cancer Research. Ed. Sparrow, S., MacMillan, London, pp. 67-79.

Riess, J.G. and Le Blanc, M. (1982). Solubility and transport phenomena in perfluorochemicals relevant to blood substitution and other biomedical applications. Pure Appl. Chem. 54, 2383-2406.

Shah, K.H., Yamamura, Y. and Usuba, A. (1984). Immunopathologic changes by artificial blood perfluorochemicals. Lab. Invest. 48, 62A.

Sharma, S.K., Davis, S.S. and Lowe, K.C. (1986). Development of stable emulsions of perfluorochemicals for biological applications. Proc. Soc. Chem. Ind., Imperial College, London.

Snedecor, G.W. and Cochran, W.G. (1980). Statistical methods. 7th Edn Iowa State College Press, Ames.

Stefaniszyn, H.J., Wynards, J.E. and Salerno, T.A. (1985). Initial Canadian experience with artificial blood (Fluorol-DA 20%) in severely anemic patients. J. Cardiovasc. Surg. 26, 337-342.

Tremper, K.K., Friedman, A.E., Levin, E.M., Lapin, R. and Camarillo, D. (1982). The pre-operative treatment of severely anemic patients with a perfluorochemical oxygen-transport fluid. N. Engl. J. Med. 307, 277-283.

Vercellotti, G.M., Hammerschmidt, D.E., Craddock, P.R. and Jacob, H.S. (1982). Activation of plasma complement by perfluorocarbon artificial blood: probable mechanism of adverse pulmonary reactions in treated patients and rationale for corticosteroid prophylaxis. Blood, 59, 1299-1304.

Waxman, K., Tremper, K.K., Cullen, B.F. and Mason, G.R. (1984). Perfluorocarbon infusion in bleeding patients refusing blood transfusions. Arch. Surg. 119, 721-724.

CEREBRAL CORTICAL OXYGENATION AND PERFUSION DURING PROGRESSIVE

NORMOVOLAEMIC HAEMODILUTION WITH HETASTARCH (VOLEX) AND FLUOSOL-DA

T. SHINOZUKA, E.M. NEMOTO AND P.M. WINTER

Department of Anesthesiology and Critical Care Medicine, University of Pittsburgh School of Medicine, Pittsburg, PA 15261, U.S.A.

INTRODUCTION

In a previous study on the effects of normovolaemic haemodilution with Hetastarch on cerebral cortical oxygenation and perfusion, we found that cortical tissue PO_2 (PtO_2) was unchanged between haematocrits (Hct) of 45 to 38%, but then fell at a rate of about 2.4% per 1% decrease in Hct between Hct of 38% to 5% (Shinozuka, Nemoto and Bleyaert, 1984). Fluosol-DA may delay the fall in cerebral cortical PtO_2 during normovolaemic haemodilution compared with haemodilution with Hetastarch.

METHODS

Thirty-four Sprague-Dawley albino rats weighing 350 to 400 g body weight were anaesthetized with sodium pentobarbital 60 mg/kg I.V., and mechanically ventilated on 60% O_2/40% N_2 via tracheostomies and immobilized with pancuronium bromide at 0.2 mg/h I.V. Anaesthesia was maintained by continuous infusion of pentobarbital at 5 mg/h I.V. Femoral artery and venous catheters were inserted for monitoring of arterial blood pressure and arterial blood gases and pH. With the rats' heads fixed in a stereotaxic apparatus, 10 μm platinum microelectrodes were inserted 500 μm into the parietal cortex for measurement of local cerebral blood flow (LCBF) while 10 μm gold microelectrodes were similarly inserted for monitoring PtO_2. Rectal temperature was controlled at 37 $\pm$ 0.5°C, mean arterial pressure (MAP) between 110 and 130 torr ($\overline{1}4.5$ and 17.1 kPa), PaO_2 about 250 torr and $PaCO_2$ between 35 and 40 torr. Local CBF was measured by H_2 clearance after equilibration on 1% H_2 and calculated by the $T_{\frac{1}{2}}$ method. PtO_2 measurements were corrected for drift and sensitivity changes based on calibrations immediately before and at the end of each study. Following control measurements at normal Hct (Fig.1), haemodilution was accomplished by withdrawing 20% of the calculated total blood volume while simultaneously infusing I.V. an equal volume of either 20% Fluosol-DA (n = 12) or 6% Hetastarch (n = 11). In this manner, Hct was progressively decreased stepwise from 45% to 40, 30, 20, 10 and 5%. Eleven rats not subjected to haemodilution served as controls. Statistical analyses were done by analysis of variance and the Student-Neumann-Keul's test. The Mann-Whitney-U test was used for statistical analysis of LCBF and PtO$_2$ as percent of control.

RESULTS

Physiological variables between the Hetastarch and Fluosol-DA haemodiluted rats were similar (Table 1). MAP was comparable between control, Hetastarch and Fluosol study groups but PaCO$_2$ and PaO$_2$ tended to be lower in the control compared with the other two groups. The differences, however, were not significant. The relationship between haemoglobin (g/100 ml) and Hct (%) during haemodilution with 6% Hetastarch was linear with a correlation coefficient of 0.96 (Fig. 2).

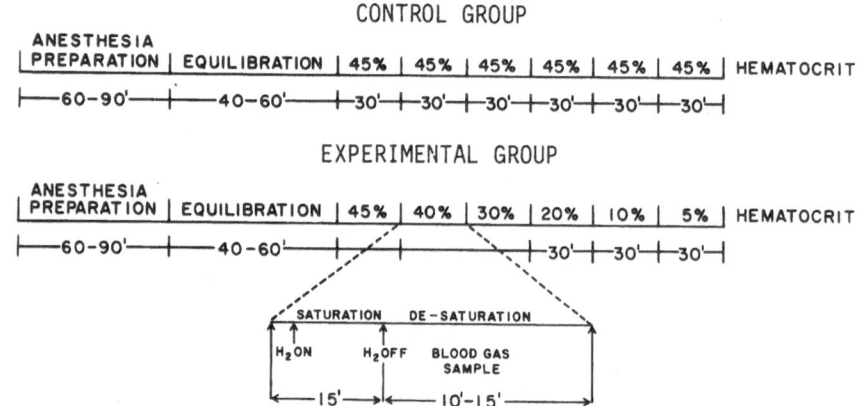

Figure 1. Experimental protocol for control and normovolaemic haemodilution (experimental) study groups in rats. Haemodilution was accomplished with either 6% Hetastarch or 20% Fluosol-DA.

In the untreated control group, LCBF was 43 $\pm$ 7 ml.100 g^{-1}.min^{-1} at 0 min (i.e., 100% of control) and remained at about that level until 120 min (Fig. 3). The rise in LCBF to 120% of control at 150 min was not significant. PtO$_2$, on the other hand, remained at about 90% of control (control PtO$_2$ = 30 $\pm$ 5 torr) between one and 2.5 h.

The increase in LCBF in percent of control during progressive normovolaemic haemodilution was similar with Hetastarch and Fluosol-DA (Fig. 4). With Hetastarch haemodilution, the relationship was described by the equation LCBF (% of control) = 95.07 $\pm$ 785/X and during Fluosol-DA haemodilution by the equation, LCBF (% of control) = 115 + 768.6/X. In both circumstances, marked increases in LCBF did not occur until Hct fell below 20%.

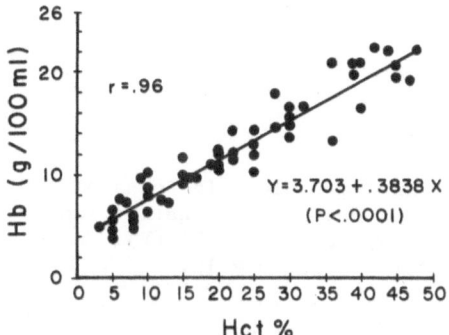

Figure 2. Relationship between changes in haematocrit (Hct) and haemo-globin (Hb) during normovolaemic haemodilution with 6% Hetastarch in rats.

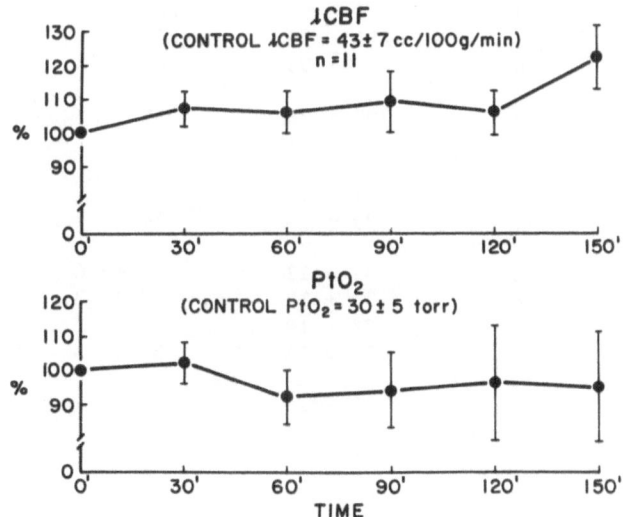

Figure 3. Local cerebral blood flow (1CBF) and tissue PO$_2$ (PtO$_2$) in the cerebral cortex of control rats during 2.5 h under pento-barbital anaesthesia.

Table 1. Physiological variables in control (C) rats and in rats subjected to progressive normovolaemic haemodilution with Hetastarch (H) or Fluosol-DA (F). Means $\pm$ S.D. For C, n = 11; for H, n = 11; for F, n = 12.

	Sample No.	C	H	F
MAP(Torr)				
	1	95 $\pm$ 39	115 $\pm$ 14	115 $\pm$ 7
	2			115 $\pm$ 8
	3	90 $\pm$ 37	119 $\pm$ 8	108 $\pm$ 13
	4	91 $\pm$ 39	121 $\pm$ 8	111 $\pm$ 16
	5	88 $\pm$ 37	119 $\pm$ 8	108 $\pm$ 11
	6	90 $\pm$ 38	116 $\pm$ 10	109 $\pm$ 7
	7			102 $\pm$ 10
	8	92 $\pm$ 39	109 $\pm$ 13	93 $\pm$ 14
PaO$_2$(Torr)				
	1	223 $\pm$ 95	263 $\pm$ 20	252 $\pm$ 33
	2			259 $\pm$ 29
	3	221 $\pm$ 94	270 $\pm$ 26	279 $\pm$ 26
	4	219 $\pm$ 94	266 $\pm$ 17	267 $\pm$ 22
	5	214 $\pm$ 95	269 $\pm$ 21	291 $\pm$ 31
	6	207 $\pm$ 93	267 $\pm$ 18	305 $\pm$ 15
	7			302 $\pm$ 22
	8	203 $\pm$ 98	259 $\pm$ 23	293 $\pm$ 32
PaCO$_2$(Torr)				
	1	29 $\pm$ 10	36 $\pm$ 4	40 $\pm$ 7
	2			42 $\pm$ 8
	3	33 $\pm$ 12	38 $\pm$ 6	40 $\pm$ 6
	4	33 $\pm$ 12	39 $\pm$ 3	41 $\pm$ 4
	5	32 $\pm$ 12	38 $\pm$ 3	39 $\pm$ 3
	6	31 $\pm$ 12	35 $\pm$ 3	38 $\pm$ 3
	7			38 $\pm$ 3
	8	31 $\pm$ 13	37 $\pm$ 2	36 $\pm$ 3

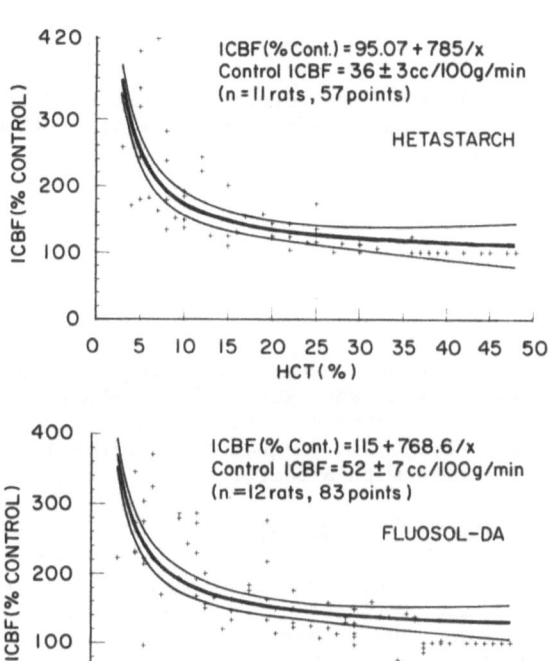

Figure 4. Local cerebral blood flow (lCBF) in the cerebral cortex of rats subjected to progressive normovolaemic haemodilution with 6% Hetastarch (top panel) or 20% Fluosol-DA (bottom panel). Ninety-five percent confidence intervals are shown. Significance levels for parameters and regression were 0.0001 for both graphs.

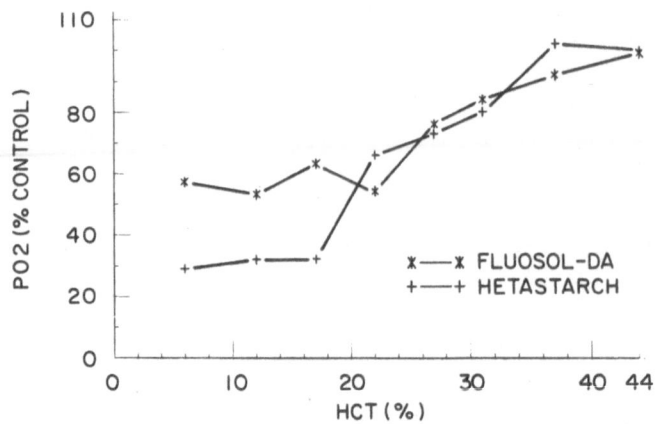

Figure 5. Rat cortical PO$_2$ in percent of control during progressive normovolaemic haemodilution with either 6% Hetastarch (n= 11) or 20% Fluosol-DA (n = 12).

Comparison of the mean changes in PtO$_2$ (Fig. 5) shows that in the Hct range between 45% and 20%, Fluosol-DA was not superior to Hetastarch in maintaining cortical tissue oxygenation whereas at Hct less than 20%, mean PtO$_2$ was higher with Fluosol-DA haemodilution compared with haemodilution with Hetastarch.

DISCUSSION

Haemodilution with Fluosol-DA does not appear to provide a clear advantage in maintaining cortical oxygenation compared with Hetastarch unless Hct falls below 20%. Changes in LCBF with haemodilution were similar with both solutions. These findings suggest that haemoglobin serves as the primary O$_2$ carrier at Hcts greater than 20 to 25% and, despite substitution of the haemodiluting agent with Fluosol-DA, cortical tissue oxygenation was not improved. However, as Hct fell below 25% and haemoglobin failed to sustain sufficient O$_2$ delivery, the additional O$_2$-carrying capacity of Fluosol-DA above that of Hetastarch was evident. Therefore, only in circumstances of severe haemodilution of Hcts below 25% would Fluosol-DA provide a clear advantage over Hetastarch. These findings are in agreement with those of Naito and Yokoyama (1978) who found that Fluosol-DA improved cortical PO$_2$ compared with Hetastarch at 10% but not at 20% Hct.

It should be mentioned that had we used 100% O$_2$ instead of 60% O$_2$, we might have found that Fluosol improves cortical PtO$_2$ compared with Hetastarch at a Hct higher than 20%. Although their studies were done in the dog myocardium, Kessler, Vogel and Gunther (1983) showed that with haemodilution to a Hct of 14% with Fluosol-DA, 100% O$_2$ resulted in a mean PtO$_2$ of 358 Torr and with 50% O$_2$, a mean PtO$_2$ of 100 Torr. Thus, a twofold increase in O$_2$ from 50% to 100% resulted in an almost fourfold increase in mean PtO$_2$.

We should also emphasize that our studies do not indicate or provide information on the potential beneficial effects of haemodilution with Fluosol-DA compared with Hetastarch in ischaemic brain injury.   Indeed, there is some evidence that Fluosol-DA was effective in treating cerebral vasospasm following subarachnoid haemorrhage in patients (Handa et al., 1983).   However, a recent study testing the efficacy of Fluosol in focal cerebral ischaemia in cats failed to reveal any beneficial effects (Kolluri et al., 1986).   Thus, the particular circumstances in which Fluosol would be of benefit must be clearly defined.

ACKNOWLEDGEMENTS

Supported in part by the American Heart Association, Dallas, Texas, Grant No. 84-1138.   Fluosol-DA was kindly provided by Alpha Therapeutic Corporation, Los Angeles, CA.

REFERENCES

Handa, H., Nagasawa, S., Yonekawa, Y., Naruo, Y. and Oda, Y. (1983). New treatment of cerebral vasospasm with Fluosol-DA 20%: protective effect on cerebral ischemia and change of cerebral blood flow (CBF). Progress in Clinical and Biological Research, Vol. 122, Alan R. Liss, Inc., New York, pp. 299-306.

Kessler, M., Vogel, H. and Gunther, H. (1983). Local oxygen supply of the myocardium after extreme hemodilution with Fluosol-DA. Progress in Clinical and Biological Research, Vol. 122, Alan R. Liss, Inc., New York, pp. 237-248.

Kolluri, S., Heros, R.C., Hedley-Whyte, E.T., Vonsattel, J. P., Miller, D. and Zervas, N.T. (1986). Effect of Fluosol on oxygen availability. Regional cerebral blood flow, and infarct size in a model of temporary focal cerebral ischemia. Stroke, 17, 976-980.

Naito, R. and Yokoyama, K. (1978). Fluosol-DA. In: Perfluorochemical Blood Substitutes. Technical Information Series No. 5. The Green Cross Corporation, Osaka.

Shinozuka, T., Nemoto, E.M. and Bleyaert, A.L. (1984). Cerebral cortical oxygenation and perfusion during Hetastarch hemodilution. In: Oxygen Transport to Tissue-VI.   Eds Bruley, D., Bicher, H.I. and Reneau, D., Plenum Press, New York and London, (Adv. Exp. Med. Biol. 180, 853-860).

OXYGEN TRANSPORT TO RENAL TISSUE: EFFECT OF OXYGEN CARRIERS[**]

G. GRONOW, Th. KELTING, Ch. SKREZEK, J.v.d. PLAS[*] and
J.C. BAKKER[*]

Department of Physiology, University of Kiel, D-2300 Kiel,
F.R.G. and [*]Central Laboratory, Netherlands Red Cross,
1066 CX Amsterdam, The Netherlands

INTRODUCTION

The isolated perfused rat kidney introduced by Weiss and colleagues in 1959 has become a commonly used tool in the field of renal physiology and pharmacology (Weiss, Passow and Rothstein, 1959; Little and Cohen, 1974; Ross, 1978; Maack, 1986). In view of technical complications such as blood clotting and the release of vasoactive factors most authors preferred a hyperoxygenated (PO_2~660 mmHg) balanced salt solution instead of blood as a perfusate. However, recent experiments with the isolated Ringer-perfused rat kidney indicate that oxygen transport to renal tissue has become a central question: due to a steep gradient of oxygen partial pressure in the outer medullary region of mammalian kidneys (Leichtweiss et al., 1969; Baumgartl et al., 1972) the poor oxygen binding capacity of hyperoxygenated salt solutions induced functional and morphological lesions in distinct renal tissue zones (Alcorn et al., 1981; Brezis et al., 1984; Schurek and Kriz, 1985). The aim of the present experiments was to compare the effects of three different oxygen carriers on function and tissue integrity of the isolated perfused rat kidney. The following served as oxygen carriers: a) coupled haemoglobin, b) washed erythrocytes, and c) perfluorocarbons. Perfusions performed with hyperoxygenated Ringer solutions (PO_2~660 mmHg) served as a control. Our data indicate that renal function (perfusion flow rate, glomerular filtration rate, absolute and fractional Na^+ reabsorption) as well as parameters of tissue integrity (i.e. the loss of enzymes, and tissue water content) are maintained best with an erythrocyte suspension and, for a limited time period, with coupled haemoglobin in the perfusate. They are reasonably maintained during Ringer perfusion, but are severely impaired in the presence of a perfluorocarbon emulsion.

[**]Dedicated to Professor Ch. Weiss in his 60th year,
Medical University of Lubeck, F.R.G.

MATERIALS AND METHODS

Preparations

Isolation and perfusion of the rat kidney have been described elsewhere (Gronow and Cohen, 1984; Gronow and Kossmann, 1985). Briefly, the right kidney of fed male Sprague-Dawley rats (body weight 350-420g) was cannulated under pentobarbital anaesthesia and transferred to a perfusion chamber which was gassed with humidified $O_2:CO_2:N_2$ mixtures and warmed by a thermostatically controlled water bath (38°C). In all experiments a balanced salt solution served as the basic medium or control: Ringer-bicarbonate, pH 7.4, completed with 6g/dl bovine albumin as a colloid, and physiological substrates such as 5 mM/l glucose, 10 mM/l l-glycine, 2 mM/l proline, and 2 mM/l l-aspartate. The perfusates recirculated at a mean 'arterial' pressure of 100 ± 4 mmHg through 1) the kidney, 2) a venous reservoir, 3) a Millipore filter, 4) an artificial lung, and 5) an arterial reservoir. Parameters of renal function and of tissue integrity were measured as described previously (Gronow and Cohen, 1984; Gronow and Kossmann, 1985). The activity of N-acetyl- -D-glucosaminidase was measured according to the method described by Maruhn (1976).

Isolated tubular segments from rat kidney cortex were prepared as reported earlier (Gronow, Benk and Franke, 1984; Gronow, Meya and Weiss, 1984). Briefly, both kidneys of Sprague-Dawley rats treated as above were flushed with ice-cold saline, filled with 0.4% collagenase solution, and, after separation and mechanical treatment of the cortical zones, suspended in oxygenated and buffered 0.2% (w:v) collagenase solution (38°C) containing physiological substrates as described for the basic medium (see above). After 15 min of incubation the isolated tubular segments of rat kidney cortex were washed twice at 4°C and resuspended at 38°C in the incubation medium under study. In all statistical comparisons of functional parameters and/or tissue integrity P-values of 0.5 or less were considered to indicate a significant difference (Student's t-test, analysis of variance).

Oxygen carriers

Oxygen carriers were added to the basic Ringer medium (see above) in the following final concentrations: 1.7% (w:v) coupled haemoglobin (1.7% Hb), 5 vol.% washed human erythrocytes (5% Ery), and 20% (w:v) perfluorocarbon (FC-43 = 20 g/dl perfluorotributylamine plus 2.56 g/dl Pluronic F-68; Green Cross Corp., Osaka, Japan). Prior to the experiments purified human haemoglobin had been coupled to 2-nor-2-formyl-pyridoxal-5'-phosphate (NFPLP) as described earlier (Bakker et al., 1985). The haemoglobin solution (> 80% NFPLP-Hb, P_{50} = 38 mmHg) was diluted with basic medium to 1.7% (or 1/9 of the haemoglobin content of blood) to meet the 1/9 haematocrit of recently tested 5 vol.% erythrocyte perfusion media (Swanson et al., 1981; Brezis et al., 1984; Schurek and Kriz, 1985).

Oxygen transport by different perfusion media was kept comparable at ~2.2 vol.% O_2 by adjusting oxygen partial pressure in the artificial lung to 300 mmHg in the case of 1.7% Hb, 5% Ery and 20% FC-43 or to 660 mmHg in the case of Ringer. Colloid osmotic pressure was measured in a membrane osmometer (Wescor, Utah, U.S.A.) and adjusted to 25 mmHg (Ringer, 1.7% Hb, 5% Ery, 20% FC-43) by varying bovine albumin concentration (Serva, Heidelberg, F.R.G.) in the basic medium. Isolated tubular segments from rat kidney cortex were tested in Ringer incubation media (PO_2~660 mmHg) containing either 1 g/dl bovine albumin or 1 g/dl Pluronic F-68 (Wyandotte, Michigan, U.S.A.).

Ringer medium

Early experiments performed with the isolated perfused rat kidney provided evidence that even hyperoxygenation ($PO_2 \sim 660$ mmHg) of a balanced salt solution does not supply sufficient oxygen to support normal renal function (Franke and Weiss, 1976). On one hand, oxygen consumption (calculated on a wet weight basis) in renal cortical tissue is comparable or even higher than in brain tissue (Cohen, 1979), on the other hand, oxygen transport to renal tissue zones below the cortical regions is limited by the special arrangement of the renal vascular tree (Fig. 1, left scheme): long bundles of the arterial system cross, as descending vasa recta, the outer and inner zone of the medulla; the venous blood is drained in ascending vasa recta which run in close proximity to the arterial system (Kriz, 1981). In consequence, a countercurrent exchange diffusion system of oxygen from the arterial to the venous system (Fig. 1, arrows) may explain the high renal venous PO_2 _in vivo_ (60–70 mmHg) and the extremely low medullary oxygen tension (10–30 mmHg). This steep corticomedullary oxygen gradient has been demonstrated in renal tissue by Weiss and colleagues by measurements with O_2-microelectrodes _in vivo_, as well as in the Ringer-perfused isolated rat kidney (Leichtweiss et al., 1969; Baumgartl et al., 1972).

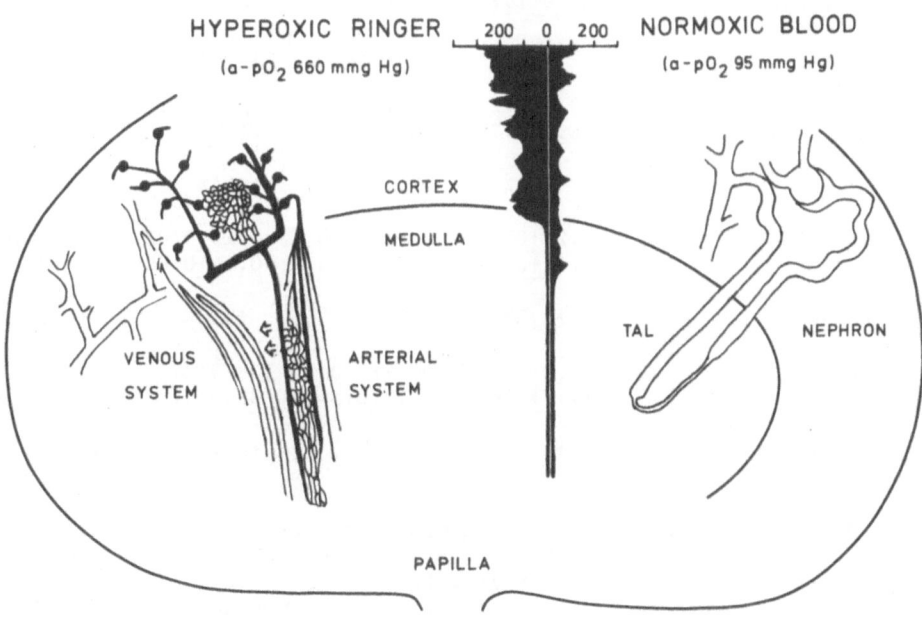

Figure 1. Left: organization of the vascular system in rat kidney (arrows, postulated arteriovenous oxygen shunting). Middle: tissue oxygen tension in rat kidney _in vivo_ (to the right) and _in vitro_ (to the left). Note the steep decrease in PO_2 near the corticomedullary border zone of the Ringer-perfused kidney (adapted from Baumgartl et al., 1972). Right: nephron structure (TAL = thick ascending limb of Henle).

Table 1. Effect of oxygen carriers on parameters of renal function and tissue integrity

Parameter	Perfusion time (min)	Ringer $\bar{x}$+SEM ($\bar{n}$=9)	5% ERY $\bar{x}$+SEM ($\bar{n}$=6)	1.7% Hb $\bar{x}$+SEM ($\bar{n}$=7)	FC-43 $\bar{x}$+SEM ($\bar{n}$=8)
Perfusion flow rate (ml/g·min)	20	31.8+1.8	19.1+2.0	20.2+1.9	19.5+1.6
	120	33.2+1.1	17.4+1.6	10.0+2.1*	3.2+1.4*
Change		+4.4%	−8.9%	−50.5%	−83.7%
Glomerular filtration rate (ml/g·min)	20	0.88+0.14	0.91+0.15	0.80+0.18	0.74+0.19
	120	0.83+0.12	0.97+0.12	0.75+0.15	0.38+0.15*
Change		−5.7%	+6.6%	−6.3%	−48.6%
Na^+-reabsorption (ueq/g·min)	20	118+27.1	126+16.5	110+14.9	100+12.4
	120	108+22.6	141+18.4	108+21.3	55+15.2*
Change		−8.5%	+11.9%	−1.8%	−45.3%
Fractional Na^+ reabsorption (% filtered Na)	20	92.2+2.0	95.5+1.7	94.9+1.5	91.5+0.9
	120	85.0+1.2*	92.1+2.1	84.3+1.8*	75.9+1.1*
Change		−7.8%	−3.4%	−10.6%	−15.6%
Urinary γGT loss (mU/g·min)	20	3.0+1.2	2.1+0.9	2.2+1.1	3.5+1.3
	120	16.6+2.7*	3.6+1.5	24.5+4.7*	55.8+6.2*
Change		+453%	+71.4%	+1014%	+1494%
Urinary NAG loss (mU/g·min)	20	0.23+0.09	0.16+0.08	0.15+0.02	0.23+0.09
	120	0.41+0.11*	0.14+0.03	0.54+0.08*	1.28+0.42*
Change		+78.3%	−12.5%	+350%	+456%
Tissue water content	20	104.0+1.7	100.9+1.3	104.3+2.1	104.0+2.9
	120	112.7+2.0*	102.2+0.5	132.9+4.5*	126.1+3.5*
% of control**		+8.7%	+1.3%	+28.6%	+22.1%

Ringer = balanced salt solution; 5% ERY = 5 vol.% erythrocyte suspension; 1.7% Hb = NFPLP-coupled haemoglobin; FC-43 = perfluorocarbon emulsion; γGT = γ-glutamyltransferase, NAG = N-acetyl-β-D-glucosaminidase. For further explanations see legend to Figure 2 and text.

* significant change ($p < 0.05$) in respect of 20 min observation.

** 100% = water content of unperfused contralateral kidney.
   (0.746 ± 0.093 g H_2O/g wet weight)

When perfusing with hyperoxygenated Ringer (Fig. 1, middle panel, left trace) the steep PO_2-gradient was clearly localized closer to the corticomedullary border, thus leaving the mitochondria-packed cells of the thick ascending limb of Henle (TAL, Fig. 1, right scheme) at a markedly reduced oxygen tension. Accordingly, distinct morphological lesions have been demonstrated recently in the TAL-region of the hyperoxygenated, Ringer-perfused rat kidney (Alcorn et al., 1981; Brezis et al., 1984; Schurek and Kriz, 1985).

In the present experiments, a more pronounced hypoxic vasodila-tation in the corticomedullary region of the Ringer-perfused rat kidney (see Fig. 1, middle scheme) may have raised perfusion flow rate (PFR) above the level observed in the presence of oxygen carriers (Fig. 2). The reduced absolute, (T-Na$^+$; Table 1) as well as fractional, Na$^+$ reabsorption (%-T-Na$^+$; Fig. 3) may then be the consequence of O_2-limited Na$^+$ transport in the TAL of the Ringer-perfused rat kidney. Concomitantly (see Table 1), in the course of perfusion a hypoxic swelling in this region (Brezis et al., 1984; Schurek and Kriz, 1985) may have 1) reduced glomerular filtration rate by increasing intratubular pressure (Gronow and Cohen, 1984), 2) increased tissue water content by the hypoxic swelling of tubular cells, and finally, 3) liberated membrane-bound γ-glutamyl-transferase (γGT, Fig. 4) as well as intracellular lysosomal N-acetyl-β-D-glucosaminidase (NAG) by the disruptive swelling of cellular membranes. Accordingly, the observed ~9% decrease in fractional Na$^+$ reabsorption in the Ringer-perfused kidney was not related to substrate depletion in our recirculating perfusion system. This point of view is supported by observations of a similar decrease in %T-Na$^+$ in a Ringer-perfused rat kidney inserted into a perfusion set-up equipped with continuous substrate replacement (Schurek and Alt, 1981; Schurek and Kriz, 1985).

Haemoglobin medium

Forced haemolysis in vivo, i.e. the intravascular release of large amounts of haemoglobin into the circulation is an established cause of acute tubular necrosis and, finally, of acute renal failure (Levinski, Alexander and Venkatachalam, 1981). However, the application of haemoglobin as a plasma expander with O_2-binding capacity includes the risk of nephrotoxic effects and of coagulation problems. This risk is diminished with purified, stroma-free haemoglobin which has been thoroughly cleared of toxic substances probably liberated from erythrocyte membranes (Kahn, Allen and Baldassare, 1985). After the infusion of large amounts of haemoglobin in vivo, when the binding capacity of plasma haptoglobins is exhausted, the renal load due to the 'escape' of free, dissociated haemoglobin subunits (through the glomer-ular membrane) can be reduced by coupling dimers to tetramers with 2-nor-2-formyl-pyridoxal-5'-phosphate. This type of coupled haemo-globin (NFPLP-Hb) not only offers the advantage of an increased vascular retention time in vivo but also avoids the unfavourable effects of a high oxygen binding capacity (Bakker et al., 1985).

Accordingly, in the present experiments a high content of coupled haemoglobin molecules in the perfusate (> 80%) was indicated by a reduced urinary excretion of free haemoglobin (0.28 $\pm$ 0.017% with NFPLP-Hb instead of ~3% with free haemoglobin). As was also observed with the other physiological oxygen carrier in the perfusate (5% Ery), and in respect of the perfusions performed with Ringer medium, the addition of 1.7% haemoglobin (1.7% Hb) to the perfusate significantly 1) reduced perfusion flow rate (Fig. 2); 2) raised absolute as well as fractional sodium reabsorption (Table 1; Fig. 3), and 3) lowered γGT-leakage (Fig. 4) and NAG-liberation (Table 1) in the course of a 60 min perfusion. These beneficial effects of 1.7% coupled haemoglobin (or 5% erythrocytes, see below) in the perfusate were probably mediated via a less pronounced arteriovenous oxygen shunting (due to the binding capacity of haemoglobin or erythrocytes) in the descending vasa recta of outer medullary tissue (Fig. 1, left scheme). As a consequence, the steep PO_2 gradient in the corticomedullary region (Fig. 1, middle scheme) would have been shifted in the presence of oxygen carriers into deeper zones of the medulla, as has been demonstrated by Weiss and co-workers in the rat kidney (Baumgartl et al., 1972).

PERFUSION FLOW RATE

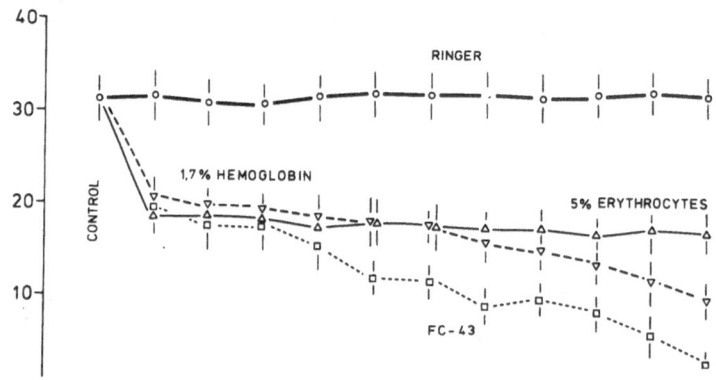

Figure 2. Perfusion flow rate ($\overline{x} \pm$ S.E.M.) with different oxygen carriers in the perfusate of the isolated rat kidney: 1.7% (w:v) NFPLP-coupled haemoglobin (n = 7); 5 vol.% human erythrocytes (n = 6); and 20% (w:v) perfluorocarbon emulsion (FC-43, n = 8), all 3 at PO_2 = 300 mmHg. Control = oxygen carrier-free Ringer solution (n = 9) at PO_2 = 660 mmHg. Ordinate; $ml \cdot gWW^{-1} \cdot min^{-1}$.

After ~1 hour of perfusion with 1.7% Hb, however, functional stability of the isolated rat kidney decreased: PFR fell continuously (Fig. 2), and absolute as well as fractional Na^+ reabsorption (Fig. 3) fell significantly, to the level found with Ringer perfusion. Accordingly, enzyme losses of ƔGT (Fig. 4) and NAG (Table 1) increased, and the water content of the 1.7% Hb-perfused kidney was even higher than after Ringer perfusions (Table 1). Light microscopic examination of Hb-perfused kidneys revealed disseminated intravascular Hb-precipitates. This disturbance of renal microcirculation may have finally impaired renal function by local effects of ischaemia, i.e. the observed decrease in glomerular filtration rate (Table 1) and sodium reabsorption (Fig. 3; Table 1) as well as an increased liberation of cellular enzymes (Fig. 4, Table 1). Similar effects have been demonstrated in the substrate-depleted and hypoxic rat kidney (Gronow and Kossmann, 1985). The present investigations do not allow conclusions about the exact mechanism of the observed formation of haemoglobin precipitates. However, a concomitant development of small haemoglobin clots at the glass surfaces of the arterial and venous reservoirs of the perfusion set-up supports the idea of an increasing haemoglobin instability/condensation during contact with unphysiological surface areas (Bakker et al., 1984).

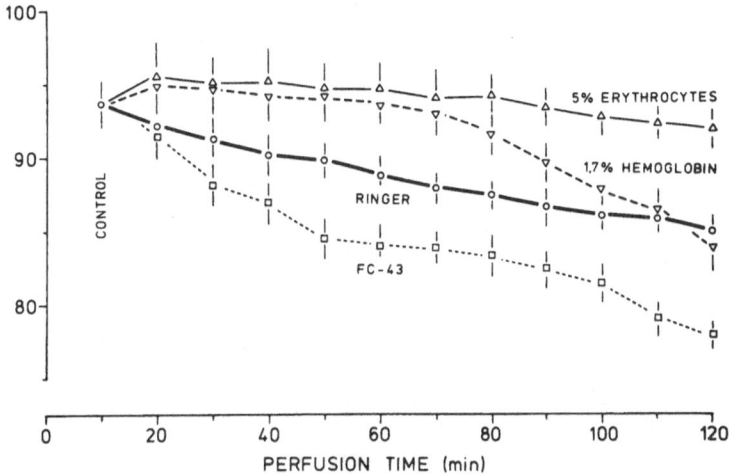

FRACTIONAL Na$^+$-REABSORPTION

Figure 3. Fractional Na$^+$ reabsorption ($\bar{x} \pm$ S.E.M.) with different oxygen carriers in the perfusate of the isolated rat kidney. FC-43 = perfluorocarbon emulsion, for further explanations see legend to Fig. 2. Ordinate; per cent of filtered Na$^+$.

Erythrocyte suspension
        In accordance with the observations of Swanson et al. (1981) the addition of 5 vol.% erythrocytes to the perfusate of the isolated rat kidney stablized renal function remarkably. Only small, insignificant alterations in renal function and tissue integrity occurred in the course of 100 min of perfusion with 5 vol.% erythrocytes (Table 1): perfusion flow rate did not decrease, glomerular filtration rate as well as Na$^+$ reabsorption remained high, and the urinary losses of tissue enzymes (γGT, NAG) were lowest in the presence of 5 vol.% erythrocytes.

        Thus, in accordance with the early observations of Weiss and co-workers (Leichtweiss et al., 1969; Baumgartl et al., 1972) the hypoxic limitation of renal function by a corticomedullary PO$_2$-

gradient (Fig. 1) appeared to be less pronounced in the presence of 5 vol.% erythrocytes (PO_2 = 300 mmHg) than in hyperoxygenated Ringer medium (PO_2 = 660 mmHg). These results support the hypothesis that it is not the oxygen transport capacity of the perfusates (which was similar under both experimental conditions) but the smaller driving force of the lower 'arterial' PO_2 in the erythrocyte suspension which may have reduced arteriovenous oxygen diffusion in the isolated rat kidney.

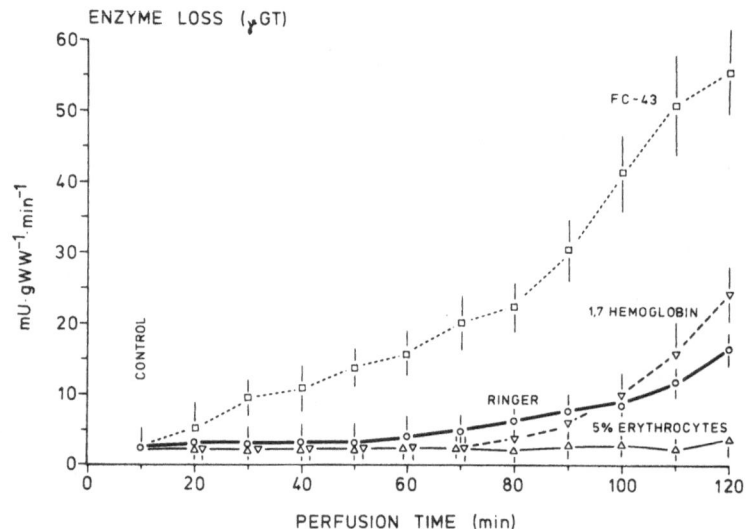

Figure 4. Urinary loss of γ-glutamyltransferase (γGT, $\bar{x}$ + S.E.M.) with different oxygen carriers in the perfusate of the isolated rat kidney. FC-43 = perfluorocarbon emulsion; for further explanations see legend to Fig. 2.

Perfluorocarbon emulsion
    Chemical inertness and a relatively high oxygen binding capacity (~6 vol.% O_2/atm O_2) commended perfluorocarbon emulsions as artificial blood substitutes in the field of experimental physiology and pharmacology (Faithfull et al., 1984). Beneficial effects of perfluorocarbons on oxygen transport to tissue were observed under conditions of low haemoglobin concentration and/or limited organ perfusion, but adverse effects on alveolar and phagocytic cell function limited their clinical application (Kahn et al., 1985); Lowe and Bollands, 1985). Data about the effect of perfluorochemicals on renal function are also conflicting: the perfusion of canine kidneys prior to transplantation with FC-43 'Oxypherol' (10% perfluorotributylamine with 1.28% Pluronic F-68) yielded good functional and morphological results (Ruland et al., 1983). However, morphological lesions in the thick ascending limb of Henle (TAL, see Fig. 1) of isolated rat kidney

preparations were more pronounced after perfusion with 'Oxypherol' than with oxygen carriers such as haemoglobin or erythrocytes in the perfusate (Brezis et al., 1984). FC-43 also induced 'foamy' vacuolization in the cytoplasm of proximal tubular cells of the rabbit kidney (Nanney, Fink and Virmani, 1984).

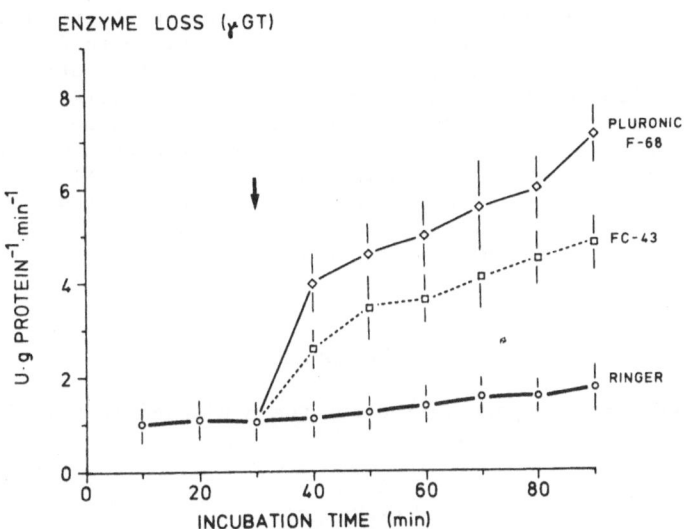

Figure 5. Loss of γ-glutamyltransferase (γGT) by isolated tubular cells from rat kidney cortex ($\overline{x}$ ± S.E.M., n=10) in the presence of 1%(w:v) Pluronic F-68 or FC-43 (20% perfluorotributylamine + 2.56% Pluronic F-68), or in Ringer control (PO_2 = 660 mmHg).

When Franke and Weiss (1976) added a 7.9 vol.% mixture of FC-43 'Fluosol' (perfluorotributylamine plus Pluronic F-68) to the Ringer perfusate of an isolated rat kidney preparation a tremendous increase in the rates of absolute and fractional Na^+ reabsorption ($T-Na^+$ up to 149 µeq $Na^+/g \cdot min^{-1}$, $\%T-Na^+$ up to 88.5%) indicated improved O_2 transport into O_2-deprived renal tissue regions. This beneficial effect of FC-43 on kidney function, however, was only transient. During the course of perfusion $T-Na^+$ was unstable and (indicating a severe impairment of active Na^+ transport) declined steeply by ~85%. Similarly, in the present experiments functional stability proved to be worse in the presence of FC-43 (Table 1 and Figs 2-4). Thus, perfusion flow rate (-83.7%), glomerular filtration rate (-48.6%), and Na^+ reabsorption ($\%T-Na^+$: -15.6%; $T-Na^+$: -45.3%) declined markedly, and (indicating a disruptive swelling of cellular membranes) urinary enzyme losses (γGT: +1494%; NAG: +456%) as well as tissue water content (+22.1%) increased significantly.

In view of some recent reports about potentially cytotoxic effects of the emulsifier Pluronic F-68 (Bucala, Kawakami and Cerami, 1983; Lane and Lamkin, 1984) we did a second series of experiments with isolated tubular segments from rat kidney cortex in which we compared the effect of 1% (w:v) Pluronic F-68 (as the only additive) with the effect of FC-43 (20% perfluorotributylamine plus 2.56% Pluronic F-68) on cellular losses of γ-glutamyltransferase. Isolated cells incubated in Ringer medium lost up to 2 units γGT/g protein·min^{-1} over the time of incubation (Fig. 5). After the addition of FC-43 to the incubation medium (arrow), however, cellular losses of γGT increased more than two fold (~5 U γGT/g·min^{-1}). When 1% Pluronic was the only additive, cellular losses of γGT into the incubation medium accelerated even more and reached a final level up to 7 U γGT/g·min^{-1}. Thus, in the present experiments the beneficial effect of oxygen-carrying perfluorotributylamine may have been overlayed by a membrane-disrupting effect (as indicated by the liberation of membrane-bound γGT from brushborder microvilli) of the detergent Pluronic F-68. Further work will be necessary to clarify the role of solubilizers in adverse effects of perfluorocarbon emulsions, and the isolated kidney appears to be a sensitive model to test oxygen transport capacity as well as the biological compatibility of perfluorocarbon emulsions and/or their additives.

ACKNOWLEDGEMENT

The authors express appreciation for the dedicated technical assistance of Mrs R.Bock.

REFERENCES

Alcorn, D., Emslie, K.R., Ross, B.D., Ryan, G.B. and Tange, J.D. (1981). Selective distal nephron damage during isolated kidney perfusion. Kidney Int. 19, 638-647.

Bakker, J.C., Plas, J. v.d., Bleeker, W.K., De Vries-Van Rossen, A., Schoester, M., Brummelhuis, H.G.J. and Loos, J.A. (1984). Oxygen affinity of hemoglobin solutions modified by coupling to PLP or NFPLP and the effects on tissue oxygenation. In: Oxygen Transport to Tissue-VI. Eds Bruley, D., Bicher, H.I. and Reneau, D., Plenum Press, New York and London, (Adv. Exp. Med. Biol. 180, 345-356).

Baumgartl, H., Leichtweiss, H.-P., Lubbers, D.W., Weiss, Ch. and Huland, H. (1972). The oxygen supply of the dog kidney: measurements of intrarenal pO_2. Microvasc. Res. 4, 247-257.

Brezis, M., Rosen, S., Silva, P. and Epstein, F.H. (1984). Selective vulnerability of the medullary thick ascending limb to anoxia in the isolated perfused rat kidney. J. Clin. Invest. 73, 182-190.

Bucala, R., Kawakami, M. and Cerami, A. (1983). Cytotoxicity of a perfluorocarbon blood substitute to macrophages in vitro. Science, 220, 965-969.

Cohen, J.J. (1979). Is the function of the renal papilla coupled exclusively to an anaerobic pattern of metabolism? Am. J. Physiol. 236, F423-F433.

Faithfull, N.S., Fennema, M., Erdmann, W., Lapin, R., Smith, A.R., Van Alphen, W., Essed, C.E. and Trouwborst, A. (1984). Tissue oxygenation by fluorocarbons. In: Oxygen Transport to Tissue-VI. Eds Bruley, D., Bicher, H.I. and Reneau, D., Plenum Press, New York and London, (Adv. Exp. Med. Biol. 180, 569-580).

Franke, H. and Weiss, Ch. (1976). The O_2 supply of the isolated cell-free perfused rat kidney. In: Oxygen Transport to Tissue-II. Eds Grote, J., Reneau, D. and Thews, G., Plenum Press, New York and London, (Adv. Exp. Med. Biol. 75, 425-431).

Gronow, G.H.J., Benk, P., Franke, H. (1984). Effect of anaerobic substrates on post-anoxic cellular functions in isolated tubular segments of rat kidney cortex. In: Oxygen Transport to Tissue-VI. Eds Bruley, D., Bicher, H.I. and Reneau, D., Plenum Press, New York and London, (Adv. Exp. Med. Biol. 180, 403-410).

Gronow, G.H.J. and Cohen, J.J. (1984). Substrate support for renal functions during hypoxia in the perfused rat kidney. Am. J. Physiol. 247, F618-F631.

Gronow, G.H.J. and Kossman, H. (1985). Perfusate oxygenation and renal function in the isolated rat kidney. In: Oxygen Transport to Tissue-VII. Eds Kreuzer, F., Cain, S.M., Turek, Z. and Goldstick, T.K., Plenum Press, New York and London, (Adv. Exp. Med. Biol. 191, 675-682).

Gronow, G., Meya, F. and Weiss, Ch. (1984). Studies on the ability of kidney cells to recover after periods of anoxia. In: Oxygen Transport to Tissue-V. Eds Lubbers, D.W, Acker, H., Leniger-Follert, E. and Goldstick, T.K., Plenum Press, New York and London, (Adv. Exp. Med. Biol. 169, 589-595).

Kahn, R.A., Allen, R.W. and Baldassare, J. (1985). Alternate sources and substitutes for therapeutic blood components. Blood, 66, 1-12.

Kriz, W. (1981). Structural organization of the renal medulla; comparative and functional aspects. Am. J. Physiol. 241, R3-R16.

Lane, T.A. and Lamkin, G.E. (1984). Paralysis of phagocyte migration due to an artificial blood substitute. Blood, 64, 400-405.

Leichtweiss, H.P., Lubbers, D.W., Weiss, Ch., Baumgartl, H. and Reschke,W. (1969). The oxygen supply of the rat kidney: measurements of intrarenal pO_2. Pflugers Arch. 309, 328-349.

Levinski, N.G., Alexander, E.A. and Venkatachalam, M.A. (1981). Acute renal failure. In: The Kidney, vol. 1. Eds Brenner, B.M. and Rector, F.C., W.B. Saunders, Philadelphia, pp. 1181-1236.

Little, J.R. and Cohen, J.J. (1974). Effect of albumin concentration on function of isolated perfused kidney. Am. J. Physiol. 226, 512-517.

Lowe, K.C. and Bollands, A.D. (1985). Physiological effects of perfluorocarbon blood substitutes. Med. Lab. Sci. 42, 367-375.

Maack, Th. (1986). Renal clearance and isolated kidney perfusion technique. Kidney Int. 30, 142-151.

Maruhn, D. (1976). Rapid colorimetric assay of β-galactosidase and N-acetyl-β-glucosaminidase in human urine. Clin. Chim. Acta, <u>73</u>, 453–461.

Nanney, L., Fink, L.M. and Virmani, R. (1984). Perfluorochemicals: morphologic changes in infused liver, spleen, lung and kidney of rabbits. Arch. Pathol. Lab. Med. <u>108</u>, 631–637.

Ross, B.D. (1978). The isolated perfused rat kidney. Clin. Sci. Molec. Med. <u>55</u>, 513–521.

Ruland, O., Hauss, J., Spiegel, H.-U. and Schoenleben, F.H. (1983). Comparative pO_2 histograms and other parameters in canine kidneys during perfusion with Oxypherol. In: Advances in Blood Substitute Research. Eds Bolin, R.B., Geyer, R.P. and Nemo, G.J., Alan R. Liss, New York, pp. 221–225.

Schurek, H.-J. and Alt, J.M. (1981). Effect of albumin on the function of perfused rat kidney. Am. J. Physiol. <u>240</u>, F569–F576.

Schurek, H.-J. and Kriz, W. (1985). Morphologic and functional evidence for oxygen deficiency in the isolated perfused kidney. Lab. Invest. <u>53</u>, 145–155.

Swanson, J.W., Besarab, A., Pomerantz, P.P. and De Guzman, A. (1981). Effect of erythrocytes and globulin on renal functions of the isolated rat kidney. Am. J. Physiol. <u>241</u>, F139–F150.

Weiss, Ch., Passow, H. and Rothstein, A. (1959). Autoregulation of blood flow in isolated rat kidney, in the absence of red cells. Am. J. Physiol. <u>196</u>, 1115–1118.

HIGHLY POLYMERIZED HUMAN HAEMOGLOBIN AS AN OXYGEN-CARRYING BLOOD SUBSTITUTE

W.K.R. BARNIKOL and O. BURKHARD[*]

Physiologisches Institut und [*]Haematologische Abteilung, Johannes Gutenberg-Universitaet, Saarstr. 21, D-6500 Mainz, F.R.G.

INTRODUCTION

There are many situations where a quick substitution of erythrocytes is warranted. However the need for blood typing generally causes considerable delay in getting the matched erythrocytes. Furthermore, in the case of a catastrophe or war there may be a great demand for blood which cannot be delivered by the blood donors available. In these cases an artificial oxygen-carrying blood substitute which does not require blood typing and which can be stored for a considerable time may save lives. For example, in the case of an accident such a blood substitute is available at once. Another field of application for such solutions lies in chronic peripheral arterial occlusion. It has been shown that in these cases an oxygen-carrying blood substitute is superior to simple isoncotic volume substitution (Moss et al., 1953).

Many laboratories are therefore trying to develop an artificial oxygen-carrying blood substitute. One approach is the use of perfluorochemicals (Clark and Gollan, 1966; Pabst, 1977; Hirlinger et al., 1982). However animal experiments have shown that at least the commercially available perfluorochemical emulsions are not inert substances but cause considerable side effects (Lowe and Bollands, 1985; Sharma et al., 1987). In a recent paper Gould et al. (1986) demonstrated that after blood loss Fluosol-DA is unnecessary in moderate anaemia and ineffective in severe anaemia.

On the other hand one may ask whether haemoglobin (Hb) solutions may be used as a red cell substitute. Three main problems arise in this area. First the Hb solutions should not change the oncotic pressure of blood significantly, to ensure that the oncotic balance is kept up. Second, the artificial blood substitute must have a certain residence time in the animal or man. Low residence times require repeated transfusions which impose an overload of protein on the organism especially on the kidneys. Third, the artificial blood substitute should not increase the blood viscosity. In nature those problems are solved in two different ways.

In higher animals the Hb solution is contained in specific compartments, the erythrocytes. Following this example some scientists have tried to use microencapsulated Hb solutions to mimic the erythrocyte situation (Nicolau and Gersonde, 1977; Hunt et al., 1985) but the microcapsules are quite unstable and have a low residence time. An alternative solution is employed in lower animals. For example the earthworm uses a large oxygen-carrying molecule, erythrocruorin, with a molecular weight of $3.4 \cdot 10^6$ dalton. The modified Hb solutions produced so far have mean molecular weights of about $0.2 \cdot 10^6$ dalton (Kothe and Bonhard, 1985).

EXPERIMENTS, RESULTS AND DISCUSSION

Our approach is to mimic the situation in the earthworm and to use Hb polymers with a very large molecular weight. To get such a product we polymerized haemoglobin in highly concentrated solutions (> 30 g/dl). The key feature of our procedure is to use a two-phase reaction system. The linking agent is dissolved in an organic phase which is mixed with the Hb solution. The linking agent diffuses slowly from the organic phase into the Hb solution.

In a typical experiment 3 ml of the highly concentrated human Hb solution are placed in a 25 ml flask and 120 μl of a saturated solution of glutardialdehyde in chloroform (made by thoroughly mixing equal volumes of chloroform and 25% aqueous aldehyde solution) are added.

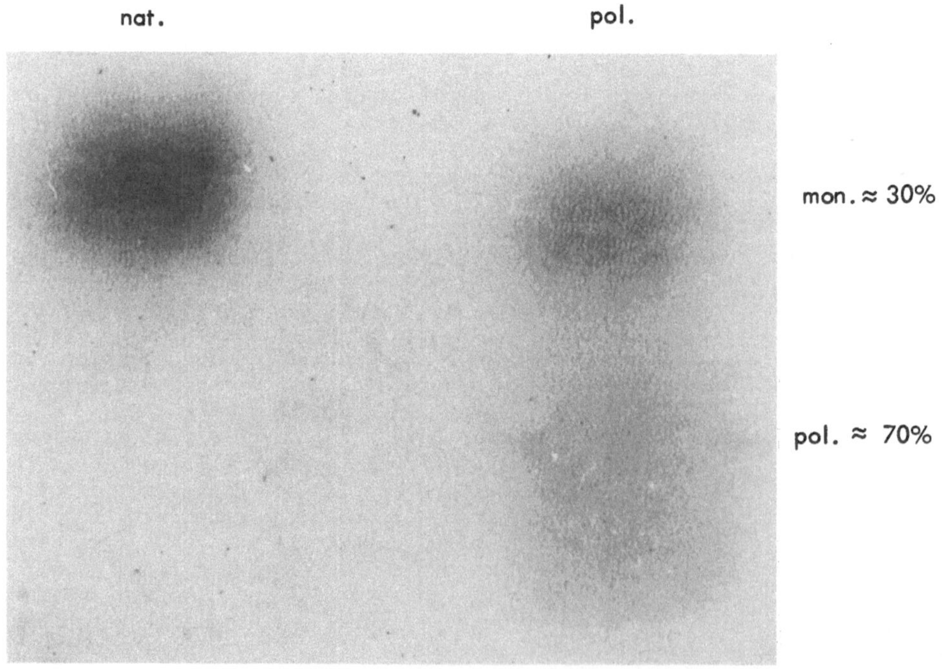

Figure 1. Sephadex G150 thin layer gel chromatography of the reaction product described in the text. Monomer is separated from polymers and oligomers. nat., native Hb; pol., polymerized; mon., monomer.

The mixture is stirred at room temperature and after approximately 90 min the reaction is stopped by dilution. The product is chromatographed (thin layer) with the aid of Sephadex G150 gel (Deutsche Pharmacia GmbH). The chromatogram is given in Figure 1. Since Sephadex gel does not retain proteins of a molar weight greater than 150 kdalton the portion of monomer and polymerized Hb can be determined. The polymer yield is about 70% in mass.

In a second experiment we tested whether high molecular weights are actually reached. For this investigation thin layer chromatography with Sephacryl S400 (Deutsche Pharmacia GmbH) was performed. As molecular weight markers we used erythrocruorin and Blue Dextran (Deutsche Pharmacia GmbH). The Blue Dextran has an average molecular weight of $2 \cdot 10^6$ dalton.

Figure 2 shows the chromatogram with Sephacryl S400. We obtained

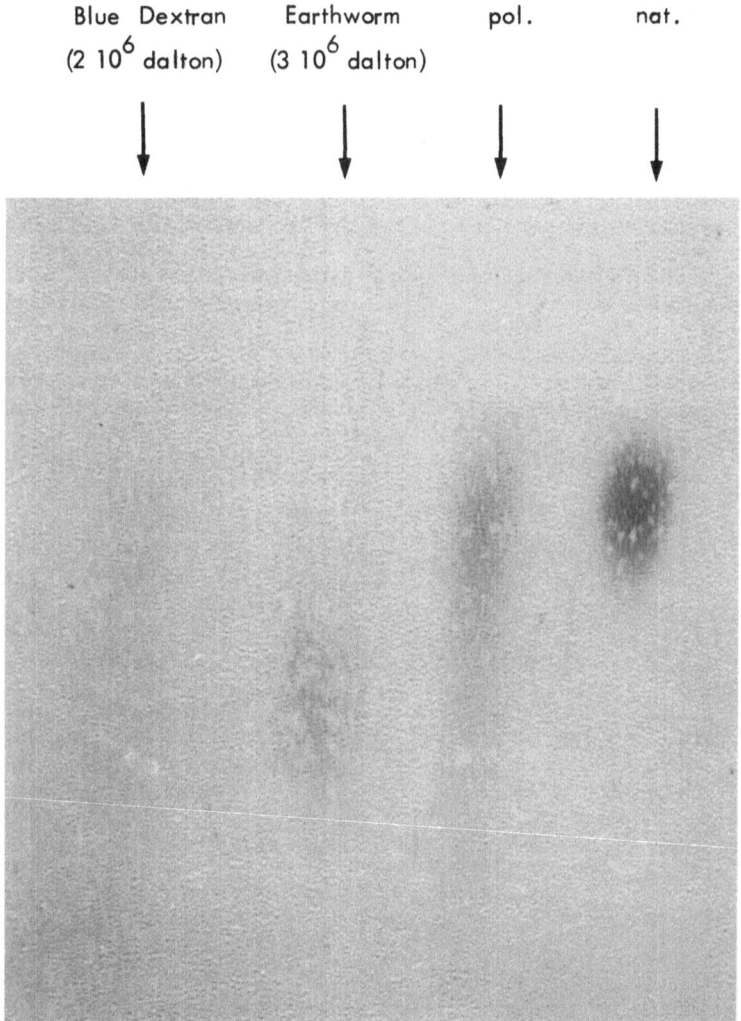

Figure 2. Sephacryl S400 thin layer gel chromatography of the reaction product as described in the text. Earthworm erythrocruorin, and Blue Dextran (average molecular weight $2 \cdot 10^6$ dalton) serve as molecular weight markers. Abbreviations as Fig. 1.

a broad distribution of molecular weights, some components having a molecular weight which is comparable to earthworm erythrocruorin and some having an even higher molecular weight.

The oxygen pressure at half saturation (P_{50}) is about 13 mmHg which is approximately half the P_{50} of blood under normal conditions. For comparison the binding curve of native Hb under the same conditions is shown; it has a P_{50} of 18 mmHg.

Some researchers believe that the P_{50} is the essential parameter which largely governs the O_2 transport to tissue. In the context of this thinking some experiments of Moss et al. (1984) are interesting. These authors demonstrated that it is more important to have a sufficiently high oxygen transport capacity than to have a high P_{50}. So it may be possible to supply adequate oxygen to tissue even with a Hb solution having a P_{50} value of about 13 mmHg. Further experiments are warranted to clarify this point.

In Figure 3 the O_2 binding curve of our product is shown.

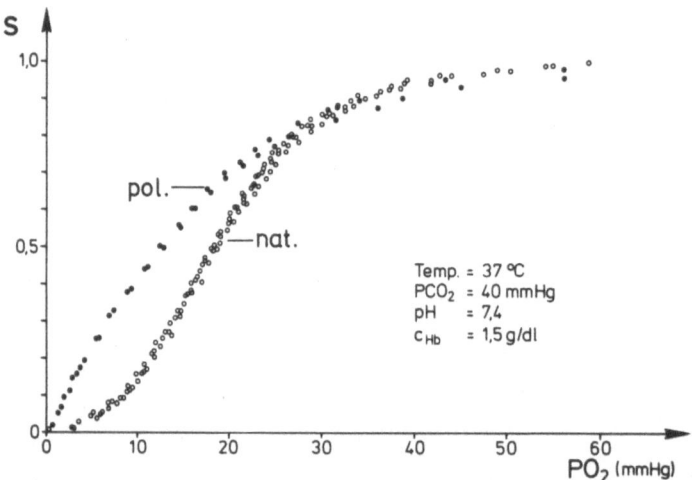

Figure 3. O_2 binding curve of polymerized Hb and native Hb under conditions given. S, O_2 saturation; PO_2, O_2 partial pressure.

In Figure 4 the Hill coefficient (n) is plotted versus O_2 saturation. It is well known that the Hill coefficient can be interpreted as a measure of cooperativity (Barnikol, 1982a,b; Barnikol and Burkhard, 1983; Burkhard and Barnikol, 1983). The cooperativity of polymerized Hb is clearly diminished compared with native Hb.

To summarize, our results demonstrated that it is possible to get polymerized human Hb with a molecular weight of $3 \cdot 10^6$ dalton or higher, in a reasonable yield. The P_{50} is not optimal but may be high enough for clinical purposes. In addition it is possible to shift the P_{50} up by further chemical modification (e.g. pyridoxylation) of our product (see for instance, Kothe and Bonhard, 1985). Further experiments with animals are warranted to measure residence time and viscosity of the highly polymerized Hb solution. Preliminary observations indicate that the viscosity of our polymerized Hb solution is comparable to that of blood.

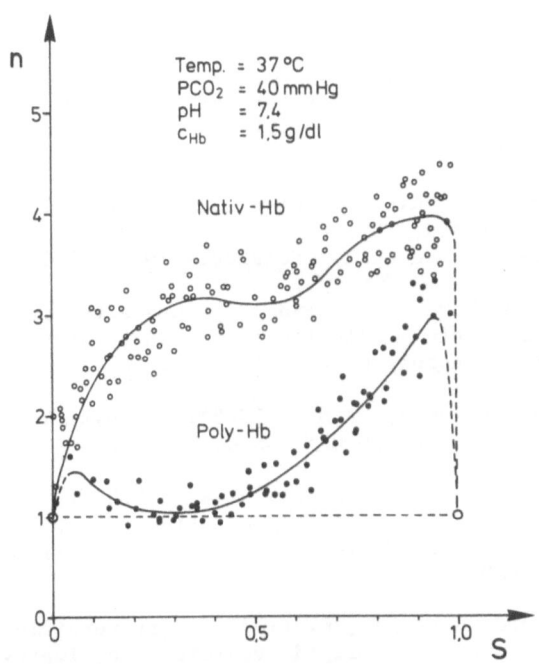

Figure 4. Dependence of Hill's coefficient (n) on O_2 saturation of poly-Hb and native Hb.

REFERENCES

Barnikol, W.K.R. (1982a). Die Feinstruktur der Sauerstoff-Haemoglobin-Bindungskurve. Funkt. Biol. Med. 1, 46–52.

Barnikol, W.K.R. (1982b). Der Einfluss der Effektoren auf die Feinstruktur der Sauerstoff-Haemoglobin-Bindungskurve. Funkt. Biol. Med. 1, 214–220.

Barnikol, W.K.R. and Burkhard, O. (1983). Feinstruktur der Sauerstoff-Bindung und intermolekulare Wechselwirkung des Haemoglobins im Tierreich: Regenwurm, Frosch, Ente und Forelle. Funkt. Biol. Med. 2, 53-59.

Burkhard, O. and Barnikol, W.K.R. (1983). Die Konzentrations-abhangigkeit molekularspezifischer physikalisch-chemischer und biologischer Eigenschaften des humanen Haemoglobins als Beweis fur intertetramere Wechselwirkungen. Funkt. Biol. Med. 2, 185-193.

Clark, L.C., Jr and Gollan, F. (1966). Survival of mammals breathing organic liquids equilibrated with oxygen at atmospheric pressure. Science, 152, 1755-1756.

Gould, S.A., Rosen, A.L., Sehgal, L.R., Sehgal, H.L., Langdale, L.A., Krause, L.M., Rice, C.L., Chamberlin, W.H. and Moss, G.S. (1986). Fluosol-DA as a red cell substitute in acute anemia. N. Engl. J. Med. 314, 1653-1656.

Hirlinger, W.K., Grunert, A., Herrmann, M., Petutschnigk, D. and Langer, K. (1982). Auswirkungen eines teilweisen Blutaustausches mit Fluosol-DA 20% auf den intakten Organismus des Schweines. Anaesthesist, 31, 660-666.

Hunt, C.A., Burnette, R.R., MacGregor, R.D., Strubbe, A.E., Lau, D.T., Taylor, N. and Kawada, H. (1985). Synthesis and evaluation of a prototypal artificial red cell. Science, 230, 1165-1168.

Kothe, N. and Bonhard, K. (1985). Characterization of a modified stroma-free hemoglobin solution as an oxygen-carrying plasma substitute. Surg. Gynecol. Obstet. 161, 563-569.

Lowe, K.C. and Bollands, A.D. (1985). Physiological effects of perfluorocarbon blood substitutes. Med. Lab. Sci. 42, 367-375.

Moss, G.S., Gould, S.A., Sehgal, L.R., Sehgal, H.L. and Rosen, A. (1984). Hemoglobin solution - from tetramer to polymer. In: The Red Cell, Sixth Ann Arbor Conference. Ed. Brewer, G., Alan R. Liss Inc., New York, pp. 191-207.

Nicolau, Y.C. and Gersonde, K. (1977). Irreversible incorporation of allosteric effectors into lipid vesicles to improve unloading by hemoglobin in erythrocytes. Deutsches Patent 2,740053.

Pabst, R. (1977). Sauerstofftransport mit stromafreien Haemoglobin-losungen und Fluorocarbonen. Med. Klinik, 72, 1555-1562.

Sharma, S.K., Bollands, A.D., Davis, S.S. and Lowe, K.C. (1987). Emulsified perfluorochemicals as physiological oxygen-transport fluids: assessment of a novel formulation. This volume.

OXYGEN BOUNDARY CROSSING PROBABILITIES

N. A. BUSCH and I. A. SILVER

Department of Pathology, University of Bristol
Bristol BS8 1TD, U.K.

SUMMARY

The probability that an oxygen particle will reach a time dependent boundary is required in oxygen transport studies involving solution methods based on probability considerations. A Volterra integral equation is presented, the solution of which gives directly the boundary crossing probability density function. The boundary crossing probability is the probability that the oxygen particle will reach a boundary within a specified time interval. When the motion of the oxygen particle may be described as strongly Markovian, then the Volterra integral equation can be rewritten as a generalized Abel equation, the solution of which has been widely studied.

INTRODUCTION

The most convenient setting for the study of oxygen transport in tissue is in the theory of stochastic processes. Here, one postulates a set of basic rules for the behaviour of a particle based on the local environment, and then uses the tools of mathematical statistics to ascertain the global behaviour of the system. Since even simple analytical results are difficult to obtain, quantitative studies using what has become known as Monte Carlo simulation have been extensively used. The Monte Carlo simulations are essentially numerical experiments, where a sequence of random numbers is generated, and a set of rules are applied for each number in the sequence, based upon the magnitude of the random number. The general analytic solution methods permit minimal use of numerical experiments to study the behaviour of the system since the behaviour is defined to conform to a set of explicit equations.

The Monte Carlo, or 'random walk' studies involve the repeated application of the local environment behaviour rules to a particle, until either the particle ceases to exist (in the more general setting of birth/death processes), leaves the domain of interest, or a time interval has passed. When the particle ceases to exist, it is subtracted from the total number of particles which are studied for the next time interval, and its effect on the reduced number accounted for in the statistics for the current time interval. If the particle still exists within the domain of interest after the current time interval has passed, then the particle is included in the statistics for the time interval. In the case that the particle leaves the domain of interest during the current time interval, then the behaviour rules for the local environment which applied to the particle on the interior of the domain no longer apply. Instead, the behaviour of the particle at the boundary must be defined according to a second set of rules. As with deterministic analysis of oxygen transport, the definition of the behaviour of the species at the boundary can be difficult, and the computational aspects of the application of the rules can be expensive.

The objective of formulating an analytic solution to the boundary crossing probability density function problem is to avoid the computational strategies of the Monte Carlo simulation studies of the phenomena of oxygen transport. The background to the deterministic solution of the problem of transport phenomena in living tissue is given by Lightfoot (1974). For deterministic analysis of the transmembrane transport phenomena, the reader is directed to the works of Lakshminarayanaiah (1969,1984). A large amount of work has centred on the use of Monte Carlo simulation of transport phenomena in living tissue. However, no concise treatise exists which is of general use in the area of oxygen transport.

The mathematical statistics of the problem of oxygen transport in tissue in the setting of stochastic processes can be written analytically, once the basic rules for the local behaviour of the species have been defined. Of particular interest is the determination of the probability of crossing a time dependent boundary, when the oxygen particle motion has been defined. The required probability is called a boundary crossing probability, and is necessary to the general solution of the diffusion, convection, and reaction problem of oxygen transport. The general solution may be written as the sum of three integrals: the first accounts for the motion of the particles originally in the domain of movement, the second accounts for birth/death processes within the domain, and the third is for behaviour at the boundary of the domain. The integrand of the third integral involves the first passage time probability density function.

THEORY

Following Fortet (1943), let the local behaviour of the oxygen particle be as $X(t)$, with time bounded by zero, and infinity; that is let $X(t)$ be a Brownian motion process which is almost surely at the spatial origin at time zero. The covariance in time of the process is defined to be equal to the minimum of the end of the time interval. Let the boundary of the domain accessible to the Brownian particle be $x(t)$, which is bounded and continuous for the entire time interval. The problem is to determine the probability that the process $X(t)$ will cross the boundary $x(t)$ before some time $t = T$. The resultant probability is then used in the solution of the general analytic diffusion, convection and reaction problem.

Let $U(t,x(t); \tau, \xi)$ be the conditional density of the process $X(t')$ at $\xi = x(t) + \epsilon$ at the time $\tau > t$, given that the process was at $x(t)$ at time t. $U(t,x; \tau, \xi)$ is the density of the process $X(t')$ at $\xi = x(\tau) + \epsilon$ at the time $\tau > t$, given that $X(t) = x$. It is clear that $U(t,x; \xi, \tau)$ is the solution of the backward Kolmogorov equation on an unbounded medium, and $U(t,x(t); \xi, \tau)$ is the solution of the forward Kolmogorov equation on the same unbounded medium. Denote $\Phi(t,x; \tau)$ as the probability that the process $X(t')$ crosses the curve $C:[x=x(t)]$ in the domain D:[(t,x)]. That is, Φ is the required boundary crossing probability, and the time derivative is the boundary crossing probability density function.

The probability that X(t') is at $\xi = x(\tau) + \epsilon$ at time $\tau > t$ given that X(t)=x, [x<x(t)] is,

$$Pr\{X(\tau)>x(\tau);X(t)=x\}=\int_{x(\tau)}^{\infty}U(t,x;\tau,\xi)d\xi, \tag{1}$$

also

$$Pr\{X(\tau)>x(\tau);X(t)=x\}= \tag{2}$$

$$\int_{t}^{\tau}[\int_{x(\tau)}^{\infty}U(t',x(t');\tau,\xi)d\xi]\{d\Phi(t,x;t',\xi(t'))/dt'\}dt'.$$

Equating Equations (1) and (2) gives

$$\int_{x(\tau)}^{\infty}U(t,x;\tau,\xi)d\xi=$$
$$\int_{t}^{\tau}[\int_{x(\tau)}^{\infty}U(t',x(t');\tau,\xi)d\xi]\{d\Phi(t,x;t',\xi(t'))/dt'\}dt'. \tag{3}$$

The function $\Phi(t,x;\tau)$ also satisfies the following:

$$\int_{z_0}^{z_1}U(t,x;\tau,\xi)d\xi=$$
$$\int_{t}^{\tau}[\int_{z_0}^{z_1}U(t',x(t');\tau,\xi)d\xi]\{d\Phi(t,x;t',\xi(t'))/dt'\}dt'. \tag{4}$$

provided that $z_0 < z_1$ and $z_0,z_1 > x(\tau)$. If $z_0 > x(\tau)$, and $U(t'x(t'); \tau, \xi) > 0$ and bounded, then the order of integration in Equation (4) may be interchanged. As the result is valid for any z_0, and z_1, then the boundary crossing probability density function satisfies the following:

$$U(t,x;\tau,\xi)= \tag{5}$$

$$\int_{t}^{\tau}U(t',x(t');\tau,\xi)\{d\Phi(t,x;t',\xi(t'))/dt'\}dt'.$$

When the motion of the oxygen particle is defined to be Brownian, then $U(t,x; \tau, \xi)$ and $U(t,x(t); \tau, \xi)$ are given by the following two equations:

$$U(t,x;\tau,\xi)= \tag{6a}$$

$$[4\pi(\tau-t)]^{-1/2}exp\{-[x(\tau)-x]^2/4(\tau-t)\},$$

and

$$U(t,x(t);\tau,\xi)=$$

$$(6b)$$

$$[4\pi(\tau-t)]^{-1/2}\exp\{-[x(\tau)-x(t)]^2/4(\tau-t)\},$$

Equation (4) is a Volterra integral equation of the first kind and when the oxygen motion is Brownian, it is easily transformed into the generalized Abel equation. The required transformation is made as follows, (cf. Smith, 1972), let

$$U(t,x;\tau,\xi)=g(t)=$$

$$(7a)$$

$$[4\pi(\tau-t)]^{-1/2}\exp\{-[x(\tau)-x]^2/4(\tau-t)\},$$

and

$$k(\tau,t)=$$

$$(7b)$$

$$[4\pi]^{-1/2}\exp\{-[x(\tau)-x(t)]^2/4(\tau-t)\},$$

and finally,

$$f(t')=d\Phi(t,x;t',\xi(t'))/dt'.$$

$$(7c)$$

The generalized Abel equation is simply then

$$g(\tau)=\int_t^\tau k(\tau,t')[\tau-t]^{-\alpha}f(t')dt'.$$

$$(8)$$

There are many existing methods for solving the generalized Abel equation, any one of which may be suitable for obtaining the boundary crossing probability density function from Equation (8). Equation (5) for the first passage time probability density function was transformed into an Abel equation simply to demonstrate that solution techniques for such equations have been known for some time. As Equation (5) is a Volterra integral equation of the first kind, any solution technique for the particular type of equation may be employed. The first known specific solution to Equation (5) for a strong Markov process was given by Durbin (1971), and has since been extended by Smith (1972), Anderssen, De Hoog and Weiss (1973) and Ricciardi and Sato (1983). Equation (5) can be easily extended to include the existence of an absorbing boundary; again, the first known such result is that given by Cuzik (1981), extended by Ricciardi, Sacerdoti and Sato (1984), and Durbin (1985). The resulting equation is a Volterra integral equation of the second kind which is easily solved using the method of Bownds and coworkers (Bownds and Wood, 1976, 1979; Bownds, 1982; Bownds and Applebaun, 1985). Of particular interest is the use of Equation (5) for the first passage time probability density function for an

Ornstein-Uhlenbeck process (Uhlenbeck and Ornstein, 1930; extended by Uhlenbeck and Wang, 1945). The reader is directed to the work of Chandrasekhar (1943) for a detailed investigation of the Ornstein-Uhlenbeck process, and to the work of Durbin (1971, 1985) for the solution of the first passage probability.

CONCLUSION

The analytic formulation of the solution to the oxygen transport problem in tissue requires knowledge of the first passage time probability density function to the domain boundary. In the case of a single spatial dimension, and time, the required probability density function has been given as the solution of a particular Volterra integral equation of the first kind. The transformation of the Volterra integral equation to a generalized Abel equation has been demonstrated. The objective of defining an analytic formulation for the boundary crossing probability density function is to avoid the uneconomic computational strategies inherent in Monte Carlo simulation studies of the phenomena of oxygen transport.

REFERENCES

Anderssen, R.S., De Hoog, F.R. and Weiss, R. (1973). On the numerical solution of Brownian motion processes. J. Appl. Probability, 10, 409-418.

Bownds, J.M. (1982). Theory and performance of a sub-routine for solving Volterra integral equations. Computing, 28, 317-332.

Bownds, J.M. and Applebaum, L. (1985). Algorithm 627, a Fortran subroutine for solving Volterra integral equations. ACM Trans. Math. Software, 11, 58-65.

Bownds, J.M. and Wood, B. (1976). On numerically solving nonlinear Volterra integral equations with fewer computations. SIAM J. Numerical Analysis, 13, 705-519.

Bownds, J.M. and Wood, B. (1979). A smoothed projection method for singular nonlinear Volterra equations. J. Approxim. Theory, 25, 120-141.

Chandrasekhar, S. (1943). Stochastic problems in physics and astronomy. Reviews of Modern Physics, 15, 1-89.

Cuzik, J. (1981). Boundary crossing probabilities for stationary gaussian processes and Brownian motion. Trans. Am. Math. Soc. 263, 469-492.

Durbin, J. (1971). Boundary-crossing probabilities for the Brownian motion and poisson processes and techniques for computing the power of the Kolmogorov-Smirnov test. J. Appl. Probability, 8, 431-453.

Durbin J. (1985). The first passage density of a continuous gaussian process to a general boundary. J. Appl. Probability, 22, 99-122.

Fortet, R. (1943). Les fonctions aleatoires du type de Markoff associees a certaines equations lineaires aux derivees partielles du type parabolique. J. Math. Pures Appl. 22, 177-243.

Lakshminarayanaiah, N. (1969). Transport Phenomena in Membranes. Academic Press, New York.

Lakshminarayanaiah, N. (1984). Equations of Membrane Biophysics. Academic Press, New York.

Lightfoot, E.N., Jr (1974). Transport Phenomena and Living Systems. New York.

Ricciardi, L.M., Sacerdote, L. and Sato, S. (1984). On an integral equation for first passage time probability densities. J. Appl. Probability, 21, 302-314.

Ricciardi, L.M. and Sato, S. (1983). A note on the evaluation of first passage time probability densities. J. Appl. Probability, 20, 197-201.

Smith, C.S. (1972). A note on boundary-crossing probabilities for the Brownian motion. J. Appl. Probability, 9, 857-861.

Uhlenbeck, G.E. and Ornstein, L.S. (1930). On the theory of Brownian motion. Physical Review, 36, 823-841.

Uhlenbeck, G.E. and Wang, M.C. (1945). On the theory of the Brownian motion II. Reviews of Modern Physics, 17, 323-342.

IN VIVO RECRUITMENT OF MITOCHONDRIAL $\dot{V}O_2$: TEST OF CURRENT MODELS USING TISSUE DATA

R.J. CONNETT

The University of Rochester, School of Medicine and Dentistry, 601 Elmwood Avenue, Rochester, NY 14642 U.S.A.

INTRODUCTION

One of the most diligently studied and least understood biological processes is the control of $\dot{V}O_2$ and oxidative phosphorylation in vivo. Currently two basic conceptual frameworks are being used to analyse $\dot{V}O_2$. One approach is based on understanding the biochemical mechanisms and is focused on the identification of the rate-limiting step or steps in the coupling between demand-induced changes in the cytosolic energy state and the rate of oxidative phosphorylation. The other approach is derived from non-equilibrium thermodynamics.

Mechanistic models describe the rates in terms of an enzyme capacity and the concentration of substrate or substrates. The form of the resulting expressions is basically that of the standard Michaelis-Menten equation. The proposed rate-limiting steps are currently in dispute. Evidence has been presented to support a major role for the adenine nucleotide translocase in isolated mitochondria (Davis and Lumeng, 1975; Kuster, Bohnensack and Kunz, 1976; Letko et al., 1980; Kuntz et al., 1981; Williamson et al., 1981). This concept has been expanded in those tissues containing significant amounts of creatine phosphate (PCr) to include the function of a 'creatine shuttle' in delivering the ADP to the mitochondrial membrane (Jacobus, 1985). Under this model $\dot{V}O_2$ would have a Michaelis-Menten dependence on [ADP] or [ADP]/[ATP] and would be relatively independent of phosphate concentration changes. An alternative model is that derived by Wilson and coworkers which implies that the turnover rate of electron transport, and hence $\dot{V}O_2$, is determined by the amount of cytochrome c oxidase, the level of reduction of cytochrome c and the extramitochondrial [ATP]/[ADP]·[Pi] (Wilson et al., 1974; Wilson, Owen and Erecińska, 1977; Wilson, Owen and Holian, 1977). This model requires that adenine nucleotide translocase should not be rate-limiting but may include a strong dependence on [Pi] (Wilson, Erecińska and Schramm, 1983).

The non-equilibrium thermodynamic description of $\dot{V}O_2$ involves the two driving forces (phosphorylation potential and redox potential) for phosphorylation and electron transport and the coupling between them. The descriptive equation derived from this type of analysis is not Michaelis-Menten but involves the dependence of the $\dot{V}O_2$ on the log ([ATP]/[ADP]·[Pi]) and log (redox ratio). Application of this type of analysis to isolated mitochondria and to liver cells has shown that the linear form of the equation gives a good fit to the data and is consistent with the best current estimates of the standard free energies and the stoichiometry. The observation of linearity requires that the coupling between $\dot{V}O_2$ and phosphorylation not be fixed (Rottenberg, 1973; Van Der Meer et al., 1978; Stucki, 1980b; Van Dam et al., 1980; Pietrobon et al., 1982; Pietrobon, Zoratti and Azzone, 1983). A theoretical analysis of the linear model demonstrates that optimal efficiency could not occur with a coupling of 1 but that optimal efficiency for power output or ATP flow occurs with a coupling near 0.95 (Stucki, 1980b). Soboll and Stucki (1985) and O'Shea and Chappell (1984) have shown that the coupling, i.e. P/O ratio, may vary with the conditions both in intact liver and in isolated mitochondria.

The principal goal of this report is to use data obtained during the rest-work transition in dog gracilis muscle to test the ability of the various models to predict in vivo oxygen consumption.

DATA

The data to be used for model testing are all derived from vascularly-isolated, in situ, dog gracilis muscles studied during a rest-work transition at about 65% of $\dot{V}O_{2max}$. The muscles were stimulated via the nerve to develop isometric contractions at 4 Hz. Sampling by fast freezing was carried out at 5, 10, 15, 30, 60 and 180 sec after the onset of stimulation. All the data and detailed methods have been reported previously (Connet, Gayeski and Honig, 1985a; Gayeski, Connett and Honig, 1985; Connett, 1986; Olgin, Connett and Chance, 1986). A summary of the data is shown in Table 1. It includes:

1. Directly measured rates and contents of metabolic intermediates

2. Values directly computed from the measured variables using well defined enzyme equilibrium conditions. These include cytosolic pH, free cytosolic [ATP] (Connett, 1986), and mitochondrial redox ratios (Olgin et al., 1986). Correction for binding of Mg, K and H has been taken into account in the computations. It was assumed that [K^+] and [Mg^{2+}] were constant at 0.1 M and 1 mM respectively. All the calculation methods have been reported previously (Connett, 1985, 1986; Olgin et al., 1986).

TESTING OF MODELS

Mechanistic models of rate-limiting steps
All mechanistic models must fit a variant of the Michaelis-Menten equation, i.e. a hyperbolic relationship between the substrate concentration and the rate.

Table 1. Data used in analysis

Stimulation duration (sec):	0	5	10	15	30	60	180
$\dot{V}O_2$ $\mu mol/g \cdot min^{-1}$	0.079 +0.006	0.905 +0.07	1.60 +0.20	1.96 +0.16	3.16 +0.55	4.69 +0.39	4.76 +0.24
ATP mM	6.92 +0.37	7.02 +0.31	6.08 +0.52	6.44 +0.39	5.43 +0.39	5.10 +0.19	5.74 +0.34
ADP µM	65.3 +3.9	274.5 +12.1	136.3 +16.7	127.2 +3.2	54.0 +3.0	57.0 +4.5	44.7 +13.0
Pi mM	1.32	3.03	5.03	4.04	10.65	15.02	15.40
PCr mM	20.5 +0.8	18.7 +0.6	17.7 +0.9	18.3 +0.7	12.7 +0.6	8.7 +0.6	7.7 +1.5
Cr mM	16.4 +0.8	18.4 +0.6	19.5 +0.9	18.8 +0.7	24.4 +0.6	28.5 +0.6	29.5 +1.5
pH	7.28 +0.07	7.82 +0.19	7.52 +0.13	7.49 +0.12	6.86 +0.13	6.64 +0.11	6.40 +0.17
PO_2 torr	16.8	15.1	13.0	13.0	6.7	8.0	10.3
NADH/NAD	0.18	0.13	0.13	0.03	0.09	0.17	0.16

1. Adenine nucleotide translocase

The kinetics of the adenine nucleotide translocase reaction are generally formulated as shown in Equation (1) below (Davis and Lumeng, 1975):

$$V = V_{max}[ADP]/\{(1+[ATP]/K_i)K_m + [ADP]\} \tag{1}$$

During the initial period where $\dot{V}O_2$ is changing most rapidly there is little change in the ATP concentration so the inhibitory term, $(1 + [ATP]/K_i)$, is nearly constant. During the first 15 sec there is a sharp rise and fall in [ADP] (Table 1). Note, there is no comparable change in $\dot{V}O_2$. This implies that Equation (1) cannot describe the change in $\dot{V}O_2$. A more quantitative test using published values of $K_m = 2.9$ µM, $K_i = 30$ mM (Jacobus, 1985) and a V_{max} that forces a fit at steady-state is shown in Figure 1. The results show, as expected from the low K_m for this process, that the translocase is operating near V_{max} under the conditions studied. Thus the adenine nucleotide translocase cannot be functioning as a rate-limiting step in these muscles. This conclusion is independent of the value of K_m used since the lowest [ADP] occurs at the highest $\dot{V}O_2$ and the small changes in [ATP] are not correlated with $\dot{V}O_2$.

2. Pi translocase

A phosphate limit on mitochondrial function is much harder to evaluate. There are at least two phosphate translocators in the mitochondrial membrane: a Pi/OH exchanger and a Pi/organic acid exchanger. It has been demonstrated that the Pi/OH exchanger can be rate-limiting at low phosphate concentrations and the rate is dependent on the extramitochondrial pH (Tobin, Mackerer and Mehlman, 1972; Coty and Pedersen, 1974). The [Pi] is initially at ~ 1 mM which is about the K_m for the Pi/OH exchange system (Coty and Pedersen, 1974). Thus during the initial 5-10 sec this step could be limiting. This limitation would be compounded by the alkaline pH change during the initial seconds of stimulation. If the substrate is $H_2PO_4^-$, the substrate for the exchange will be decreased with an increase in pH

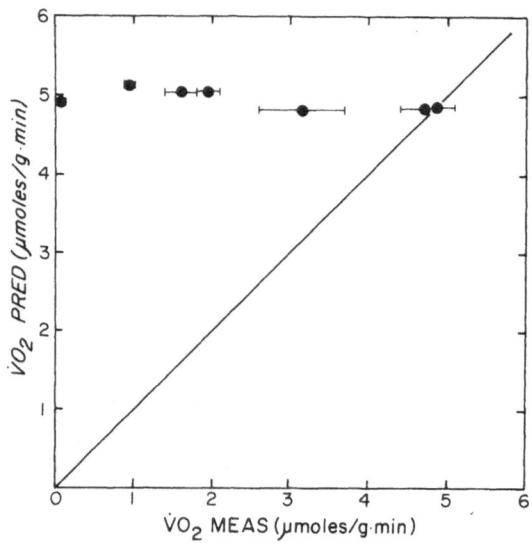

Figure 1. Fit to adenine nucleotide translocase model. The concentrations for each time point were inserted into Equation (1). K_m = 2.9 μm; K_i = 30 mM and V_{max} = 34 μmoles/min. The error bars indicate the S.E.M. of the measured $\dot{V}O_2$. PRED, predicted; MEAS, measured.

even though the total [Pi] increases. Alternatively, the increase in [OH^-] associated with the increase in pH could function as a competitive inhibitor comparable to the effect of ATP on the adenine nucleotide translocase. This would lead to a signficant phosphate limit on the increase in $\dot{V}O_2$ during the early phases of the rest-work transition. Application of a model comparable to Equation (1) with ADP and ATP replaced with Pi and OH^- respectively will describe the data for the first three sampling times but cannot describe the full time course of the rest-work transition.

In summary, we can conlude that no single rate-limiting translocase step can account for the kinetics of $\dot{V}O_2$ change observed during the rest-work transition.

LINEAR THERMODYNAMIC MODELS

The basic equation for this model is shown in Equation (2) below:

$$\dot{V}O_2 = L_o \, A_o + L_{po} \, A_p \qquad\qquad (2)$$

where L_o is the phenomenological coupling coefficient for the electron flow between NADH and oxygen, L_p is the cross-coupling coefficient between the redox gradient and the phosphorylation potential,

and   the A_o and A_p terms refer to the oxidation and the phosphorylation affinity terms respectively. These are the free energy changes in the reactions (Van Der Meer et al., 1978; Stucki, 1980b; Van Dam et al., 1980; Pietrobon et al., 1982).

The data in Table 1 indicate that there is a small increase in the apparent oxidation of the mitochondria with the onset of work which is, with the exception of the 15 sec point, essentially constant throughout the rest-work transition. On the log scales used in this analysis even the peak in redox after 15 seconds of stimulation is relatively small compared with the changes in $\dot{V}O_2$ or the phosphorylation potential and therefore will be neglected in the initial analysis. Further, cellular PO_2 always remained above that thought to be limiting to $\dot{V}O_2$ even at $\dot{V}O_2$ max (Connett, Gayeski and Honig, 1985b). Thus oxygen should be a zero order reactant and not lead to changes in A_o under these conditions. Since the change in A_o is very small, Equation (2) can be approximated by Equation (3) below:

$$\dot{V}O_2 = A_o' + L_{po} \, A_p \qquad\qquad (3)$$

or $\dot{V}O_2 = A_o' + A_1' \cdot \log(f(R_p))$

where $R_p = [ATP]/[ADP] \cdot [Pi]$ (using the total concentrations)

$A_o' = L_o \, A_o + L_{po} \cdot G^o{}_{phos}$

and $A_1' = L_{po} \cdot RLn(10)$

The nature of $f(R_p)$ as well as G^o depend on the forms of the phosphorylated compounds involved in the reaction. The measured values are approximately the total free concentrations. These include not only the charged forms but also the forms chelated with Mg^{2+}, and K^+ as well as the fraction titrated with H. In Table 2, Equations (2.1) to (2.4) are the equations used for interconverting the various forms as proposed by Rosing and Slater (1972).

While the equation for the hydrolysis of ATP is usually written in the form of Equation (4) this does not take into account charge balance and substrate forms. It should be more properly expressed as shown in Equations (5) to (7).

$$ATP + H_2O = ADP + Pi \qquad\qquad (4)$$

$$ATP^{4-} + H_2O = ADP^{3-} + H_2PO_4^- \quad (or \; HPO_4{}^{2-} + H^+) \qquad\qquad (5)$$

$$ATPH^{3-} + H_2O = ADPH^{2-} + H_2PO_4^- \quad (or \; HPO_4{}^{2-} + H^+) \qquad\qquad (6)$$

$$ATPMg^{2-} + H_2O = ADPMg^- + H_2PO_4^- \quad (or \; HPO_4{}^{2-} + H^+) \qquad\qquad (7)$$

Table 2.   Equations for binding corrections

$$[ATP^{4-}] = [ATP]/b_t \tag{2.1}$$
$$[AD^{3-}] = [ADP]/b_d \tag{2.2}$$
$$[HPO_4{}^{2-}] = [Pi]/b_p \tag{2.3}$$
$$[PCr^{2-}] = [PCr]/b_c \tag{2.4}$$

where   [ATP]. [ADP] and [Pi] are the total free concentrations of ATP, ADP and phosphate respectively,

and b_x are the correction terms for K^+, Mg^{2+} and H^+ binding to the compounds and take the form shown in Equation (2.5) below:

$$b_x = 1 + K_x^{Mg}[Mg^{2+}] + K_x^K[K^+] + K_x^H[H^+](1 + K_x^{HMg}[Mg^{2+}]) \tag{2.5}$$

where K_x^y are binding constants given in the form: $[X][Y]/[XY]$

Table 3.   Definitions of $f(R_p)$

From Equations (4) to (7) given in the text we can define R_p^i, where i refers to the forms of the nucleotides, as follows:

$$R_p \qquad = [ATP]/[ADP][Pi] \tag{3.1}$$

$$R_p^- \qquad = [ATP^{4-}]/[ADP^{3-}][H_2PO_4{}^-] \tag{3.2}$$

$$R_p^H \qquad = [ATPH^{3-}]/[ADPH^{2-}][H_2PO_4{}^-] \tag{3.3}$$

$$R_p^{Mg} \qquad = [ATPMg^{2-}]/[ADPMg^-][H_2PO_4{}^-] \tag{3.4}$$

These can be expressed in terms of $\log(f(R_p))$ as follows:

$$\log(R_p)^- = \log(R_p) + pH + \log(b_d \cdot b_p/b_t) - \log(K_p^H) \tag{3.5}$$

$$\log(R_p^H) = \log(R_p) + pH + \log(b_d \cdot b_p/b_t) - \log(K_p^H \cdot K_d^H/K_t^H \tag{3.6}$$

$$\log(R_p^{Mg}) = \log(R_p) + pH + \log(b_d \cdot b_p/b_t) - \log(K_p^H \cdot K_d^{Mg}/K_t^{Mg}) \tag{3.7}$$

If the reaction is with $HPO_4{}^{2-}$ rather than $H_2PO_4{}^-$ then the term $-\log(K_p^H)$ is eliminated from each equation.

For creatine the equation equivalent to (3.6) is:

$$\log(R_p^H) = \log([PCr]/[Cr] \cdot [Pi]$$
$$+ \log(b_p/b_c) \; \log(K_p^H \cdot K_d^H/K_t^H \cdot K_{cpk} \tag{3.8}$$

where $K_{cpk} = [ATP^{4-}][Cr]/\{[ADP^{3-}][PCr][H^+]\} = 3.6 \times 10^8$

and   PCr, Cr are the measured values of total phosphocreatine and creatine respectively.

These charge balanced equations lead to three definitions of $\log(f(R_p))$ shown in Table 3, Equations (3.1) to (3.3). Note that in order for charge balance to be maintained either $H_2PO_4^-$ (or $HPO_4^{2-} + H^+$) must be included in the definition of $f(R_p)$. This leads to the inclusion of a pH term in the $\log(f(R_p))$ term as shown in Equations (3.5) to (3.7). Thus if one carries out a linear regression to test the fit of the data expressed in terms of R_p the fit should be very poor unless pH were constant or were included in the fitting equation. Further, the regression coefficients for both pH and R_p should be almost identical. The difference will be due to the pH dependence of the $\log(b_d \cdot b_p/b_t)$ term. Over the pH range of interest this term is linearly dependent on pH with a slope of 0.34. Thus the coefficient for the pH term should be ~ 0.34 different from that for $\log(R_p)$. In Table 4, Fits 1 and 2 to Equations (4.1) and (4.2) show the results of a linear regression of $\dot{V}O_2$ on $\log(R_p)$ alone, and with pH included. Very clearly the prediction from the analysis above is supported.

From the creatine kinase equilibrium an equivalent expression for $f(R_p)$ can be derived using the measured PCr, Cr and Pi. This is given in Table 3, Equation (3.8). A fit using these parameters should be identical to that obtained with R_p and pH with the same slope and an intercept that differs by $a_1 \cdot \log(K_{cpk})$ or ~ 24. The result of fitting with $\log([PCr][Cr] \cdot [Pi])$ is given in Table 4, Fit 3. Again the prediction is supported.

As a further test of the original assumption that redox changes and $[O_2]$ do not contribute significantly to the variation in $\dot{V}O_2$ under these conditions, the mitochondrial redox estimated from the fluorescence measurements with and without O_2 were included in the regression equation (Eqn 4.4, Table 4). The results of the linear regression on this equation is shown in Table 4, Fit 4. Although the inclusion of the redox does increase the amount of variation accounted for by the equation the coefficient is not signficiantly different from 0. Inclusion of oxygen had no effect on the fit and the results are not shown.

Figure 2 illustrates the efficacy of these equations in predicting $\dot{V}O_2$. The large circles indicate the measured values and 1 S.E.M. The residuals of three fits are plotted with the symbols, as defined in the figure legend, indicating Equations (4.2), (4.3) and (4.4). Given the errors in the measurements all the fits are excellent. The theoretically best model, that including the mitochondrial redox, does give values that place all points within 1 S.E.M. of the measured $\dot{V}O_2$. However, all fits are within the confidence intervals of the measurements.

DISCUSSION

With access to a uniquely complete set of data on the early period of the rest-work transition in red muscle we have examined the ability of a number of quantitative models to account for the time course of the change in $\dot{V}O_2$. A limiting role for either adenine nucleotide translocase or phosphate translocase cannot account for the full transition. This is not necessarily in conflict with the results obtained in isolated mitochondria since there are a number of simultaneous changes occurring in the intact muscle that have not been examined in the in vitro studies. The phosphate levels in vivo start at lower levels than those used in most in vitro studies.

Table 4. Fits to thermodynamic equations

Fit (Eqn)	Fitting Parameters				
	a_0	a_1	a_2	a_3	R^2
1. (Eqn 4.1)	6.02 ± 1.54*	−3.12 ± 1.69			0.40
2. (Eqn 4.2)	24.5 ± 0.3	−2.41 ± 0.35	−2.70 ± 0.25		0.98
3. (Eqn 4.3)	−0.14 ± 0.31	−2.74 ± 0.19			0.98
4. (Eqn 4.4)	24.32 ± 0.29	−2.31 ± 0.32	−2.75 ± 0.24	−0.62 ± 0.46	0.99

Equations

$$\dot{V}O_2 = a_0 + a_1 \cdot \log(R_p) \tag{4.1}$$

$$\dot{V}O_2 = a_0 + a_1 \cdot \log(R_p) + a_2 \cdot pH \tag{4.2}$$

$$\dot{V}O_2 = a_0 + a_1 \cdot \log([PCr]/[Cr][Pi]) \tag{4.3}$$

$$\dot{V}O_2 = a_0 + a_1 \cdot \log(R_p) + a_2 \cdot pH + a_3 \cdot \log(NADH/NAD) \tag{4.4}$$

* $\pm$ standard error; R_2, squared correlation coefficient.

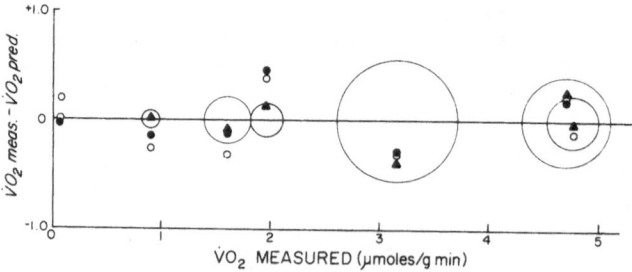

Figure 2. Fits of thermodynamic model to $\dot{V}O_2$ measured. The abscissa shows the time course of $\dot{V}O_2$ from left to right with the S.E.M. indicated by the circles. The ordinate is the residuals of the model fit. The symbols indicate: o, fit with adenine nucleotides and pH, Eqn (4.2), Table 4; •, fit with phosphocreatine, Eqn (4.3), Table 4; ▲, fit with adenine nucleotides, pH and mitochondrial redox, Eqn (4.4), Table 4. pred., predicted; meas., measured.

Most importantly there are large changes in the cytosolic pH which lead to significant redistribution of the adenine nucleotides and phosphate between the various charged forms. Studies with isolated mitochondria are generally carried out at near constant pH. We can, however, conclude that none of the single enzymatic steps tested is rate- limiting throughout the rest-work transition in vivo.

Application of the linear thermodynamic approach with the charge-balanced phosphorylation potential gives an excellent descriptive equation. This approach requires knowledge of the free concentrations of the adenine nucleotides and phosphate as well as the cytosolic pH and estimates of $[K^+]$, $[Mg^{2+}]$ and binding constants. The analysis can also be applied with estimates of phosphocreatine, creatine and phosphate alone and obviates the necessity of pH measurements. This latter approach has a number of advantages. All the components are in relatively high concentration and can be determined accurately either by direct chemical methods on frozen tissue or non-invasively by NMR. Further the only binding constants contributing significantly are those for phosphate. As suggested previously (Pietrobon et al., 1983; Connett et al., 1985b; Olgin et al., 1986), neither the observed changes in $[O_2]$ nor the small changes in mitochondrial redox contribute significantly to the recruitment of $\dot{V}O_2$ in this pure, red skeletal muscle. This again is in contrast to the results obtained in studies with isolated mitochondria. However, as suggested in an elegant theoretical analysis by Stucki (1980a), the most efficient operation of the system requires a buffering of driving forces. He has termed this a thermodynamic buffering. Cytosolic redox appears to change in parallel with the changes in $\dot{V}O_2$ in these muscles (Olgin et al., 1986). Similar results have been demonstrated in an in vitro system which included both functional mitochondria and the glycolytic pathway (Jong and Davis, 1983). This change in the cytosolic redox may function to buffer any changes in mitochondrial redox and result in an essentially constant driving force for oxidative phosphorylation. The result is a system that is driven directly by the changes in cytosolic phosphorylation potential. The net effect is a direct quantitative coupling between demand and supply of phosphorylation energy in the cytosol. This, in thermodynamic terms, is the strict conductance matching required for the efficient linear operation of a coupled system.

ACKNOWLEDGEMENT

This work is supported in part by a USPHS Grant No. AM 36154.

REFERENCES

Connett, R.J. (1985). In vivo glycolytic equilibria in dog gracilis muscle. J. Biol. Chem. 260, 3314-3320.

Connett, R.J. (1987). Factors governing the recruitment of glycolysis and oxygen consumption during the rest-work transition in red skeletal muscle. Symposium, Gubbio, Italy, Nov. 1985, Elsevier, Amsterdam (in press).

Connett, R.J., Gayeski, T.E.J. and Honig, C.R. (1985a). Energy sources in fully aerobic rest-work transitions: a new role for glycolysis. Am. J. Physiol. 248, H922-H929.

Connett, R.J., Gayeski, T.E.J. and Honig, C.R. (1985b). An upper bound on the the minimum PO_2 for O_2 consumption in red muscle. In: Oxygen Transport to Tissue-VII. Eds Kreuzer, F., Cain, S.M., Turek, Z. and Goldstick, T.K., Plenum Press, New York and London, (Adv. Exp. Med. Biol. 191, 291-300).

Coty, W.A. and Pedersen, P.L. (1974). Phosphate transport in rat liver mitochondria. Kinetics and energy requirements. J. Biol. Chem. 249, 2593-2598.

Davis, E.J. and Lumeng, L. (1975). Relationships between the phosphorylation potentials generated by liver mitochondria and respiratory state under conditions of adenosine diphosphate control. J. Biol. Chem. 250, 2275-2282.

Gayeski, T.E.J. Connett, R.J. and Honig, C.R. (1985). Oxygen transport in rest-work transition illustrates new functions for myoglobin. Am. J. Physiol. 248, H914-H921.

Jacobus, W.E. (1985). Respiratory control and the integration of heart high-energy phosphate metabolism by mitochondrial creatine kinase. Ann. Rev. Physiol. 47, 707-725.

Jong, Y.A. and Davis, E.J. (1983). Reconstruction of steady state in cell-free systems. Interactions between glycolysis and mitochondrial metabolism: regulation of the redox and phosphorylation states. Arch. Biochem. Biophys. 222, 179-183.

Kunz, W., Bohnensack, R., Bohme, G., Kuster, U., Letko, G. and Schonfeld, P. (1981). Relations between extramitochondrial and intramitochondrial adenine nucleotide systems. Arch. Biochem. Biophys. 209, 219-229.

Kuster, U., Bohnensack, R. and Kunz, W. (1976). Control of oxidative phosphorylation by the extramitochondrial ATP/ADP ratio. Biochim. Biophys. Acta, 440, 391-402.

Letko, G., Kuster, U., Duszynski, J. and Kunz, W. (1980). Investigation of the dependence of the intramitochondrial [ATP]/[ADP] ratio on the respiration rate. Biochim. Biophys. Acta, 593, 196-203.

Olgin, J., Connett, R.J. and Chance, B. (1986). Mitochondrial redox changes during rest-work transition in dog gracilis muscle. In; Oxygen Transport to Tissue-VIII. Ed. Longmuir, I.S., Plenum Press, New York and London, (Adv. Exp. Med. Biol. 200, 545-554).

O'Shea, P.S. and Chappell, J.B. (1984). The relationship between the rate of respiration and the protonmotive force. The role of proton conductivity. Biochem. J. 219, 401-404.

Pietrobon, D., Zoratti, M., Azzone,, G.F., Stucki, J.W. and Walz, D. (1982). Non-equilibrium thermodynamic assessment of redox-driven H^+-pumps in mitochondria. Eur. J.Biochem. 127, 483-497.

Pietrobon, D., Zoratti, M. and Azzone, G.F. (1983). Molecular slipping in redox and ATPase H^+ pumps. Biochim. Biophys. Acta, 723, 317-321.

Rosing, J. and Slater, E.C. (1972). The value of ΔG^o for the hydrolysis of ATP. Biochim. Biophys. Acta, 267, 275-290.

Rottenberg, H. (1973). The thermodynamic description of enzyme-catalyzed reactions. Biophys. J. 13, 503–511.

Soboll, S. and Stucki, J. (1985). Regulation of the degree of coupling of oxidative phosphorylation in intact rat liver. Biochim. Biophys. Acta, 807, 245–254.

Stucki, J.W. (1980a). The thermodynamic buffer enzymes. Eur. J. Biochem. 109, 257–267.

Stucki, J.W. (1980b). The optimal efficiency and the economic degrees of coupling of oxidative phosphorylation. Eur. J. Biochem. 109, 269–283.

Tobin, R.R., Mackerer, C.R. and Mehlman, M.A. (1972). pH effects on oxidative phosphorylation of rat liver mitochondria. Am. J. Physiol. 223, 83–88.

Van Dam, K., Westerhoff, H.V., Krab, K., Van Der Meer, R. and Arents, J.C. (1980). Relationship between chemiosmotic flows and thermodynamic forces in oxidative phosphorylation. Biochem. Biophys. Acta, 591, 240–250.

Van Der Meer, R., Akerboom, T.P.M., Groen, A.K. and Tager, J.M. (1978). Relationship between oxygen uptake of perifused rat-liver cells and the cytosolic phosphorylation state calculated from indicator metabolites and a redetermined equilibrium constant. Eur. J. Biochem. 84, 421–428.

Williamson, J.R., Steinman, R., Coll, K. and Rich, T.L. (1981). Energetics of citrulline synthesis by rat liver mitochondria. J. Biol. Chem. 256, 7287–7297.

Wilson, D.F., Erecińska, M. and Schramm, V.L. (1983). Evaluation of the relationship between the intra- and extramitochondrial [ATP]/[ADP] ratios using phosphoenolpyruvate carboxykinase. J. Biol. Chem. 258, 10464–10473.

Wilson, D.F., Owen, C.S. and Erecińska, M. (1977). Quantitative dependence of mitochondrial oxidative phosphorylation on oxygen concentration: a mathematical model. Arch. Biochem. Biophys. 195, 494–498.

Wilson, D.F., Owen, C.S. and Holian, A. (1977). Control of mitochondrial respiration: a quantitative evaluation of the roles of cytochrome c and oxygen. Arch. Biochem. Biophys. 182, 749–762.

Wilson, D.F., Stubbs, M., Oshino, N. and Erecińska, M. (1974). Thermodynamic relationships between the mitochondrial oxidation-reduction reactions and cellular ATP levels in ascites tumour cells and perfused rat liver. Biochemistry, 13, 5305–5312.

A COMPARISON OF A FOUR WAVELENGTH ANALYSIS AND MULTICOMPONENT WAVE-

LENGTH ANALYSIS APPLIED TO DETERMINATION OF HAEMOGLOBIN SATURATION

F. DEGNER and T.E.J. GAYESKI

University of Rochester, 601, Elmood Avenue, Rochester,
NY 14620, U.S.A.

INTRODUCTION

Cryomicrospectrophotometry has been used to study haemoglobin (Hb) and myoglobin (Mb) saturation in living tissue (Figulla, Hoffman and Lubbers, 1983). This paper evaluates two possible artefacts in cryomicrospectrometry applied to Hb measurements in small blood vessels. First, microcirculatory vessels may differ in haematocrit compared with systemic vessels (Hoffman et al., 1984). The resulting difference in Hb concentration might influence the spectroscopic signal and hence the calculation of Hb saturation. Second, the freezing rate of the tissue during the sample preparation influences the ice crystal size. Hence, light scattering and measured spectra may vary as a function of depth from a frozen surface. The present study determines the influence of different haemoglobin concentrations and freezing rates on the cryomicrospectrophotometric measurements and calculated saturations. In addition, we compared mouse, man and dog Hb using dog Hb as the standard. Finally, this paper compares the linear, multicomponent, wavelength analysis of Hoffman and Lubbers (1985) with a four wavelength method of Gayeski (1981). The latter method requires intensity measurements at four different wavelengths per saturation calculation over a 'narrow' spectral range (547–588 nm) as compared with many wavelengths (up to 500) over a 'broad' spectral range (500–600 nm). A simple method of determining haemoglobin saturation allows easier implementation.

METHODS

Microspectrophotometer

Figure 1 shows the microspectrophotometer schematically. A one kW xenon lamp in series with a grating monochromator (Schoeffel Instrument Corp., Model GM 250, 1200 lines/mm blazed at 300 μm) is the illumination source. The lamp is ignited by a Schoeffel Instrument Corp., Model 2551 SP Lamp Igniter and is powered by a Hewlett Packard Model 6269-B-DC Power Supply. A computer-controlled stepping motor (A.W. Hayden Co., Model 7532) moves the monochromator's wavelength micrometer to predetermined wavelengths. Light transmitted through a Leitz Orthoplan microscope, a Mirror Housing 500, a Pol-Vertical Illuminator and a Leitz 25x objective (519535: numerical aperture of 0.22 and free working distance of 14.8 mm) epi-illuminates specimens. The long working distance and a heating coil wrapped around the lens prevents accumulation of condensed water on lens elements. A Leitz MPV I photometer body images the object plane onto an aperture of variable diameter. Light passing through the aperture is measured via a Centronic photomultiplier tube (Model P4283 TIR) in a Peltier-device-cooled housing (Schoeffel Instrument Corp., Model D500T). The cooled photomultiplier tube housing is powered by a Schoeffel Instrument Corp. Model DP 5001-2T power supply, and the photomultiplier tube, by a Keithley Model 244 high voltage power supply. The output current of the PMT was measured on a Keithley Model 414J picoammeter the analog output of which was converted to a digital signal and stored in an IMSAI 8080 microcomputer. Typical light signals are at least three orders of magnitude greater than the dark current of the photomultiplier tube.

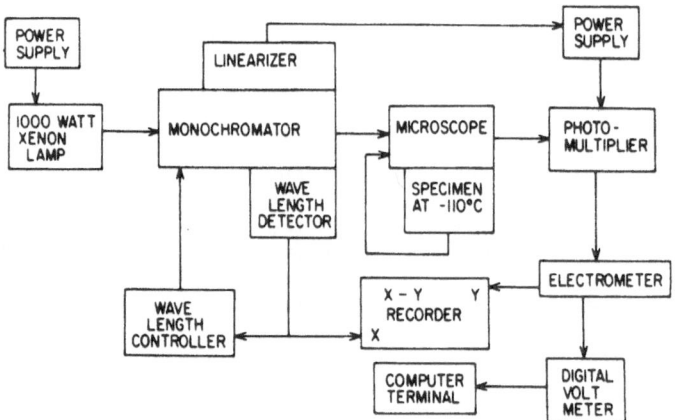

Figure 1.   Schematic representation of the cryomicrospectrophotometer.

Microscope cold stage

A styrofoam container 20 x 12 x 8 cm, sealed with GE RTV615A silicone rubber compound, serves as a reservoir for liquid nitrogen. An 8 cm diameter x 5 cm high, hollow brass cup closed at its lower end separates the sample from the liquid nitrogen. A sample holder is mounted within this brass cup. To minimize sample preparation time, easy mounting of the sample on the cold stage is necessary. The sample holder constructed to achieve this objective comprises two parts: a base and a specimen holder. The base consists of a solid brass rod, 5 cm diameter by 3 cm high. The top surface of the rod is milled to form a centred well 4.5 cm in diameter by 1.3 cm deep. In the centre of the well a second well 1 cm square by 1 cm deep is formed. On the sides of this inner well, slots are milled to facilitate transfer of the specimen holder into this base. The specimen holder consists of a 1 cm^3 brass block milled to fit snugly into the second milled cubic hole in the sample holder. In the centre of this holder, a 6 x 8 x 8 mm deep hole is milled. A set screw on the long side of the holder secures the specimen in place. The bottom surface of the sample holder is permanently fixed' to the aforementioned hollow brass cup. Thus, heat is removed from the sample holder's base via the liquid nitrogen. In the milled wells of the holder, 95% ethanol is placed. To regulate the temperature of this ethanol at -110°C (freezing point of 95% ethanol is -120°C), a heating coil is wrapped around the external diameter of the base. Ethanol, being hydroscopic, prevents frost formation on the specimen surface. The depth of the ethanol is adjusted so that the ethanol barely covers the surface of the specimen under observation.

To transfer the sample from storage under liquid nitrogen onto the cold stage, the sample is mounted in the specimen holder while both are under liquid nitrogen. This mounted specimen is transferred through air into a precooled ethanol-dry ice bath (-75°C). A precooled (-75°C) scalpel blade is used to cleave a fresh surface on the specimen for observation. The mounted, freshly-cleaved specimen is then transferred to the sample holder which is maintained at -110°C. Total immersion time of specimens in the ethanol-dry ice bath is less than two minutes. Mb and Hb saturations are stable for at least 5 minutes at -75°C and three hours at -110°C. Needles containing the quick-frozen samples were mounted on the microscope stage after preparation in a similar fashion.

Sample collection

Venous blood was drawn from humans, C3H mice and mongrel dogs. All blood was tonometered in an Instrumentation Laboratories (IL) tonometer Model 237. To achieve 100% and 0% saturation, we tonometered samples with 20% O_2 - 5.6% CO_2, balance N_2 or 5.6% CO_2, balance N_2 for ten minutes. Intermediate saturations were obtained from the subjects or by use of an IL Tonometer Control module 208 gas mixer. Saline was used to dilute appropriate blood samples before tonometering. The blood specimens were analysed on an IL Model 182 Co-oximeter for saturation and haemoglobin concentration. The blood samples were frozen in one of two ways. To maximize freezing rate, scored 20 gauge metal needles were filled with the blood sample and transferred into a N_2-cooled Freon 13 bath. Heat transfer was high because of the high heat conductivity of the metal and the absence of vaporization of the Freon 13. In the second method, standard plastic 3 ml syringes were filled with the blood sample and immersed in liquid N_2. Heat transfer was slower because of the lower heat conductivity, greater diameter of the syringes compared with the needles, and the presence of a N_2 vaporization layer.

Calculations and analysis

Four wavelength analysis
The Lambert Beer law for homogeneous media states that

$$I(L) = Io(L)*EXP(-a(L)*C*d) \qquad (1)$$

where L = wavelength

        I(L) = light intensity at wavelength L after passing through the medium

        Io(L) = incident light at wavelength L

        a(L) = extinction coefficient of the pigment in the medium at wavelength L

$$C = \text{concentration of the pigment in the medium}$$

$$d = \text{pathlength of the light in the medium}$$

In our system, I, Io(L), a(L) are typically wavelength dependent. Taking the natural log of both sides, the optical density D(L) is defined as:

$$D(L) = -\ln(I(L)/Io(L)) = a(L)*C*d \qquad (2)$$

If one adds a scattering term at wavelength $\underline{L}$ (s(L)) to the argument of the exponential, Equation (2) becomes

$$D(L) = a(L)*C*d+s(L) \qquad (3)$$

This 'linearization of scattering' has been utilized by Pittman and Duling (1975) and Gayeski (1981) after the modelling of Twersky (1962, 1964). This linearization approximates the non-linear relationship between absorption and scattering in reflection of Hoffman and Lubbers (1985). It applies when the range of absorption change is small.

If the sample contains two non-interacting pigments, oxygenated and deoxygenated haemoglobin, the concentrations of oxygenated Hb (C1) and deoxygenated haemoglobin (C2) can be related to their respective extinction coefficients at wavelength $\underline{L1}$,

$$D(L1) = a1(L1)*C1*d+a2(L1)*C2*d+s(L1) \qquad (4)$$

where a1(L1) = oxygenated haemoglobin extinction coefficient
a2(L1) = deoxygenated haemoglobin extinction coefficent.

The total concentration of haemoglobin is defined by the sum of its components,

$$C = C1+C2 \qquad (5)$$

The Hb saturation (S) is defined as $\quad S = C1/C \qquad (6)$

Hence,

$$D(L1) = (a1(L1)S+a2(L1)(1-S))Cd+s(L1) \qquad (7)$$

Similarly, assuming equal pathlengths for both moieties, one can take the difference between optical densities at two wavelengths (L1 and L2):

$$D(L1)-D(L2)=((S(a1(L1)-a1(L2))+(1-S)(a2(L1)-a2(L2)))Cd+s(L1)-s(L2) \qquad (8)$$

If s(L1) = s(L2), the difference between two optical densities is independent of the scattering term. This assumption appears to be valid from at least 540 to 590 nm in our system. The difference between another pair of optical densities D(L1) and D(L3) is:

$$D(L1)-D(L3)=((S(a1(L1)-a1(L3))+(1-S)(a2(L1)-a2(L3)))cd \qquad (9)$$

$$(\text{assuming } s(L1)=s(L3))$$

Taking the ratio (R) of Equations (8) to (9) eliminates the dependence of R on concentration and pathlength.

$$R = \frac{D(L1)-D(L2)}{D(L1)-D(L3)} = \frac{(I(L1)Io(L2)/Io(L1)I(L2)}{(I(L1)Io(L3)/(L1)I(L3))} \qquad (10)$$

$$= \frac{S(a1(L1)-a1(L2))+(1-S)(a2(L1)-a2(L2))}{S(a1(L1)-a1(L3))+(1-S)(a2(L1)-a2(L3))}$$

If L1 and L3 are isosbestic wavelengths i.e. a1(L1) = a2(L1) and a1(L3) = a2(L3), the saturation can be derived from Equation (10) in the form:

$$S = a*R+b \qquad (11)$$

where

$$a = \frac{a2(L1)-a2(L3)}{a2(L2)-a1(L2)}$$

$$b = \frac{a2(L2)-a2(L1)}{a2(L2)-a1(L2)}$$

Empirical determination of slope and intercept

To determine a and b in Equation (11), we determined R for two known Hb saturations. We chose saturations of 100% and 0% as determined by the IL Co-oximeter. Using intensity measurements at three wavelengths, we determined the ratio R in Equation (10). Note that the ratio is dependent on the ratio of the intensities of incident light. As we cannot measure these directly, we use the relative intensity of our light source measured at the appropriate wavelength via reflection from an aluminium oxide mirror. The slope a and intercept b were then calculated by solving the resulting simultaneous equations for 0% and 100% Hb saturation.

Isosbestic wavelengths for haemoglobin at −110 ^{o}C are located at 586, 569 and 545 nm. Non-isosbestic wavelengths of 558 and 576 nm were selected. One can define ratios R with combinations of three wavelengths as follows:

R558 as R with L1 = 586 nm, L2 = 569 nm and L3 = 558 nm,

R576 as R with L1 = 586 nm, L2 = 569 nm and L3 = 576 nm,

RISO as R with L1 = 586 mn, L2 = 569 nm and L3 = 545 nm.

RISO is the ratio R calculated with all isosbestic wavelengths. Hence, RISO is independent of S. It serves as a check on the choice of the 'linearization' of the relationship between light absorption and light scattering. Variations in RISO in situ are interpreted as changes in light scattering conditions due to changes in ice crystal size or to changes in Hb concentrations. Approximations to light scattering theory and time-dependent system noise could lead to errors in saturation calculations. To minimize the effect of these errors, Hb saturation was determined as the average saturation (SAV) of the two saturation estimates (S558, S576). A least square linear regression between SAV and the known co-oximeter saturation was performed.

Multicomponent wavelength analysis

Lubbers and associates (Lubbers, 1973; Hoffman and Lubbers, 1985) have demonstrated that a multicomponent wavelength analysis can accurately reconstruct complex spectra with contributions of up to eight pigments plus light scattering. For our two-pigment case we compared the results of their analysis with the four wavelength analysis from haemoglobin spectra recorded on our cold stage. A least square linear regression between SAV and the multicomponent wavelength analysis was performed.

Data collection
    A spectrum from 537 to 588 nm was recorded in one nm steps. Each intensity measurement consisted of the average of ten light intensity measurements for each wavelength. The image of the measuring diaphragm projected on the sample was varied from 3 x 3 µm to 20 x 20 µm.

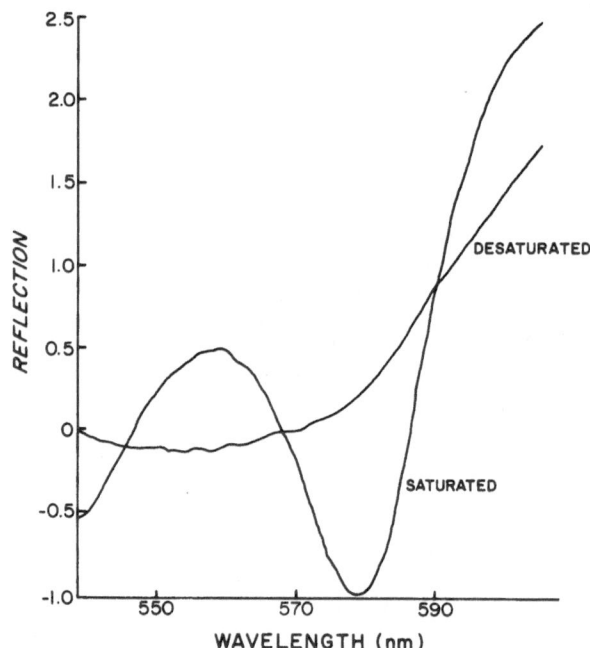

Figure 2. Spectra of fully saturated and desaturated haemoglobin. Spectra recorded in 1 nm increments.

RESULTS

    Figure 2 shows the mean spectrum (n = 10) for desaturated and saturated human blood with a haematocrit of 45%. The ordinate shows the extinction relative to the extinction at 560 nm. One can locate the isosbestic points at 545, 569 and 586 nm. The nonisosbestic points at the extremes of the difference spectrum between saturated and desaturated haemoglobin in blood were found at 558 nm and 576 nm.

Figure 3 shows the relationship between SAV and the saturation determined by the co-oximeter. The least square linear regression estimate and the 95% confidence interval were determined for human blood quick frozen with a haematocrit of 45%. Each data point represents the mean saturation values of 10 calculations. Values for a haematocrit of 10% do not differ from this regression line (filled circles). The values for mouse and dog blood (haematocrit 45%) also fall within the 95% confidence interval.

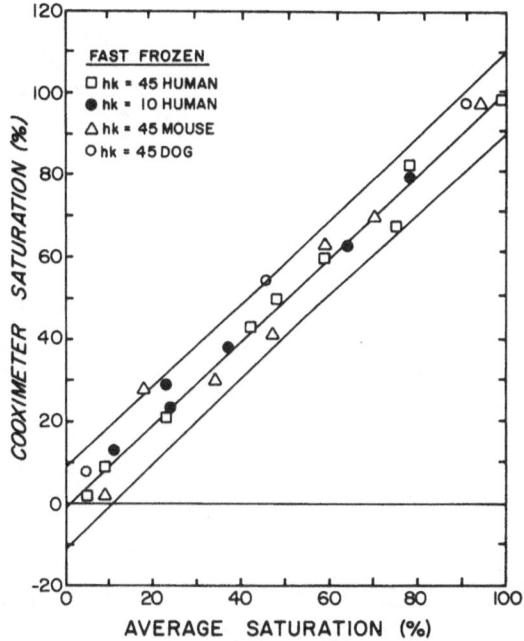

Figure 3. A comparison of Hb saturation as calculated by the Instrumentation Laboratories Co-oximeter and the four wave length analysis for haemoglobin, haematocrits shown.

Figure 4 shows the influence of the freezing rate on the relationship between SAV and co-oximeter saturation. The lower line (filled circles) represents results for rapidly frozen human blood (haematocrit 45%) as shown in Figure 3. The upper line (open circles) is the linear regression for slow frozen human blood (Hct 45%). The two least square regression estimates are almost parallel and differ in the cordinate intercept by about 15% saturation. The two points at 41% and 43% co-oximeter saturation represent measurements on blood frozen by immersion in liquid N_2 in a 20 ml syringe and a 1 ml syringe

respectively.   No effect on the results is apparent. These in vitro
results suggest that differences between in situ freezing rates within
1 mm of the frozen surface and freezing rates at the surface are much
less than between the fast and slow freezing methods employed in these
experiments (personal observation).

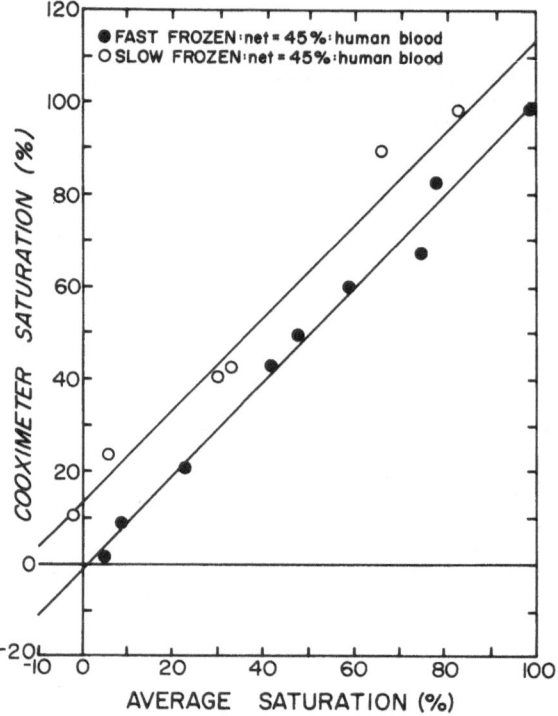

Figure 4.   The effect of freezing rate on Hb saturation using the four
wavelength analysis.

There was no significant relationship between RISO and SAV for any
haemoglobin at any saturation or freezing rate. Figure 5 shows the
relationship between RISO and SAV for quick vs slow frozen human blood
with a haematocrit of 45%. In contrast to the rapidly frozen blood the
RISO for the slowly frozen blood decreases slightly with increasing
co-oximeter saturation. The variation observed in RISO represents less
than a 5% variation in SAV. There is no saturation dependence in RISO
for fast frozen blood. In dog gracilis muscle frozen in situ, Gayeski
and Honig (1986) have reported no change in RISO for myoglobin within 1
mm of the freezing surface.

160

Figure 6 demonstrates agreement between the four wavelength analysis and the multicomponent wavelength analysis. This agreement confirms that for a two-pigment system in frozen tissue visualized in reflection, the methods yield identical results.

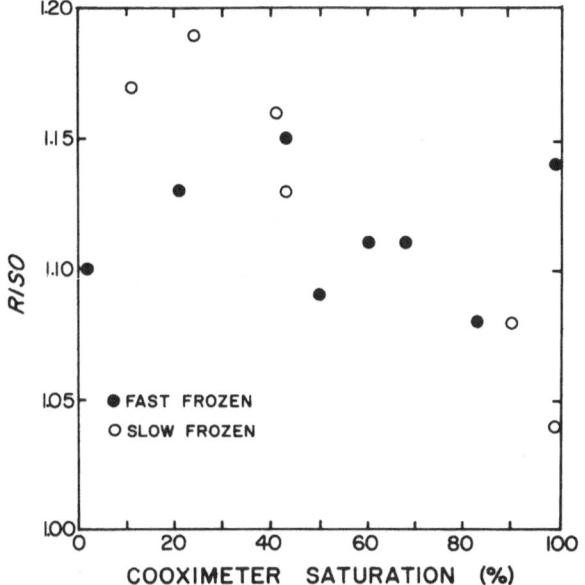

Figure 5. The effect of freezing on the relationship between RISO and SAV. The change in RISO observed for slow frozen blood is the equivalent of approximately a 5% change in SAV.

DISCUSSION

The variability in capillary haematocrit reported by others (Fung, 1973; Schmid-Schonbein et al., 1980) raised concern about measuring Hb saturation in frozen tissue in small veins. The small effect of changes in haematocrit on calculated haemoglobin saturation in vitro suggests that haematocrits between 10% and 45% will not greatly affect these calculations at comparable freezing rates. Thus, construction of frequency histograms of haemoglobin saturation from small veins in frozen tissue is possible. These frequency histograms will determine if heterogeneities of oxygen supply and demand exist under a variety of conditions.

Since a variety of animals are used in oxygen delivery and utilization experiments, we compared mouse, dog and man to see if there were any measurable differences in haemoglobin saturation calculations amongst these species. Using the four wavelength method, none was found. Thus, our results have general applicability to these species.

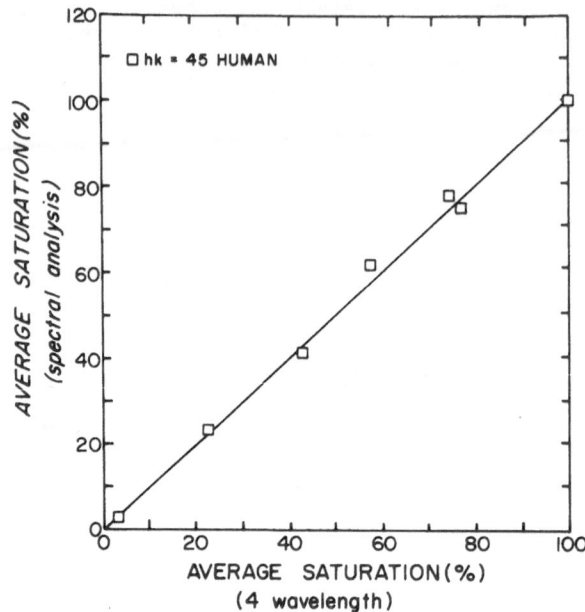

Figure 6.   The   agreement   between   the multicomponent method of Hoffman
and Lubbers (1985) and the four wavelength method is clear
for the tested condition.

Finally, we compared a four wavelength spectral analysis with a
multicomponent wavelength analysis. The former method had been applied
to calculating haemoglobin saturation (haematocrit 45%) and myoglobin
saturation in dog gracilis (Gayeski, 1981). This work extends its
applicability to haemoglobin of varying haematocrits if freezing rates
are relatively constant. Our results suggest that gross differences in
freezing rate will affect saturation calculations. The results of this
simpler method agree with those of the elegant, and more general method
of Hoffman and Lubbers (1985). Hence, analysis of reflected spectra
without complex computer algorithms are feasible if the spectra consist
of absorption by two pigments and light scattering has minimal wavelength dependence over the wavelength range used. Both of these requirements are met in the case of haemoglobin with a haematocrit range of
10% to 45% in the 540 to 590 nm range.

ACKNOWLEDGEMENTS

Supported in part by CA20329 and CA11198 from the National
Institutes of Health, the Max-Planck Gesellschaft and Grant HB03290
from the United States Public Health Service. The authors wish to
express their gratitude to Drs Carl Honig and Robert Sutherland for
their stimulation during these experiments. In addition, the technical
assistance of Mr W. Reeves and J. Frierson was greatly appreciated.

REFERENCES

Figulla, H.R., Hoffman, J. and Lubbers, D.M. (1983). Coronary conductivity and tissue oxygenation as measured by the myoglobin O_2 saturation and the cytochrome aa_3 redox state in the Langendorff guinea pig heart preparation. In: Oxygen Transport to Tissue-IV. Eds Bicher, H.I. and Bruley, D.F., Plenum Press, New York and London, (Adv. Exp. Med. Biol. 159, 579-585).

Fung, Y.C. (1973). Stochmastic flow in capillary blood vessels. Microvasc. Res. 5, 34-38.

Gayeski, T.E.J. (1981). A cryogenic microspectrophotometric method for measuring myoglobin saturation in subcellular volumes; application to resting dog gracilis muscle. Ph.D. Thesis, University of Rochester, Rochester, U.S.A.

Gayeski, T.E.J. and Honig, C.R. (1986). O_2 gradients from sarcolemma to cell interior in red muscle at maximal $\dot{V}O_2$. Am. J. Physiol. 251, H789-H799.

Hoffman, J., Heinrich, U., Ahmad, H.R. and Lubbers, D.W. (1984). Analysis of tissue reflection spectra obtained from brain or heart, using the two-flux theory for non-constant light scattering. In: Oxygen Transport to Tissue-VI. Eds Bruley, D.F., Bicher, H.I. and Reneau, D., Plenum Press, New York and London, (Adv. Exp. Med. Biol. 180, 555-563).

Hoffman, J. and Lubbers, D.W. (1985). Quantitative analysis of reflection spectra: evaluation of simulated reflection spectra. In: Oxygen Transport to Tissue-VII. Eds Kreuzer, F., Cain, S.M., Turek, Z. and Goldstick, T.K., Plenum Press, New York and London, (Adv. Exp. Med. Biol. 191, 85-90).

Lubbers, D.W. (1973). Spectrophotometric examination of tissue oxygenation. In: Oxygen Transport to Tissue. Eds Bicher, H.I. and Bruley, D.F., Plenum Press, New York and London, (Adv. Exp. Med. Biol. 37A, 45-54).

Pittman, R. and Duling, B.R. (1975). A new method for the measurement of percent oxyhaemoglobin. J. Appl. Physiol. 38, 321-327.

Schmid-Schonbein, G.W., Skalak, R., Usami, S. and Chien, S. (1980). Cell distribution in capillary networks. Microvasc. Res. 19, 18-44.

Twersky, V. (1962). Multiple scattering of waves and optical phenomena. J. Opt. Soc. Am. 52, 145-171.

Twersky, V. (1964). On propagation in random media of discrete scatterers. Am. Math. Soc. 16, 84

TIME COURSES OF ERYTHROCYTIC OXYGENATION IN CAPILLARIES OF THE LUNG:

LOWER AND UPPER BOUNDS ON RED CELL TRANSIT TIMES

K. GROEBE and G. THEWS

Physiologisches Institut der Universitat Mainz, Saarstrasse 21, D-6500 Mainz, F.R.G.

With the aid of a 2-dimensional computer simulation (Groebe and Thews, 1986), time courses of erythrocytic oxygen uptake in the lung were calculated. A number of different conditions i.e. normoxia, hypoxia and varying rates of oxygen uptake ($\dot{V}O_2$) were considered. Bounds on red cell transit times through the capillaries of the lung were inferred for different values of cardiac output ($\dot{Q}$) from the time courses. Furthermore, the effects of certain simplifications made in former model calculations were studied, e.g. neglecting reaction kinetics.

Table 1. Basic data used in model calculations

Erythrocyte volume	92.0 fl	
Erythrocyte radius	2.2 μm ⎫	
Erythrocyte length	6.0 μm ⎬ Miyamoto and Moll (1971)	
Intererythrocyte gap	1.8 μm ⎭	
Capillary radius	2.7 μm	Weibel (1963, 1964)
		Gehr et al. (1978)
Harmonic mean thickness of alveolar-capillary membrane	0.62 μm	Gehr et al. (1978)
Erythrocyte haemoglobin concentration	$5.23 \cdot 10^{-3}$ mol/1	
O_2 diffusion coefficient in the erythrocyte	$0.95 \cdot 10^{-5}$ cm^2/s	Clark et al. (1985)
O_2 diffusion coefficient in plasma	$2.18 \cdot 10^{-5}$ cm^2/s	Goldstick et al. (1976)
O_2 diffusion coefficient in endothelium	$1.2 \cdot 10^{-5}$ cm^2/s	Vaupel (1976)
HbO_2 diffusion coefficient in the erythrocyte	$1.44 \cdot 10^{-7}$ cm^2/s	Clark et al. (1985)

Table 2a. Basic physiological data, alveolar-arterial PO_2 differences (AaD), and O_2 loading times in the case of hypoxia. Data were taken from Bartels et al. (1955). Left column represents mean values of all participants in the investigation whereas in the right column the data of a single individual with a lower cardiac output were used

Oxygen uptake	$\dot{V}O_2$ (1/min)	0.3	0.3	0.3	0.3	0.3	0.3
Cardiac output	$\dot{Q}$ (1/min)	9					6
Shunt fraction	$\dot{Q}_{sh}/\dot{Q}$ (%)	1.33					2
Base excess	BE (mmol/1)	0					0
Mixed venous O$_2$ saturation	$S_{\bar{v}}O_2$ (%)	58	58	58	59	62	57
Alveolar PO$_2$	P_AO_2 (mmHg)	47			47		47
Total AaD	AaD_{tot} (mmHg)	8.7	8.7	9*	8*	6*	9.3
AaD caused by diffusion limitation	AaD_{diff} (mmHg)	max. 8.5 min. 5.7	7*	6.5*	5*	8*	6*
O$_2$ loading time	t_k (ms)	min. 251 max. 320	284	286	310	258	326

* values assumed within the range of experimentally determined data.

Table 2b. Basic physiological data, alveolar-arterial PO_2 differences (AaD), and O_2 loading times in the case of normoxia at different $\dot{V}O_2$. Data were taken from Thews (1984)

Oxygen uptake	$\dot{V}O_2$ (1/min)	0.3	1	2	3
Cardiac output	$\dot{Q}$ (1/min)	6	11.4	16.8	20.8
Shunt fraction	$\dot{Q}_{sh}/\dot{Q}$ (%)	2	1.05	0.71	0.58
Base excess	BE (mmol/l)	0	0	-3	-10
Mixed venous O_2 saturation	$S_{\bar{v}}O_2$ (%)	73	57	44	33
Alveolar PO_2	PAO_2 (mmHg)	104	106	107.5	113
Total AaD	AaD_{tot} (mmHg)	11.8	13.5	14.5	19.5
AaD caused by diffusion limitation	AaD_{diff} (mmHg)	$\leq$5.8 0.1*	$\leq$8.8 1.0*	$\leq$9.8 $\geq$5	$\leq$14.9 $\geq$7.5
O_2 loading time	t_k (ms)	$\geq$133 197	$\geq$163 202	$\geq$197 $\leq$212	$\geq$205 $\leq$224

* values assumed within the range of experimentally determined data.

The O_2 dissociation curves used are based on the data of Severinghaus (1965) modified with the aid of nomograms given by Thews (1979) according to the acid-base-status for the respective conditions and the actual PCO_2 at different locations along the capillary. Initial and boundary conditions were chosen according to data given by Bartels et al. (1955) for the case of hypoxia and by Thews (1984) for the case of normoxia and are listed in Tables 2a and 2b respectively.

Our findings suggest:
1. Transit times at $\dot{Q}$ = 6-10 1/min calculated for hypoxia experiments are about 295 ms ($\geq$251 ms and $\leq$326 ms). At normoxia at $\dot{V}O_2$ = 0.3-1.0 1/min, oxygen uptake is almost complete after two-thirds of the capillary transit time. Therefore, no estimates of transit time can be derived from these calculations. In the case of normoxia at higher performance levels ($\dot{V}O_2$ = 2-3 1/min, $\dot{Q}$ = 16.8-20.8 1/min), oxygen uptake is diffusion-limited. Therefore, we were able to specify transit times of about 210 ms ($\geq$197 ms and $\leq$225 ms). In all cases investigated, the respective times obtained for the completion of O_2-uptake were significantly shorter than pulmonary capillary transit times published by Johnson et al. (1960). The mechanisms of capillary recruitment and dilatation of lung capillaries keep transit times remarkably constant over a wide range of $\dot{V}O_2$ values, thus preventing severe restriction of O_2 supply to the organism.

2. Saturation times calculated with or without consideration of Hb-O_2-reaction kinetics differed by less than 5%, whereas neglecting facilitated diffusion and diffusion in the longitudinal direction introduced errors of about 20% and 15%, respectively.

3. Time courses of O_2 uptake under conditions of hypoxia agree well with former calculations in which an exponential increase of O_2-saturation inside the erythrocyte was assumed. The shape of the time courses calculated for normoxic conditions, however, departs distinctly from the exponential course.

REFERENCES

Bartels, H., Beer, R., Fleischer, E., Hoffheinz, H.J., Krall, J., Rodewald, G., Wenner, J. und Witt, I. (1955). Bestimmung von Kurzschlussdurchblutung und Diffusionskapazitat der Lunge bei Gesunden und Lungenkranken. Pflugers Arch. 261, 99-132.

Clark, A., Jr, Federspiel, W.J., Clark, P.A.A. and Cokelet, G.R. (1985). Oxygen delivery from red cells. Biophys. J. 47, 171-181.

Gehr, P., Bachofen, M. and Weibel, E.R. (1978). The normal human lung: ultrastructure and morphometric estimation of diffusion capacity. Respir. Physiol. 32, 121-140.

Goldstick, T.K., Ciuryla, V.T. and Zuckerman, L. (1976). Diffusion of oxygen in plasma and blood. In: Oxygen Transport to Tissue-II. Eds Grote, J., Reneau, D. and Thews, G., Plenum Press, New York and London, (Adv. Exp. Med. Biol. 75, 183-190).

Groebe, K. and Thews, G. (1986). Theoretical analysis of oxygen supply to contracted skeletal muscle. In: Oxygen Transport to Tissue-VIII. Ed. Longmuir, I.S., Plenum Press, New York and London, (Adv. Exp. Med. Biol. 200, 495-514).

Johnson R.L., Jr, Spicer, W.S., Bishop, J.M. and Forster, R.E. (1960). Pulmonary capillary blood volume, flow and diffusing capacity during exercise. J. Appl. Physiol. 15, 893-902.

Miyamoto, Y. and Moll, W. (1971). Measurements of dimensions and pathway of red cells in rapidly frozen lungs in situ. Respir. Physiol. 12, 141-156.

Severinghaus, J.W. (1965). Blood gas concentrations. In: Handbook of Physiology, Sect. 3, Respiration, Vol. II. Eds Fenn, W.O. and Rahn, H., American Physiological Society, Washington D.C., pp. 1475-1487.

Thews, G. (1979). Der Einfluss von Ventilation, Perfusion, Diffusion und Distribution auf den pulmonalen Gasaustausch. In: Funktionsanalyse biologischer Systeme. Bd. 5. Ed. Thews, G., Akademie der Wissenschaften und der Literatur, Mainz, Steiner, Weisbaden.

Thews, G. (1984). Theoretical analysis of the pulmonary gas exchange at rest and during exercise. Int. J. Sports Med. 5, 113-119.

Vaupel, P. (1976). Effect of percentual water content in tissues and liquids on the diffusion coefficients of O_2, CO_2, N_2, and H_2. Pflugers Arch. 361, 201-204.

Weibel, E.R. (1963). Morphometry of the Human Lung. Springer, Berlin.

Weibel, E.R. (1964). Morphometrics of the lung. In: Handbook of Physiology, Sect.3, Respiration, Vol.I. Eds Fenn, W.O. and Rahn, H., American Physiological Society, Washington D.C., pp. 285-307.

DIFFUSION PATHWAYS IN OXYGEN SUPPLY OF CARDIAC MUSCLE

L. HOOFD, Z. TUREK and K. RAKUSAN[*]

Department of Physiology, Medical Faculty, Catholic University, Nijmegen, The Netherlands and [*]Department of Physiology, School of Medicine, University of Ottawa, Ottawa, Ontario, Canada

In modelling oxygen transport in tissue, it is generally accepted that the oxygen in this medium has to be transported by diffusion. It is not quite clear, however, along which routes this O_2 diffusion occurs. Such routes can be spatial and functional. A spatially different diffusion route is an alternative pathway in the tissue, whereas facilitated diffusion via binding to myoglobin is an example of a functionally different additional transport route. Examples of both types will be considered here.

MATHEMATICAL APPROACH

The first and best-known model of oxygen diffusion is the Krogh-model (Krogh, 1919) in which the reaction-diffusion equation for transport into a tissue cylinder was solved for radial symmetry with a centrally located capillary. The basic equation linking diffusion and consumption is expressed here as:

$$\mathcal{P}\nabla^2 P = M \tag{1}$$

where $\mathcal{P}$ is the oxygen permeability ('Krogh's diffusion coefficient', or product of solubility and diffusion coefficient of oxygen), $\bar{\nabla}^2$ is the Laplace operator (second derivative with respect to spatial coordinates), P is the oxygen partial pressure and M is the oxygen consumption. The solution for the tissue cylinder is described by the Krogh-Erlang formula:

$$P = P_c - \frac{MR^2}{2\mathcal{P}}\left[\ell n(\frac{r}{r_c}) - \frac{r^2 - r_c^2}{2R^2}\right] \tag{2}$$

where P_c is the capillary PO_2, r_c is the capillary radius, and R is the outer limit, the radius of the Krogh cylinder. The boundary condition imposed by R is that the oxygen flux across R is zero: no oxygen leaves or enters the cylinder. The other boundary condition is the oxygen partial pressure in the capillary, P_c. In spite of its many assumptions (Kreuzer, 1982), the Krogh model has been very helpful in analysis of tissue PO_2.

One of the main limitations of the Krogh model is its geometry, the radial symmetry with the capillary located in the centre. We tried to investigate this problem starting from the differential Equation (1). Other solutions of this equation, pertaining to other geometries and boundary conditions, are obtained by adding to the original solution functions of the forms $r^{\pm k}\cos(k\phi)$ and $r^{\pm k}\sin(k\phi)$, in cylindrical coordinates r, ϕ. Together with the terms of the Krogh-Erlang equation, these constitute solutions for the tissue PO_2 field for other geometrical situations.

ASYMMETRIC CYLINDER

The first attempt to alter the Krogh situation was by adding $\cos(k\phi)$ terms for a single value of k to the Krogh-Erlang equation:

$$P = P_c - \frac{MR^2}{2\mathcal{P}}\left[\ell n(\frac{r}{r_c}) - \frac{r^2-r_c^2}{2R^2} + g\{(\frac{r}{r_c})^k - (\frac{r}{r_c})^{-k}\}\cos(k\phi)\right] \quad (3)$$

where g is the factor representing the weight of the alteration. The boundary conditions are the same as for the Krogh cylinder; $PO_2=P_c$ at $r=r_c$ and zero flux across the border. The latter condition now implies a non-circular region; e.g. for k = 1, $\phi = 0°$ ($\cos(\phi) = +1$) and $\phi = 180°$ ($\cos(\phi) = -1$) the flux $J = -\mathcal{P}(dP/dr)$ is found as:

$$J = \frac{MR^2}{2r}\left[1 - (\frac{r}{R})^2 \pm gk\{(\frac{r}{r_c})^k + (\frac{r}{r_c})^{-k}\}\right] \quad (4)$$

This flux is zero for a much larger value of r at $\phi = 0°$ ($r = R_+$; corresponding to the plus sign before the factor g) than at $\phi = 180°$ ($r = R_-$; corresponding to the minus sign). Thus, the diffusion region will be asymmetric, where the extent of the deformation is determined by g. So instead of the parameter g, the deformation can also be characterized by the ratio R_+/R_-. Examples of diffusion areas for different values of k and R_+/R_- are given in Figure 1.

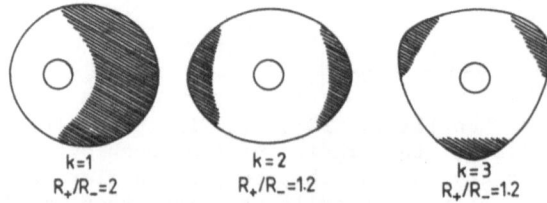

| k=1 | k=2 | k=3 |
| $R_+/R_-=2$ | $R_+/R_-=1.2$ | $R_+/R_-=1.2$ |

Figure 1. Different shapes of the diffusion area in the extensions of the Krogh model. The shapes are characterized by k, the symmetry number, and R_+/R_-, the ratio of largest to shortest distance from the capillary centre.

For all such regions, there are zones where PO_2 falls below the minimum Krogh PO_2, as indicated by the shaded areas in the figure. For $k = 1$, asymmetric region, the deviations can be quite large (Turek, Hoofd and Rakusan, 1986) but similar calculations for symmetrical deformation, $k = 2,3,4...$ show much less pronounced differences against the Krogh model.

CONCENTRIC DIFFUSION

A geometry completely different from the Krogh model is that of concentric diffusion. The diffusion region again is circular, but the capillaries are located around this region at the border, as represented in Figure 2.

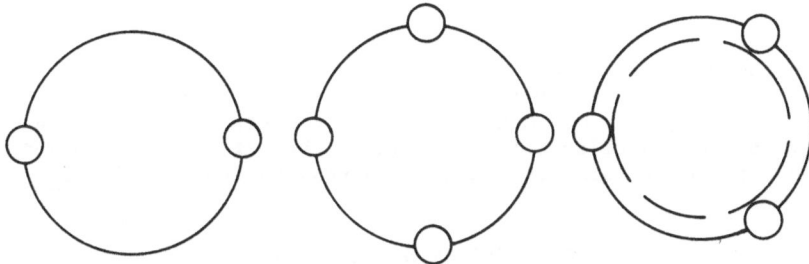

Figure 2. Concentric diffusion areas, with capillaries located outside. From left to right: homogeneous area, 2 capillaries; homogeneous area, 4 capillaries, and heterogeneous area (2 zones with different transport properties), 3 capillaries.

An approximate solution for this case can be found by summing a suitable infinite series of terms $r^{\pm k}\cos(k\phi)$ which finally leads to the expression:

$$P = P_0 - \frac{MR^2}{2\mathcal{P}}\left[\tfrac{1}{2}(\tfrac{r}{R})^2 - u_k(\tfrac{r}{R},\phi)\right] \tag{5}$$

where now P_0 is the oxygen pressure at the centre of the region and the function u_k is defined as:

$$u_k(x,\phi) = \tfrac{1}{k}\ell n\{1 - 2x^k\cos(k\phi) + x^{2k}\} \tag{6}$$

Again one of the boundary conditions was chosen such that $J \rightarrow 0$ for $r \rightarrow R$; except for the points where $\cos(k\phi) = 1$ where P approaches infinity. Therefore, the capillaries will be located at these points and capillary PO_2 is:

$$P_c = P_0 - \frac{MR^2}{2\mathcal{P}}\left[\tfrac{1}{2}(1-\tfrac{r_c}{R})^2 - \tfrac{2}{k}\ell n\{1 - (1-\tfrac{r_c}{R})^k\}\right] \tag{7}$$

If scaled properly, PO_2 profiles of these models are not much different from the Krogh values (Rakusan, Hoofd and Turek, 1984).

Contrary to the Krogh model, however, this model can allow for concentric tissue zones with different properties, as shown in the rightmost panel of Figure 2.

$$P = P_0 + \frac{1}{2\mathcal{P}_1}\left[\tfrac{1}{2}M_1 r^2 - \tfrac{1}{2}G(\mathcal{P}_1+\mathcal{P}_2)u_k(\tfrac{r}{R},\phi) - \tfrac{1}{2}G(\mathcal{P}_1-\mathcal{P}_2)u_k(\tfrac{rR}{R_1^2},\phi)\right] \qquad r \leq R_1$$

$$\text{(8)}$$

$$P = P_1 + \frac{1}{2\mathcal{P}_2}\left[\tfrac{1}{2}M_2 r^2 - (M_2 R^2 - G\mathcal{P}_2)\ell n(\tfrac{r}{R}) - G\mathcal{P}_2 u_k(\tfrac{r}{R},\phi)\right] \qquad r \geq R_1$$

where M_1, $\mathcal{P}_1$, R_1 refer to the inner zone and M_2, $\mathcal{P}_2$, R to the outer zone and G is given by:

$$G = \frac{1}{\mathcal{P}_1}(M_1 R_1{}^2 - M_2 R_1{}^2 + M_2 R^2) \qquad\qquad \text{(9)}$$

P_1 is evaluated by matching both solutions at $r = R_1$. If consumption in the outer zone, e.g. representing the connective tissue between myocytes, is taken as being much smaller than in the inner zone this affects PO_2 to only a minor extent (Rakusan et al., 1984). If, however, oxygen permeabiltiy is high in the outer zone, the influence on the PO_2 field is much more marked, increasing the overall PO_2.

FACILITATION BY MYOGLOBIN

The final pathway of O_2 considered here is a functional one: reversible binding to myoglobin (Mb). If the myoglobin is mobile (diffusion coefficient $D_{Mb} > 0$), facilitated diffusion of O_2 is possible through binding to and subsequent release from the Mb. In principle, facilitated diffusion can be accounted for quite easily, the O_2 flux being enhanced by MbO_2 diffusion:

$$J = -\mathcal{P}\frac{dP}{dx} - D_{Mb}\frac{d[MbO_2]}{dx} = -\mathcal{P}\frac{d}{dx}\{P + P_F S\} \qquad\qquad \text{(10)}$$

where

$$P_F = \frac{D_{Mb}c_{Mb}}{\mathcal{P}} \qquad\qquad \text{(11)}$$

is called the 'facilitation pressure' (de Koning, Hoofd and Kreuzer, 1981; Hoofd, 1986) and c_{Mb} is total myoglobin concentration. For rat heart, we calculated a facilitation pressure of about 15 mmHg corresponding to an Mb concentration as described by Turek et al. (1973) and $D_{Mb}/\mathcal{P}$ of Federspiel (1986). Hence, facilitation by myoglobin can add a virtual 15 mmHg driving force to the capillary PO_2 at most. This demonstrates the possibilities and also the limitations of facilitation.

Calculations for a Krogh area and an asymmetric region are shown in Figure 3 for a capillary P_c of 10 mmHg in order to elucidate the effect of facilitation. With facilitation, there is a considerable increase in the region supplied with oxygen. Furthermore, the already flat PO$_2$ profile in the outer region of the Krogh cylinder – almost 70% of the Krogh area has a PO$_2$ below 2 mmHg! – becomes even flatter with facilitation.

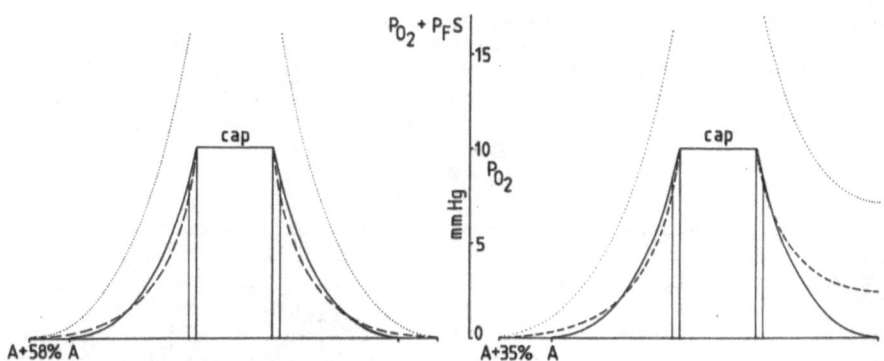

Figure 3.   PO$_2$ profiles in a Krogh (left) and an asymmetric (right) region without (solid line) and with (broken line) facilitation. Capillary wall is indicated by the thin vertical lines. The oxygen-supplied area (A) increases by an amount as indicated in the figure when there is facilitation. The dotted line represents the total driving force P + P$_F$S.

OXYGEN DISTRIBUTION IN RAT MYOCARDIUM

The general trend of the above theoretical treatment is that, with the exception of the (very) asymmetric region, there are no large deviations from the Krogh model. Therefore, we incorporated the facilitated diffusion in a computer program that calculates PO$_2$ histograms (Rakusan et al., 1984) based on the Krogh model. The measured values of rat myocardial capillary distribution, incorporating heterogeneity of capillary spacing estimated by the method of capillary domains (Hoofd et al., 1985), were determined on cross-sections of the left ventricle of normal rat hearts (Turek, Hoofd and Rakusan, 1987); the other input data were taken from Turek and Rakusan (1981). Cross-sectional capillary PO$_2$ was assumed to be homogeneous, disregarding possible effects of intraerythrocyte facilitated diffusion. Due to the heterogeneity in capillary spacing these histograms are quite broad. Examples are shown in Figure 4.

Facilitation decreases the percentage of anoxic tissue, particularly in the work situation, but cannot fully compensate for it. In the model it was assumed that the ratio between flow and O$_2$ consumption was homogeneous. So, the only heterogeneity effective here was the heterogeneity of capillary spacing. Figure 4 shows that facilitated diffusion alone cannot overcome the effect of this type of heterogeneity.

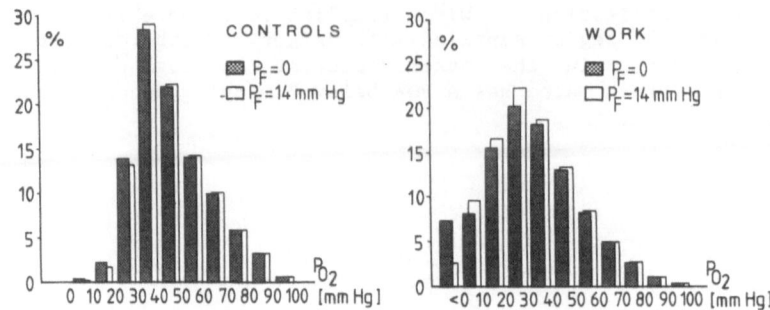

Figure 4. PO$_2$ histograms calculated with (open columns) and without (shaded columns) facilitation for rat myocardium. Left panel: resting situation; right panel: work situation, with an equal increase in oxygen consumption and blood flow (2.5x).

CONCLUSION

Although a correct description of tissue oxygenation might be quite complicated we can learn much from simple modelling with (approximate) analytical solutions for the descriptive mathematical equations. This was one of the major advantages of the Krogh model, and quite simple extensions of this model can be made as described above. Also, analytical solutions have the advantage that they can easily be incorporated into other mathematical models of tissue.

ACKNOWLEDGEMENTS

Supported in part by NATO grant number RG. 86/0073 and by the Ontario Heart Foundation.

REFERENCES

de Koning, J., Hoofd, L.J.C. and Kreuzer, F. (1981). Oxygen transport and the function of myoglobin. Theoretical model and experiments in chick gizzard smooth muscle. Pflugers Arch. **389**, 211-217.

Federspiel, W.J. (1986). A model study of intracellular oxygen gradients in a myoglobin-containing skeletal muscle fiber. Biophys. J. **49**, 857-868.

Hoofd, L.J.C. (1986). Facilitated diffusion of oxygen in tissue and model systems. Thesis, Catholic University, Nijmegen.

Hoofd, L., Turek, Z., Kubat, K., Ringnalda, B.E.M. and Kazda, S. (1985). Variability of intercapillary distance estimated on histological sections of rat heart. In: Oxygen Transport to Tissue-VII. Eds Kreuzer, F., Cain, S.M., Turek, Z. and Goldstick, T.K., Plenum Press, New York and London, (Adv. Exp. Med. Biol. **191**, 239-247).

Kreuzer, F. (1982). Oxygen supply to tissues: the Krogh model and its assumptions. Experientia, **38**, 1415-1426.

Krogh, A. (1919). The number and distribution of capillaries in muscles with calculations of the oxygen pressure head necessary for supplying the tissue. J. Physiol. **52**, 409-415.

Rakusan, K., Hoofd, L. and Turek, Z. (1984). The effect of cell size and capillary spacing on myocardial oxygen supply. In: Oxygen Transport to Tissue-VI. Eds Bruley, D., Bicher, H.I. and Reneau, D., Plenum Press, New York and London, (Adv. Exp. Med. Biol. 180, 463-477).

Turek, Z., Hoofd, L. and Rakusan, K. (1986). Myocardial capillaries and tissue oxygenation. Can. J. Cardiol. 2, 98-103.

Turek, Z., Hoofd, L. and Rakusan, K. (1987). A comparison of the methods for assessment of the heterogeneity of myocardial capillary spacing. This volume.

Turek, Z. and Rakusan, K. (1981). Lognormal distribution of inter-capillary distance in normal and hypertrophic rat heart as estimated by the method of concentric circles: its effect on tissue oxygenation. Pflugers Arch. 391, 17-21.

Turek, Z. Ringnalda, B.E.M., Grandtner, M. and Kreuzer, F. (1973). Myoglobin distribution in the heart of growing rats exposed to a simulated altitude of 3500 m in their youth or born in the low pressure chamber. Pflugers Arch. 340, 1-10.

SIMULATION OF THE POINT SPREAD FUNCTION FOR LIGHT IN TISSUE BY A MONTE CARLO METHOD

P. VAN DER ZEE and D.T. DELPY

Department of Medical Physics and Bioengineering, University College Hospital, Shropshire House, Capper Street, London WC1E 6JA, U.K.

SUMMARY

We have been able by a Monte Carlo technique to generate the point spread function (PSF) for light in tissue for a generalized range of tissue characteristics. We have demonstrated that these can be described by an equation containing a gaussian, diffusion and exponential term. The PSF equation will allow one to estimate the limits of spatial resolution achievable with near infrared (NIR) imaging systems, and may be used in image deconvolution algorithms. Additionally an equation has been derived describing the average photon pathlength through the tissue. Finally, the light transmission and reflection (backscattering) have been illustrated as functions of scattering and absorption coefficients. These results can be used in attempting to quantify data from non-invasive NIR spectroscopy systems.

INTRODUCTION

The technique of non-invasive near infrared (NIR) spectroscopy is being applied increasingly in the field of medicine (Jobsis, 1977). This technique can provide continuous information on the changes occurring in blood and tissue oxygenation. It has been applied particularly in the area of cerebral monitoring, where for obvious reasons, one cannot use invasive monitoring techniques. While information on relative changes in oxygenation is of clinical significance, quantitative data would be much more useful. If in addition one could spatially localize the signal, it would then be possible to produce an image of the tissues indicating variations in oxygenation status (Jarry et al., 1984; Cope and Delpy, in press). Both these aims can only be achieved if the nature of light transport in tissue is understood. We are currently using the NIR technique clinically to monitor oxygenation and blood volume changes in the brain of full term and premature infants (Wyatt et al., 1986). In addition, we are also developing techniques for NIR imaging across the head of the preterm infant (Arridge et al., 1986). A theoretical and experimental study of light transport in tissue is an integral part of this work.

For light in the NIR, tissue is a highly scattering material. Data obtained from post-mortem brain indicate a scattering length (length between interactions) of a few millimeters (Svaasand, Doiron and Profio, 1981). Therefore, for infant head diameters of 5 to 9 cm, light transport is dominated by multiple scattering. As a result, the optical pathlength through the brain is significantly different from the spacing between the optodes placed on the surface. For this reason, it has not normally been possible to quantitate the spectroscopic data. Knowledge of the average photon pathlength in the tissue would make this possible. Additionally, when attempting to image through the tissues, multiple scattering will cause considerable blurring of the image. If one could know the point spread function (PSF) for light travelling through the tissues it would be possible to assess the likely resolution achievable with various imaging schemes. Techniques proposed for imaging range from a simple collimated light source and camera, to CT systems and ultimately time-resolved detection. Finally, if the PSF were known, one would be able to employ deconvolution techniques to the images in order to enhance resolution.

There are various ways to calculate the transport of light through a scattering and absorbing medium. One approach is to analytically solve the wave equation or the diffusion equation (Gate, 1973; Svaasand et al., 1981). However, for these equations to be solvable even for the case of a uniform homogeneous medium, one has to make many approximations which especially for imaging make the results of doubtful value. It is also extremely difficult to incorporate any geometrical information or inhomogeneities into the tissue model. A second approach is to use the Monte Carlo technique (Wilson and Adam, 1983; Maarek et al., 1986), which can easily accommodate complex tissue models, but which requires considerable computation time. We have chosen to use this technique and have employed it to calculate the form of the point spread functions for light in tissues of varying absorption and scattering coefficients. From these, a general equation describing these PSFs has been derived using a curve fitting routine. With this equation one can rapidly calculate the point spread function for any tissue thickness. Additionally, a general equation describing the average photon pathlength through tissue has been derived. Finally, a preliminary analysis of the variation of transmitted and reflected intensities with absorption and scattering coefficients has been carried out.

THE MODEL

We have developed a full 3D model of light transport in tissue, using as a starting point the models described by Wilson and Adam (1983) and Maarek et al. (1984). In our model the following assumptions have been made about the photon interactions in tissue:

1) Scattering takes place at discrete scattering centres. Successive scattering lengths are calculated as: $L = -\ln(R')/Mus$, where R' is a random number between 0 and 1, and Mus is the tissue scattering coefficient;

2) scattering of light by the tissues is assumed to be non-isotropic. There is now considerable evidence for this supposition (Kullenberg, 1974; Maarek et al., 1984). The volume scattering function, from which the scattering angle can be derived tends to be strongly peaked in the forward direction. An extensive search of the literature has, however, failed to find

180

measured values for this function. We are currently performing experiments to measure this function in living brain tissue. In the meantime we have used a calculated form for this function. This was calculated using Mie scattering theory (Bohren and Huffman, 1983), and a range of particle sizes. Particle sizes were selected by measuring the average size of neurones and other structures in slices of brain tissue. The refractive index of water (1.33) was used for the medium, and for the cell membranes a value of 1.46. The relative refractive index was therefore 1.10. These values can be compared with reported average values of 1.49 and 1.38 for rat gut, respectively in vitro and in vivo (Gahm and Witte, 1986). The final volume scattering function was obtained by summing the individual scattering functions for each particle size. The resulting function, and the particle sizes employed are shown in Figure 1;

3) absorption takes place at a molecular level and is therefore equal to exp(−Mua*L), where Mua is the absorption coefficient and L the pathlength. Here we differ from Wilson and Adam (1983) who assume absorption to take place at the discrete scattering sites, and scattering to be isotropic.

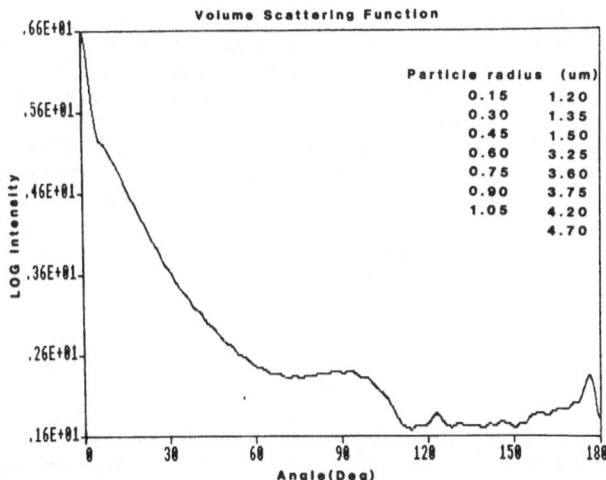

Figure 1. Average tissue volume scattering function calculated from Mie scattering theory for the range of particle sizes shown.

In the model, the path of a photon is simulated through a homogeneous tissue slab, entering it perpendicular to the lower surface. Photons are followed until they exit the tissue, when their exit coordinates, angles and total pathlength are recorded. For each PSF the model is run until 60000 photons have been transmitted through the tissue. The resulting data can be displayed as a function of any of the stored parameters (e.g. exit angle, flight time etc.).

Calculations have been performed for a fixed tissue thickness of 10 mm, but for a wide range of scattering and absorption coefficients. The range chosen (Table 1) encompasses the extremes of values quoted in the literature (Svaasand et al., 1981; Svaasand and Ellingson, 1983). Results from other tissue thicknesses can be obtained by simple scaling of the scattering and absorption coefficients.

Table 1.

Mus (mm^{-1}): 4.0; 2.0; 1.0; 0.5; 0.333; 0.25; 0.2; 0.167

Mua (mm^{-1}): 1.0; 0.4; 0.167; 0.105; 0.077; 0.061; 0.05; 0.02

A non-linear least squares curve fitting routine has been applied to the calculated PSFs and mean path lengths (Gaushouse, 1965). In order to provide a good fit over the wide dynamic range of the data, the logarithm of the chosen function has been fitted to the logarithm of the data.

The choice of function to be used in the curve fitting process is of critical importance. There are a large number of possible functions that would fit the PSFs obtained from the Monte Carlo model. One could for instance simply use a polynomial. This would, however, be difficult to relate to the known physical parameters of the model and would not provide much insight into the problem. We have therefore chosen to use functions which one would expect to obtain from a simplified analysis of the light transport problem. Whitman and Beran (1970) have calulated the beam spread of light in a scattering medium using a small angle approximation, and predict a beam profile that is gaussian in nature. At the extreme of light transport in a highly scattering medium, a photon diffusion model is applicable (Johnson, 1970) and a diffusion term would be expected. We therefore initially attempted to fit the Monte Carlo-derived PSFs using a function containing a gaussian and a diffusion term. Although providing a reasonable fit, it was found necessary to add a third term of an exponential form in order to describe fully the PSFs over the whole range of absorption and scattering coefficients.

RESULTS

Figures 2 and 3 show the point spread functions calculated from the Monte Carlo model for the extremes of the absorption and scattering coefficient range. The data plotted represent all the exiting photons, irrespective of exit angle. The figures are plotted on logarithmic as well as linear scales in order to demonstrate the extent of the 'tail' of the PSF. Figure 4 is the PSF for a mid range value of absorption and scattering coefficients. The equation to which the PSFs have been fitted is:

$$I(r) = P1exp(-r^2*P2) + P3exp(-((1+(r/d)^{2^{0.5}}))*P4)/(1+(r/d)^2) + P5exp(-r*P6) + K(0) \qquad (1)$$

where:
  d is the tissue thickness,
  $P1exp(-r^2*P2)$ is the gaussian term,
  $P3exp(-((1+(r/d)^{2^{0.5}}))*P4)/(1+(r/d)^2)$ is the diffusion term,
  $P5exp(-r*P6)$ is the exponential term, and
  K(0) represents the intensity contribution of unscattered photons.

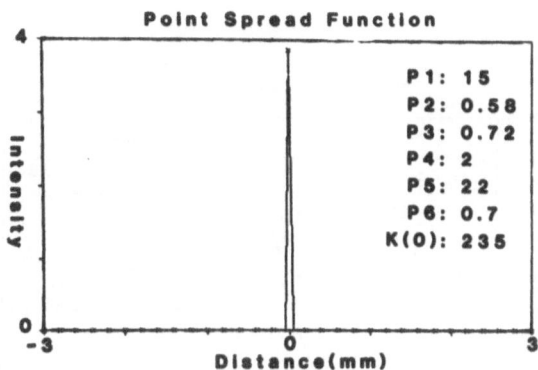

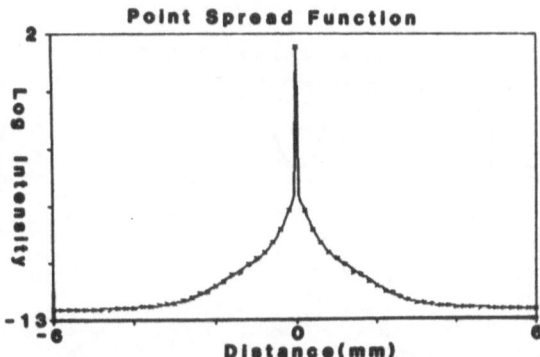

Figure 2. Point spread functions for the extreme of high absorption and low scattering (Mua = 1.0, Mus = 0.17). The functions are displayed on linear (upper figure) and logarithmic (lower figure) scales. The crosses represent the Monte Carlo values, the solid line is calculated from the fitted equation. Values for P1...P6 and K(O) are given in the upper figure.

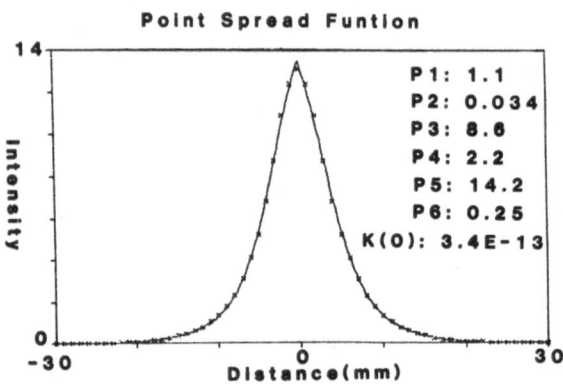

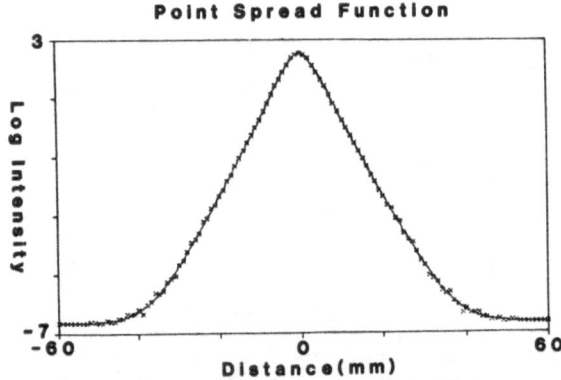

Figure 3. Point spread functions for the extreme of low absorption and
high scattering (Mua = 0.02, Mus = 4). For further
explanation see Fig. 2.

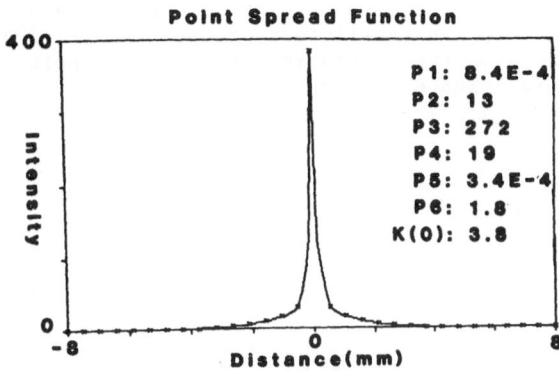

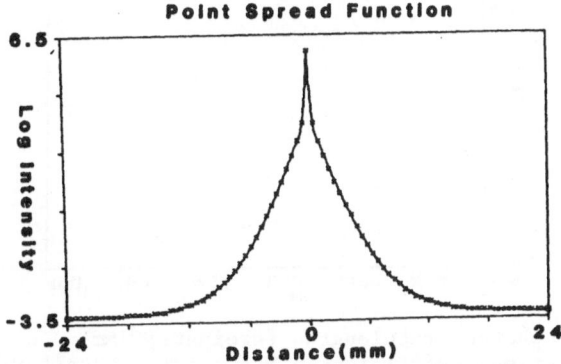

Figure 4. Point spread functions for intermediate values of absorption and scattering (Mua = 0.105, Mus = 0.5). For further explanation see Fig. 2.

The parameters Pl to P6 and K(O) can be related to tissue absorption and scattering coefficients. Preliminary work has been carried out to determine the form of these relationships. A final analysis will not be attempted until an experimentally measured volume scattering function has been obtained. Typical values for Pl...P6 and K(O) for the illustrated PSFs are shown on the figures.

The mean photon pathlength, Mp, has however been analysed as a function of Mua and Mus. It has been found to obey the following equation:

$$Mp = d(1+Mus/(3+4.2Mua(5.3+Mus)))$$ (2)

The factor d is the tissue thickness and Mus and Mua the tissue scattering and absorption coefficients respectively. Examples of this fit as a function of the scattering and the absorportion coefficient are shown in Figures 5 and 6 respectively.

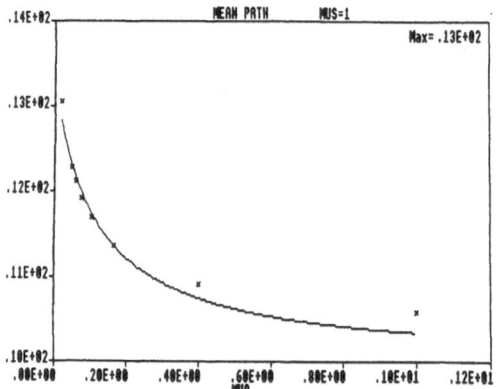

Figure 5. Mean photon pathlength (ordinate; mm) as a function of absorption coefficient, Mua, for a scattering coefficient Mus = 1.0 and a tissue thickness of 10 mm.

The model also provides us with data on the transmission and reflection coefficient (backscatter). Figure 7 shows the transmission as a function of absorption for different scattering coefficients, and in Figure 8 as a function of scattering for different absorptions. Similar results for backscatter are shown in Figures 9 and 10.

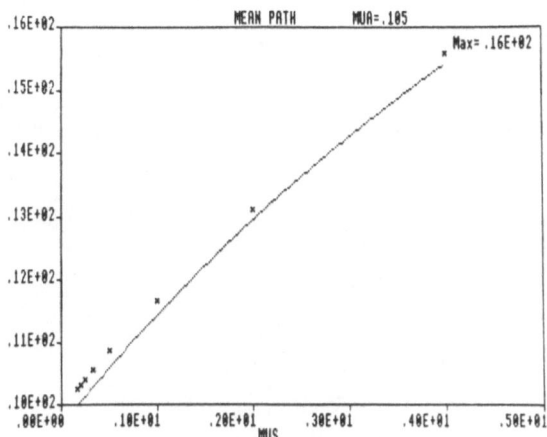

Figure 6. Mean photon pathlength (ordinate; mm) as a function of scattering coefficient, Mus, for an absorption coefficient Mua = 0.105 and a tissue thickness of 10 mm.

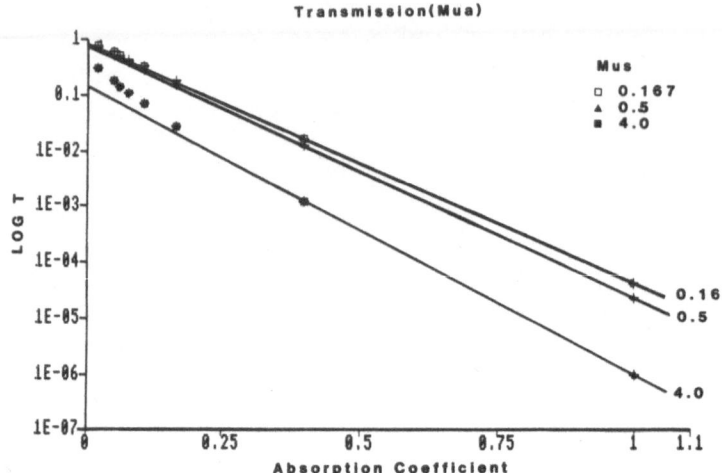

Figure 7. Total transmission as a function of absorption coefficient, Mua, for three scattering coefficients: Mus = 0.167; 0.5; 4.0.

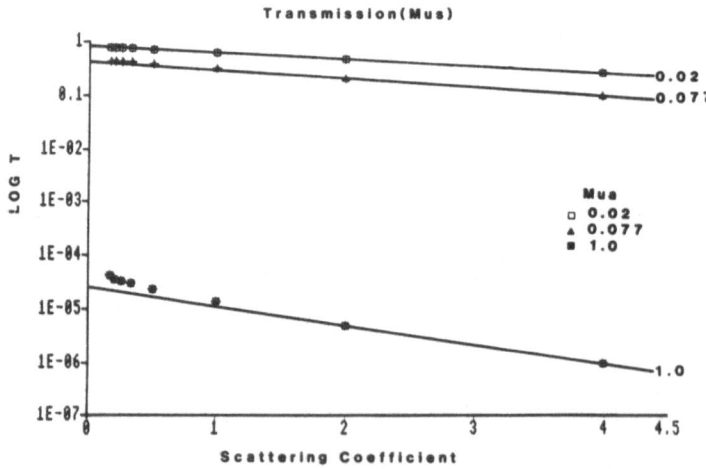

Figure 8. Total transmission as a function of scattering coefficent, Mus, for three absorption coefficients: Mua = 0.02; 0.077; 1.0.

DISCUSSION

It can be seen from Figures 2, 3 and 4 that Equation (1) gives a good fit to the Monte Carlo derived PSFs. The behaviour of P1..P6 as a function of Mua and Mus has not yet been rigorously analysed although some general trends can be seen from the results shown in Figures 2, 3 and 4. The derivation of these relationships remains a major element of future work.

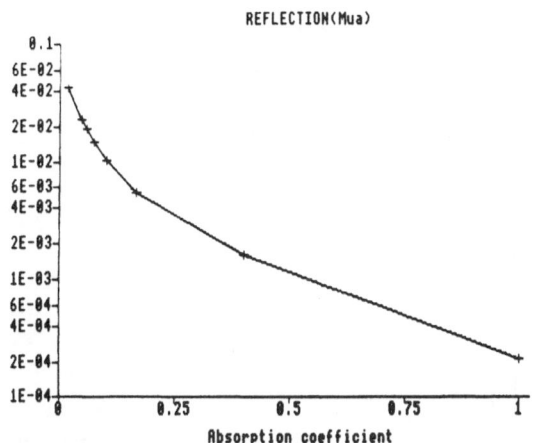

Figure 9. Total reflection (backscatter) as a function of absorption coefficient, Mua, for a scattering coefficient Mus = 0.5. Ordinate, log reflection.

The mean pathlength, as expected, increases with increasing scattering, and decreases with increasing absorption. It can be seen from Equation (2) that the mean pathlength is more sensitive to changes in scattering than in absorption. This has encouraging implications for the spectroscopic work, since scattering is likely to remain reasonably constant, being dominated by the relatively static tissues. It must be noted, however, that variations in mean pathlength with changes in absorption or scattering will also alter the effective volume from which the signal is derived. The data on light transmission as a function of absorption (Fig. 7) show that Log (T) is not a linear function of absorption. It can be seen that this non-linearity is most pronounced at high scattering coefficients. This confirms the experimental observations that Beers law does not apply in the presence of scattering. Log (T) as a function of scattering (Fig. 8) can also be seen to be a non-linear function, the rate of change being largest for low absorption.

When looking at the results for backscattering, it should be borne in mind that in the model we have assumed no reflection at the tissue boundaries due to differences in refractive index between the tissue and the surrounding medium. From the plot of reflection (backscatter) as a function of absorption (Fig. 9) it is apparent that Log(R) increases rapidly and in a non-linear way with decreasing absorption. This is because at lower absorption, scattering from greater depths starts to contribute to the reflected signal. Figure 10 shows Log(R) as a function of scattering. As expected, reflection is strongly dependent on scattering, although as scattering increases, one would expect R to approach a limiting value. Finally, it should be remembered that the reflection depends also on the shape of the volume scattering function.

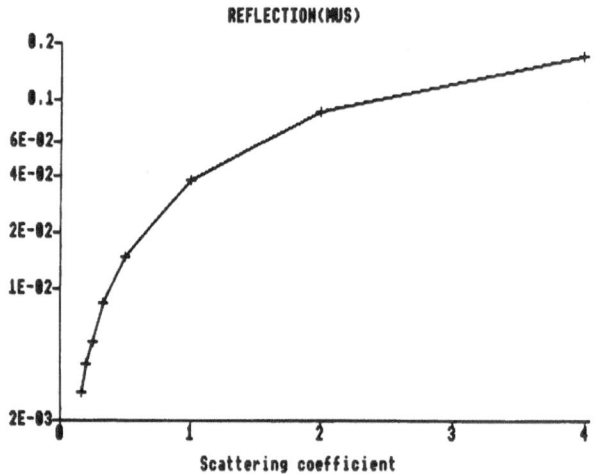

Figure 10. Total reflection (backscatter) as a function of scattering coefficient, Mus, for an absorption coefficient Mua = 0.077. Ordinate, log reflection.

ACKNOWLEDGEMENTS

The work was carried out with funding provided by the Wolfson Foundation, the Science and Engineering Research Council and Hamamatsu Photonics Ltd.

REFERENCES

Arridge, S.R., Cope, M., van der Zee, P., Hilson, P.J. and Delpy, D.T. (1986). Visualisation of the oxygenation state of the brain and muscle in newborn infants by near infrared transillumination. In: Information Processing in Medical Imaging. Ed. Bacharach, S.L., Martinus Nijhoff, Dordrecht, Holland, pp. 155-176.

Bohren, C.F. and Huffman, D.R. (1983). Absorption and Scattering of Light by Small Particles. Wiley Interscience, New York.

Cope, M. and Delpy, D.T. A system for long term cerebral blood and tissue oxygenation measurement on newborn infants by near infrared transillumination. In: Optical Monitoring in situ. Ed. Jobsis, F.F., Plenum Press, New York and London (in press).

Gahm, T. and Witte, S. (1986). Measurement of the optical thickness of transparent tissue layers. J. Microsc. 141, 101-110.

Gate, L.F. (1973). The determination of light absorption in diffusing material by a photon diffusion model. J. Phys. D: Applied Physics, 4, 1049-1056.

Gaushouse, (1965). Non linear least squares fit. University of Wisconsin Computing Center, December 1965 issue.

Jarry, G., Ghesquiere, S., Maarek, J.M., Fraysse, F., Debray, S., Biu-Mong-Hung and Laurent, D. (1984). Imaging mammalian tissues and organs using laser collimated transillumination. J. Biomed. Eng. 6, 70-74.

Jobsis, F.F. (1977). Noninvasive, infrared monitoring of cerebral and myocardial oxygen sufficiency and circulatory parameters. Science, 198, 1264-1267.

Johnson, C.C. (1970). Optical diffusion in blood. IEEE Trans. BME 17, 129-133.

Kullenberg, G. (1974). Observed and computed scattering functions. In: Optical Aspects of Oceanography. Eds Jerlov, N.G. and Nielsen, M., Academic Press, London, pp. 25-49.

Maarek, J.M., Jarry, G., Crowe, J., Bui-Mong-Hung and Laurent, D. (1986). Simulation of laser tomoscopy in a heterogeneous biological medium. Med. and Biol. Eng. and Comp. 24, 407-414.

Maarek, J.M., Jarry, G., de Cosnac, B., Lansiart, A. and Bui-Mong-Hung (1984). A simulation method for the study of laser transillumination of biological tissues. Annals Biomed. Eng. 12, 281-304.

Svaasand, L.O., Doiron, D.R. and Profio, A.E. (1981). Light distribution in tissue during photoradiation therapy. USC Institute for Physics and Imaging Science. Report MISG. 900-02.

Svaasand, L.O. and Ellingson, R. (1983). Optical properties of human brain. Photochem. Photobiol. 38, 293-299.

Whitman, A.M. and Beran, M.J. (1970). Beam spread of laser light propagating in a random medium. J. Opt. Soc. Am. 60, 1595-1602.

Wilson, B.C., Adam, G. (1983). A Monte Carlo model for the absorption and flux distribution of light in tissue. Med. Phys. 10, 824-830.

Wyatt, J.S., Cope, M., Delpy, D.T., Wray, S. and Reynolds, E.O.R. (1986). Continuous non-invasive monitoring of cerebral oxygenation in sick new born infants by near infrared spectrophotometry. Proc. Eur. Soc. Ped. Res., Groningen, Abstracts, p.37

USE OF ADAIR FOUR-STEP KINETICS IN MATHEMATICAL SIMULATION OF OXYGEN TRANSPORT IN THE MICROCIRCULATION

E.W. YAP and J.D. HELLUMS

Rice University, P.O.Box 1892, Houston, TX 77251, U.S.A.

ABSTRACT

The Adair four-step kinetic model for the reactions of haemoglobin and oxygen recognizes five haemoglobin species, corresponding to deoxyhaemoglobin and one species for each level of oxygenation of the four haem groups. Thus, an oxygen transport problem involves a system of five simultaneous non-linear partial differential equations for diffusion with chemical reaction. This mathematical complexity has impeded application of the Adair model despite its theoretical advantages over the one-step model often used in practice.

The Adair kinetic model has been incorporated into a simulation of microcirculatory oxygen transport. The results show that the usual one-step kinetic model is inaccurate in comparison with the Adair model. However, an empirical modification can be made to the one-step model to ensure compatibility with the equilibrium curve. This modified one-step kinetic model (the VRC model) is much more tractable mathematically than the Adair model. In the physiological range of fluxes, the VRC kinetic model appears to be of sufficient accuracy for most purposes, and the mathematical complexity of the Adair model is not required.

INTRODUCTION

The haemoglobin molecule has four nearly identical protein haem (Hb) groups each of which is capable of binding reversibly with one molecule of oxygen. Several oxygen reaction kinetic models have been proposed which take into account the structure of haemoglobin. The Adair reaction scheme (Adair, 1925) assumes a simple mass action kinetic model for each intermediate oxyhaemoglobin species as shown in Figure 1. It can be seen that the haemoglobin molecule, Hb_4, may react successively with four oxygen molecules to give, eventually, the fully saturated oxyhaemoglobin molecule. Gibson (1970), studied the reaction of haemoglobin with oxygen by stopped-flow methods for the purpose of estimating the numerical values of the eight rate constants. He found that he was able to obtain the best fit for the experimental kinetic and equilibrium data by setting the values of the rate constants as given in Figure 1.

The mathematical complexity of the Adair model has retarded its use in mathematical simulation of oxygen transport. The purpose of the present work is to determine if the Adair model has significant advantages over certain simpler models in the mathematical simulation of oxygen transport under conditions of physiological relevance. The simpler models are much easier to deal with mathematically, but they are on a much less satisfactory theoretical basis. The assessment of the models was carried out by direct comparison of results from mathematical simulation of oxygen release under fluxes and space and time scales characteristic of the microcirculation.

$$Hb_4 \quad + \quad O_2 \quad \underset{k_1}{\overset{k_1'}{\rightleftharpoons}} \quad Hb_4O_2 \tag{1}$$

$$Hb_4O_2 \quad + \quad O_2 \quad \underset{k_2}{\overset{k_2'}{\rightleftharpoons}} \quad Hb_4(O_2)_2 \tag{2}$$

$$Hb_4(O_2)_2 \quad + \quad O_2 \quad \underset{k_3}{\overset{k_3'}{\rightleftharpoons}} \quad Hb_4(O_2)_3 \tag{3}$$

$$Hb_4(O_2)_3 \quad + \quad O_2 \quad \underset{k_4}{\overset{k_4'}{\rightleftharpoons}} \quad Hb_4(O_2)_4 \tag{4}$$

$k_1' = 17.7 \times 10^6 \ M^{-1} \ sec^{-1}; \quad k_1 = 1900 \ sec^{-1}$

$k_2' = 33.2 \times 10^6 \ M^{-1} \ sec^{-1}; \quad k_2 = 158 \ sec^{-1}$

$k_3' = 4.89 \times 10^6 \ M^{-1} \ sec^{-1}; \quad k_3 = 539 \ sec^{-1}$

$k_4' = 33.0 \times 10^6 \ M^{-1} \ sec^{-1}; \quad k_4 = 50 \ sec^{-1}$

Figure 1. The Adair reaction scheme with values for the corresponding rate constants.

A reaction kinetic model often used in mathematical simulation, is the simple one-step mass model:

$$Hb_4 + O_2 \quad \underset{j_1}{\overset{j_1'}{\rightleftharpoons}} \quad Hb_4O_2 \tag{5}$$

This model assumes that the rate of reaction of each haem subunit, is independent of the number of oxygen molecules already bound to the haemoglobin molecule and that the rates of reaction are directly proportional to the concentration of each species. In this case, both rate coefficients, j_1 and j_1' are constants. At equilibrium, the fractional oxyhaemoglobin saturation, y, is often related to the dissolved oxygen concentration, u, by the Hill equation:

$$y = \frac{Ku^P}{1 + Ku^P} \qquad (6)$$

The constants K and p are chosen empirically to fit measurements. From Figure 2, it can be seen that the Hill equation with $p = 2.75$ and $K = 1.179 \times 10^{12}$ approximately describes the typical experimental data given in the figure.

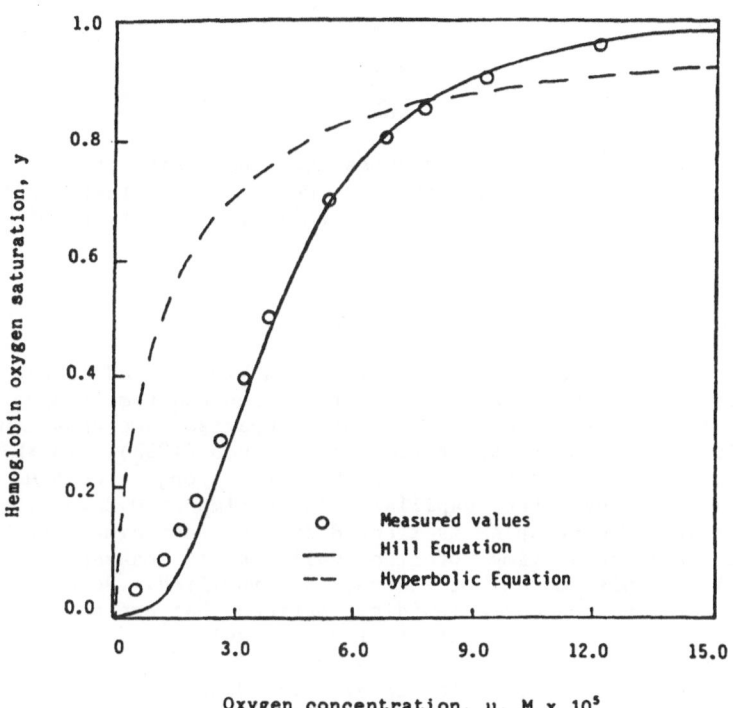

Figure 2. Oxyhaemoglobin equilibrium curves.

If the Hill equation is simplified by setting $p = 1$, the 'hyperbolic' equilibrium relationship results with the following form:

$$y = \frac{Ku}{1 + Ku} \qquad (7)$$

At equilibrium, the kinetic expression for the one-step mass action model reduces to:

$$y = \frac{(j'_1/j_1)u}{1 + (j'_1/j_1)u} \qquad (8)$$

Equation (8) is of the same form as Equation (7). Thus, it describes the hyperbolic curve shown in Figure 2. It is seen that this hyperbolic curve does not follow the equilibrium relationship accurately. This incompatibility with the equilibrium relationship is the source of error when the one-step model is used.

Moll (1969) recognized the serious error introduced by use of the simple one-step model, and proposed a modification of the model which is empirically forced to be compatible with the equilibrium curve. This model is a variable rate constant (VRC) model. The dissociation coefficient, j_1, is made a function of the oxygen concentration in such a way that at equilibrium, the kinetic expression will reduce to the Hill equation (or to some other representation of the known equilibrium curve). In this case, if j'_1 is chosen to be constant, it is easy to show that j_1 should depend on u as shown by Equation (9):

$$j_1(u) = \frac{j'_1}{KuP-1} \tag{9}$$

Another form of the VRC model assumes that the dissociation coefficient is a function of y rather than u. We have found that the two forms give essentially equivalent results under the conditions of this work (Yap, 1985).

MATHEMATICAL FORMULATION

The results of comparisons of reaction kinetic models are known to depend on the oxygen fluxes and on the space and time scales involved (Sheth and Hellums, 1980). For this comparison, we chose the simple radial diffusion model of Baxley and Hellums (1983). In this model, the erythrocyte is idealized as a cylindrical object of 6 μm diameter which moves through the capillary in a time of 0.2 to 1.0 seconds. Thus, the diffusion problem is treated in one space dimension (radial), and time (residence time of the cell as it passes through the capillary). This model, of course, is simplified. However, it gives estimates of the concentration distributions that are suitable for the present purpose, the comparison of kinetic models.

The equations for this model are discussed in more detail elsewhere (Baxley and Hellums, 1983) and are given below for both one-step kinetics and VRC kinetics.

$$\frac{\partial u}{\partial t} = D_u \frac{1}{r} \frac{\partial}{\partial r} (r \frac{\partial u}{\partial r}) + v_t(j_1 y - j'_1 u (1-y)) \tag{10}$$

$$\frac{\partial y}{\partial t} = D_h \frac{1}{r} \frac{\partial}{\partial r} (r \frac{\partial y}{\partial r}) - j_1 y + j'_1 u (1-y) \tag{11}$$

with conditions at r = R, the radius of the cell,

$$\frac{\partial y}{\partial r} = 0 \quad \text{and} \quad - D_u \frac{\partial u}{\partial r} = q_o \tag{12}$$

These equations and conditions, together with initial conditions (capillary inlet conditions) on u and y, constitute a complete problem, where

    r   denotes radial position
    t   denotes residence time of the cell in the capillary
    y   denotes haemoglobin oxygen saturation
    v_t denotes the total (oxy plus deoxy) haem concentration
       (four times the haemoglobin concentration)
    D_u denotes the diffusivity of oxygen
    D_h denotes the diffusivity of haemoglobin

Equation (10) governs oxygen transport and reaction; Equation (11), that of oxyhaemoglobin. The first of the conditions in Equation (12), corresponds to a statement of no net flux of haemoglobin at the cell membrane. The second condition designates a constant, specified flux, q_o, of oxygen through the cell membrane.

The rate coefficients, j_1 and j'_1, are constants in the simple one-step reaction kinetic model. For the VRC reaction kinetic model, all the equations and conditions are the same. The only difference is that j_1 is not constant. It is given by Equation (9).

For the case of the four-step Adair kinetic model, the full set of equations to be solved is given in Table 1. The derivation of the equations is given in detail elsewhere (Yap, 1985). It can be seen that the set of equations to be solved is much more complicated than that of the one-step or VRC models. The number of partial differential equations is increased from two to five. It should be noted that the diffusivity of all the oxyhaemoglobin species is assumed to be equal to that of haemoglobin. The haemoglobin molecule has a high molecular weight compared with oxygen.

The boundary conditions on the equations of Table 1 are the same as those in Equation (12), except we have a zero flux condition on all four haemoglobin species instead of just the one of Equation (12). The initial (capillary inlet) conditions were imposed such that the oxygen and haemoglobin species are in equilibrium. The fractional oxyhaemoglobin saturation, y, is the number of oxygenated haem groups divided by the total number of haem groups. Therefore,

$$y = 1/4 \; (y_1 + 2y_2 + 3y_3 + 4y_4) \tag{13}$$

where

$$4y_1 = [HbO_2]/v_t \tag{14}$$

$$4y_2 = [Hb(O_2)_2]/v_t \tag{15}$$

$$4y_3 = [Hb(O_2)_3]/v_t \tag{16}$$

$$4y_4 = [Hb(O_2)_4]/v_t \tag{17}$$

At equilibrium it can be shown that

$$y_1 = f_1(u) = \frac{\alpha_1 u}{(1 + \alpha_1 u + \alpha_1 \alpha_2 u^2 + \alpha_1 \alpha_2 \alpha_3 u^3 + \alpha_1 \alpha_2 \alpha_3 \alpha_4 u^4)} \tag{18}$$

TABLE 1: Differential Equations with Adair Kinetics

$$\frac{\partial u}{\partial t} = D_u \frac{1}{r} \frac{\partial}{\partial r} \left(r \frac{\partial u}{\partial r} \right) - \left(k_1' + (k_2'-k_1')y_1 + (k_3'-k_1')y_2 \right.$$

$$+ (k_4'-k_1')y_3 - k_1'y_4)uv_t + (k_1 y_1 + k_2 y_2 + k_3 y_3 + k_4 y_4)v_t \qquad (1.1)$$

$$\frac{\partial y_1}{\partial t} = D_h \frac{1}{r} \frac{\partial}{\partial r} \left(r \frac{\partial y_1}{\partial r} \right) + \left(-(k_1 + k_1'u + k_2'u)y_1 + (k_2 - k_1'u)y_2 \right.$$

$$-k_1'uy_3 - k_1'uy_4 + k_1'u) \qquad (1.2)$$

$$\frac{\partial y_2}{\partial t} = D_h \frac{1}{r} \frac{\partial}{\partial r} \left(r \frac{\partial y_2}{\partial r} \right) + \left(-(k_2 + k_2'u) \ y_2 + k_2'uy_1 + k_3 y_3 \right) \qquad (1.3)$$

$$\frac{\partial y_3}{\partial t} = D_h \frac{1}{r} \frac{\partial}{\partial r} \left(r \frac{\partial y_3}{\partial r} \right) + \left(-(k_3 + k_4'u)y_3 + k_3'uy_2 + k_4 y_4 \right) \qquad (1.4)$$

$$\frac{\partial y_4}{\partial t} = D_h \frac{1}{r} \frac{\partial}{\partial r} \left(r \frac{\partial y_4}{\partial r} \right) + \left(-k_4 y_4 + k_4'uy_3 \right) \qquad (1.5)$$

198

$$y_2 = f_2(u) = \alpha_2 u f_1(u) \tag{19}$$

$$y_3 = f_3(u) = \alpha_3 u f_2(u) \tag{20}$$

$$y_4 = f_4(u) = \alpha_4 u f_3(u) \tag{21}$$

where α_n's are the equilibrium constants i.e.

$$\alpha_n = k'_n/k_n, \qquad n = 1, \ldots, 4 \tag{22}$$

Baxley and Hellums (1983) compared the simple one-step mass action and VRC models by solving Equations (10) through (12) with and without the VRC modification of Equation (9). The oxygen capillary wall concentrations predicted by the two models deviated by over 60% at the venous end of the capillary. The large discrepancy is due to the fact that both systems are near local chemical equilibrium on average across the capillary. Therefore, the reaction path for the one-step model will approximately follow the hyperbolic equilibrium curve, while that of the VRC model will approximately follow the Hill equilibrium curve. Hence, the large difference between the results is due to the large error in the equilibrium curve compatible with the one-step model as shown in Figure 2.

Hence, for a valid comparison between the VRC and the Adair models it is necessary to ensure that both kinetic models are compatible with precisely the same equilibrium curve. The kinetic rate constants for the Adair model were obtained by Gibson (1970) for specimens with a low content of 2,3-DPG (P_{50} below physiological levels). The equilibrium curve, shown in Figure 2, is not an experimental curve. Rather it is the curve generated by the Adair model.

The requirement that the VRC method be compatible with precisely the same equilibrium curve can be satisfied if the rate constant, j_1, of the VRC method is varied with oxygen concentration, u, according to Equation (23), as shown by Yap (1985):

$$j_1 = j'_1 u \frac{4 + 3\alpha_1 u + 2\alpha_1 \alpha_2 u^2 + \alpha_1 \alpha_2 \alpha_3 u^3}{\alpha_1 u + 2\alpha_1 \alpha_2 u^2 + 3\alpha_1 \alpha_2 \alpha_3 u^3 + 4\alpha_1 \alpha_2 \alpha_3 \alpha_4 u^4} \tag{23}$$

All the partial differential equations were solved numerically by the finite element collocation method of Madsen and Sincovec (1976) which uses basic spline polynomials for the spacial discretization. This method has been shown to be much more efficient than finite difference methods (Baxley and Hellums, 1983) in problems of the type considered here.

Values of the parameters used in the solution are as in Table 2. Many of these parameters have been used by prior workers (Sheth and Hellums, 1980; Baxley and Hellums, 1983) starting with Moll (1969).

RESULTS AND DISCUSSION

Figures 3 and 4 give the radial oxygen and oxyhaemoglobin profiles at three axial positions (erythrocyte residence times). These are the solutions to the differential equations discussed earlier for comparison of the VRC and Adair kinetic models.

Table 2. Values of the parameters used in comparison of Adair and VRC kinetic models

Capillary diameter, μm	6
Haem group concentration, v_t, M	0.022
Oxygen Flux, M–cm/sec	1.5×10^{-6}
Oxygen Diffusivity, D_u, cm^2/sec	8×10^{-6}
Haemoglobin Diffusivity, D_h, cm^2/sec	6.5×10^{-8}
Association rate constant, j'_1, M^{-1},sec^{-1} (Equation (5))	3.5×10^{-6}

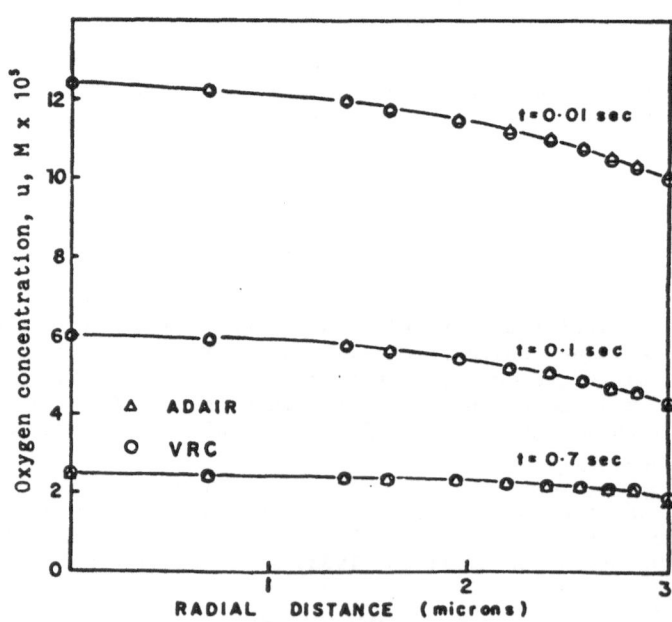

Figure 3. Radial oxygen concentration profile with VRC and Adair Kinetics. The parameter, t, denotes erythrocyte residence time in a 6 μm capillary.

200

Values of the parameters considered to be characteristic of oxygen transport to tissue were used as shown in Table 2 for a six micron diameter capillary. It is clear from the figures that the differences in results for the two kinetic models are small. The largest difference in oxygen concentration occurs near the capillary wall and is about 1.5% for an erythrocyte residence time of 0.1 sec.

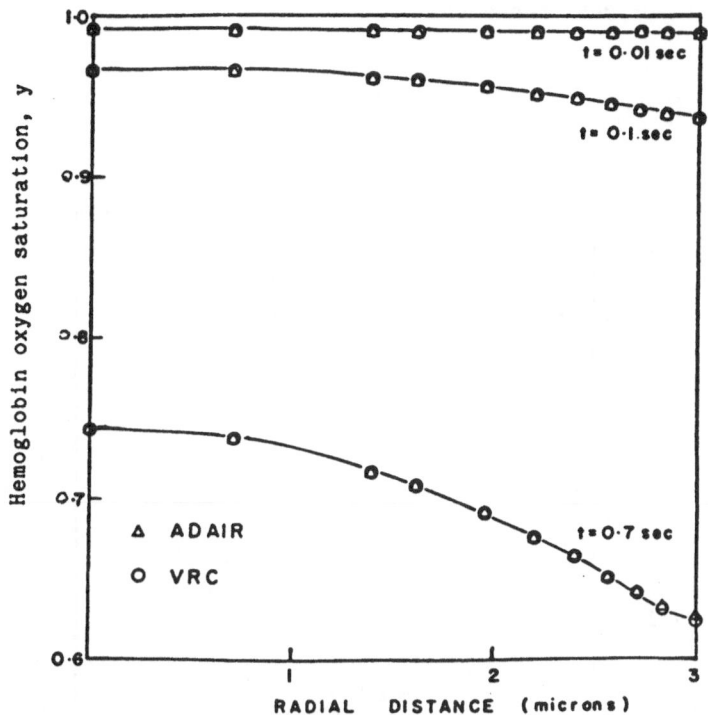

Figure 4. Radial oxyhaemoglobin concentration profile with VRC and Adair kinetic schemes. The parameter, t, denotes erythrocyte residence time in a 6 μm capillary.

The same solutions are compared in Figure 5 in terms of the capillary wall oxygen concentration. Here it can be seen that the difference increases with residence time, but remains small. There is less than 7% difference in all cases. To put this small difference into perspective, it is important to note that the overall radial concentration gradients from the red cell on through the surrounding tissue is small. Some workers estimate the gradients in the tissue to be only a few mmHg.

Another comparison of the results is given in Figure 6 in terms of the Nusselt Number. The Nusselt Number, Nu, is a dimensionless flux defined by

$$ Nu = \frac{Rq_o}{D_u(\bar{u} - u_w)} $$

where $\bar{u}$ is the radially averaged (mixed-mean) oxygen concentration and u_w is the capillary wall oxygen concentration.

In terms of the Nusselt Number, the differences appear larger, ranging up to about 28%. However, again as a matter of perspective, Sheth and Hellums (1980) have pointed out that changes in the Nusselt Number are not directly proportional to changes in overall resistance to oxygen delivery. The intracapillary resistance has been estimated to constitute about half the total resistance to oxygen transport (Hellums, 1977; Baxley and Hellums, 1983). On this basis, a 28% change in the Nusselt Number results in only a 12% change in overall resistance to oxygen transport.

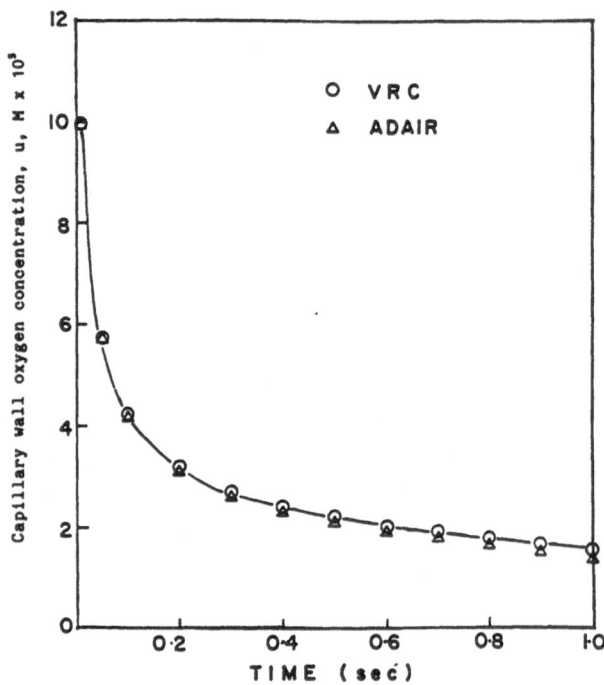

Figure 5. Comparison of capillary wall oxygen concentration as predicted by the VRC and Adair kinetic schemes.

Therefore, it may be concluded that the VRC model is adequate for many purposes, compared with the more complex Adair model. This observation needs to be qualified since it is not necessarily valid for all oxygen fluxes. The flux used in this comparison is in the range where diffusional processes play an important role. The reaction is fast enough for the oxygen and haemoglobin to be nearly in local chemical equilibrium across the capillary. In this case, the chemical kinetic model is of only secondary importance as long as it is compatible with the equilibrium relationship. This point is illustrated in Figure 7 where the radially-averaged oxyhaemoglobin fractional saturation, $\bar{y}$, is plotted against the radially-averaged oxygen concentration, $\bar{u}$, for the same solution. Expressed in this way, the points for the VRC and Adair kinetic models fall so close together that they are indistinguishable.

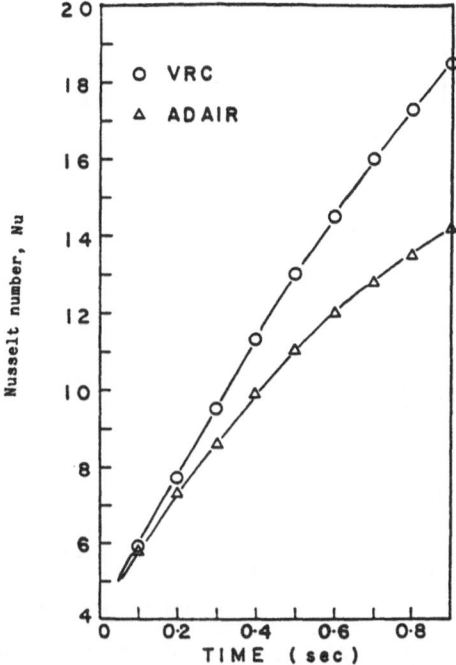

Figure 6. Comparison of the Nusselt number as predicted by the VRC and Adair kinetic schemes.

It can be seen that the reaction paths taken by the two kinetic models trace the equilibrium curve very closely. Previously, we (Baxley and Hellums, 1983) and others (Keller and Friedlander, 1966; Kreuzer and Hoofd, 1970, 1972; Gijsbers, and van Ouwerkerk, 1976; Clark et al., 1985) have shown that the deviation from local chemical equilibrium is small everywhere except in a thin 'boundary layer' very close to the capillary wall. Consequently it appears that a kinetic model more complicated than the VRC is unnecessary for most purposes, providing the VRC is compatible with the correct oxygen-oxyhaemoglobin equilibrium curve, and providing we restrict our attention to oxygen fluxes in the range of physiological interest for oxygen delivery to tissue.

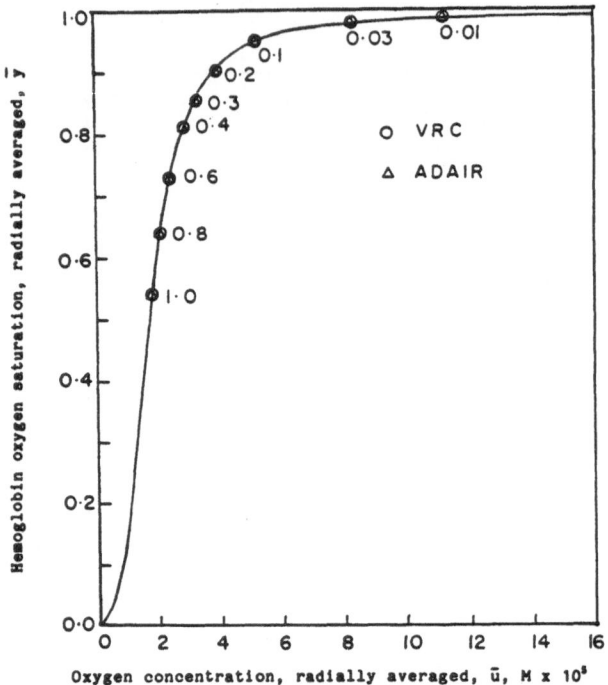

Figure 7. Reaction path diagram for Adair and VRC kinetic models. The numbers along the curve denote the residence time in seconds of the erythrocyte in a 6 μm capillary.

The choice of kinetic models can be more important at high fluxes. To study the effect of various fluxes on the choice of kinetic models the same set of equations was solved using a constant capillary wall oxygen concentration boundary condition, instead of the constant capillary wall oxygen flux boundary condition. This boundary condition yields a solution in which the flux varies with axial position. It is high near the entrance and decreases in a monotonic way with distance downstream. Figures 8 and 9 show the radial oxygen and oxyhaemoglobin profiles at two short erythrocyte residence times (very near the entrance). A capillary wall oxygen concentration of 8.0×10^{-5} M was used, because the value is close to the average value computed with the constant oxygen flux boundary condition.

Note that the deviations between the two solutions are appreciable in terms of the haemoglobin oxygen saturation (Fig.9), but only for very brief residence times. Even then, the maximum deviation shown on the figure is only about 16% of the drop in saturation from the centre of the capillary to the capillary wall.

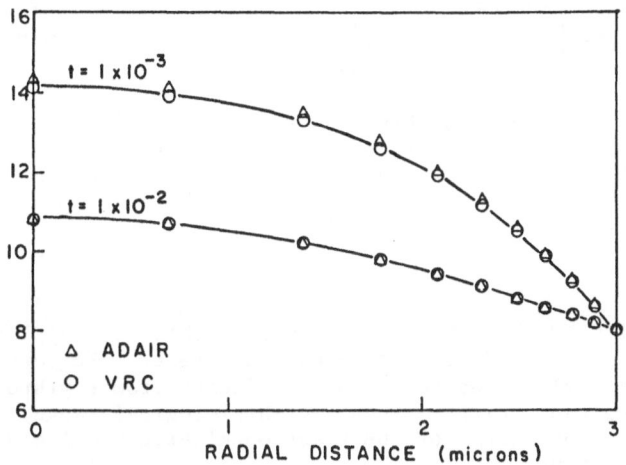

Figure 8. Radial oxygen profiles with a constant capillary wall oxygen concentration boundary condition for a 6 um capillary. The parameter, t, denotes the residence time in seconds in a 6 um capillary. The ordinate shows oxygen concentration,u, M x 10^5.

The derivations and solutions and calculations for a number of other conditions are given in much more detail by Yap (1985). All these results are consistent with our finding here that the VRC model has sufficient accuracy for most practical purposes. The mathematical complexity of the Adair model is not required for simulation of oxygen delivery to tissue. The key point is that the reaction kinetic model must be consistent with the oxygen-oxyhaemoglobin equilibrium curve.

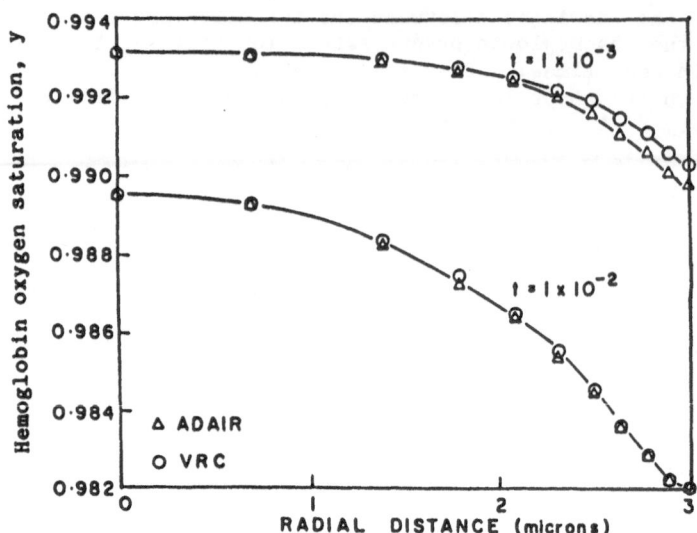

Figure 9. Radial oxyhaemoglobin saturation profiles with a constant capillary wall oxygen concentration boundary condition. The parameter, t, is the residence time in seconds in a 6 μm capillary.

It is likely that the conclusion from this work applies only over a range of fluxes of physiological interest. In certain in vitro experiments in flow reactors, oxygen fluxes from erythrocytes are an order of magnitude higher than the physiological range. Under these circumstances, the form of the reaction kinetic model may be expected to be more important.

ACKNOWLEDGEMENT

This work was supported by the National Institutes of Health under grant number 2R01 HL 19824.

REFERENCES

Adair, G.S. (1925). The hemoglobin system: VI. The oxygen dissociation curve of hemoglobin. J. Biol. Chem. 63, 529-545.

Baxley, P. and Hellums, J.D. (1983). Simulation of oxygen transport in the microcirculation. Ann. Biomed. Eng. 11, 401-416.

Clark, A., Federspiel, W.J., Clark, P.A.A. and Cokelet, G.R. (1985). Oxygen delivery from red cells. Biophys. J. 47, 171-181.

Gibson, Q.H. (1970). The reaction of oxygen with hemoglobin and the kinetic basis of the effect of salt on binding of oxygen. J. Biol. Chem. 245, 3285-3288.

Gijsbers, G.H. and van Ouwerkerk, H.J. (1976). Boundary layer resistance of steady-state oxygen diffusion facilitated by a four-step chemical reaction with hemoglobin in solution. Pflugers Arch. 365, 231-241.

Hellums, J.D. (1977). The resistance to oxygen transport in the capillaries relative to that in the surrounding tissue. Microvasc. Res. 13, 131–136.

Keller, K.H. and Friedlander, S.K. (1966). The steady-state transport of oxygen through hemoglobin solutions. J. Gen. Physiol. 49, 663–679.

Kreuzer, F. and Hoofd, L.J.C. (1970). Facilitated diffusion of oxygen in the presence of hemoglobin. Respir. Physiol. 8, 280–302.

Kreuzer, F. and Hoofd, L.J.C. (1972). Factors influencing facilitated diffusion of oxygen in the presence of hemoglobin and myoglobin. Respir. Physiol. 15, 105–124.

Madsen, N.K. and Sincovec, R.F. (1976). PDECOLD: General collocation software for partial differential equations. ACM Trans. Math. Software.

Moll, W. (1969). The influence of hemoglobin diffusion on oxygen uptake and release by red cells. Respir. Physiol. 6, 1–15.

Sheth, B.V. and Hellums, J.D. (1980). Transient oxygen transport in hemoglobin layers under conditions of the microcirculation. Ann. Biomed. Eng. 8, 183–196.

Yap, E.W.H. (1985). Mathematical modelling of oxygen transport in the microcirculation. M.S. Thesis, Rice Univ., Houston, Texas, U.S.A.

ARTERIAL OBSTRUCTION INDUCED BY PAF-ACETHER

(1-O-ALKYL-SN-GLYCERO-3-PHOSPHOCHOLINE)

R.H. BOURGAIN, R. ANDRIES, C. BOURGAIN[*] and
P. BRAQUET[**]

Laboratorium voor Fysiologie en Fysiopathologie, and
[*]Laboratorium voor Experimentele Pathologie, Vrije
Universiteit Brussel, Brussels 1090, Belgium and
[**]Institut H. Beaufour, Le Plesis Robinson, F 92350
France

INTRODUCTION

Platelet activating factor (PAF-acether) is a very potent mediator which is involved in many inflammatory and allergic reactions (Demopoulos, Pinckard and Hanahan, 1979; Vargaftig et al., 1981). Since its discovery in 1972 by Benveniste, Henson and Cochrane it has been demonstrated that this agent can induce leukopenia (O'Flaherty et al., 1981) and plasma leakage at the venular site of the vascular compartment (Bjork et al., 1983) followed eventually by arterial hypotension and death. In a previous paper (Bourgain et al., 1985), we demonstrated the effect of PAF-acether on the arterial wall in the guinea-pig following topical superfusion of the mediator in well standardized conditions. It became evident from these investigations that within two to three minutes following a three minute application of the mediator (10^{-7} M), a platelet thrombus developed at the site of superfusion. Over time, this platelet thrombus recruited leukocytes; these invaded the thrombotic mass and simultaneously underwent diapedesis. It happened that fragments of the thrombotic mass detached from the main structure and embolized; nevertheless the thrombus increased in size and occlusion of the vascular lumen occasionally occurred, leading to peripheral ischaemia and tissue hypoxia.

In order to study in detail the sequence of these phenomena, an ultrastructural analysis was performed by transmission (TEM) and scanning (SEM) electronmicroscopy on arterial segments exposed to topical challenge by the mediator. Specimens were examined at regular intervals from the instant immediately following cessation of superfusion with the mediator to 40 minutes afterwards.

Quantification of the thrombus was recorded continuously by an optoelectronic device developed in the laboratory (Bourgain et al., 1984).

METHODS

Male guinea-pigs (700–800 g) were used for this _in vivo_ investigation. The technique for the induction of anaesthesia, ventilation and preparation of the arterial segment has been published in detail elsewhere (Bourgain et al., 1984), and will be summarized. Following dissection of a branch of the mesenteric artery over a distance of approximately 2 mm, microprojection of the arterial segment to be investigated was performed in a dark room onto a series of thirty light-dependent resistances. The appearance of the white thrombus which contrasted in light intensity with the blood stream was registered as an electrical potential.

For SEM, total body perfusion fixation was performed at physiological pressure through a catheter in the left common carotid artery (1.5% glutaraldehyde in 0.1 M cacodylate buffer with 2% sucrose and 2 mM $CaCl_2$, pH 7.4, 37°C) and the mesentery excised. After overnight immersion fixation in 1.5% glutaraldehyde the arterial segment was dissected free from the surrounding tissue, postfixed in 1% OsO_4 for one hour, dehydrated in ascending ethanol concentrations and dried by the critical point method using liquid CO_2. The method of Reidy and Schwartz (1982) was used to open the arterial segment. The specimens were sputter coated with 15 nm gold and examined in a Philips SEM 505 scanning electron microscope at 30 kV. For investigation by TEM, the arterial segment was seized in forked (5 mm) forceps, excised and immediately immersed in 2.5% charcoal-treated glutaraldehyde in 0.075 M cacodylate buffered to pH 7.2, and left to fix for two hours at room temperature. After overnight rinsing in 0.2 M cacodylate buffer, specimens were post-fixed for one hour in unbuffered 1% osmium tetroxide. They were stained in bulk for one hour at 37°C with 0.5% uranyl acetate in veronal buffered to pH 5.3 and embedded in Spurr's resin. Transverse sections were cut over at least half the length of each specimen. Semi-thin sections were taken at regular intervals and stained with alkaline methylene blue for examination by light microscopy. Thin sections were stained either by potassium permanganate or uranyl acetate followed by lead citrate and examined in a Zeiss EM-9 electron microscope.

Superfusion of the arterial segment by PAF-acether (10^{-7} M) was performed for two to three minutes and followed by superfusion of isotonic Ringer solution. All solutions were kept at 37°C and administered at a flow rate of 1 ml.min^{-1}.

For ultrastructural analysis, the arterial specimens were studied 2, 5, 10, 20 and 30 minutes after superfusion with the mediator was discontinued.

RESULTS

Normal Endothelium

SEM analysis clearly demonstrated that the endothelial lining consists of closely adjoining cells with their longitudinal axes in parallel with the direction of the blood stream except at the site of a bifurcation where the distribution of the direction of the long axes was more random, probably due to the turbulence and formation of eddies in this area. The nuclei of the endothelial cells bulged slightly into the vascular lumen but no gaps were seen between the cells. In normal arterial segments which were investigated, no signs of desquamation of the endothelial cells could be observed.

Effect of superfusion with PAF-acether

Following superfusion with PAF-acether for two to three minutes at 10^{-7} M, retraction of the endothelial cells occurred in the area corresponding to the site of application of the mediator. As a result of this retraction, sub-intimal tissue became exposed to the blood stream and platelets adhered to these exposed areas between the endothelial cells. Within three to four minutes after the end of topical application of the agent, some platelets started to cover the retracting and desquamating endothelium. Leukocytes were invariably recruited by the thrombus; they consisted mainly of neutrophils which surroundeed and penetrated the thrombotic mass. No fibrin deposition was visible. TEM photographs clearly demonstrated that the leukocytes invaded the thrombus and performed diapedesis. The platelets became degranulated and at the de-endothelialized site, diapedesis of some thrombocytes became evident.

Within six to seven minutes, the thrombotic mass increased in size due to simultaneous extension of the thrombus over the surrounding endothelial cells and the apposition of newly formed layers of plate-lets onto the original layer (Fig. 1).

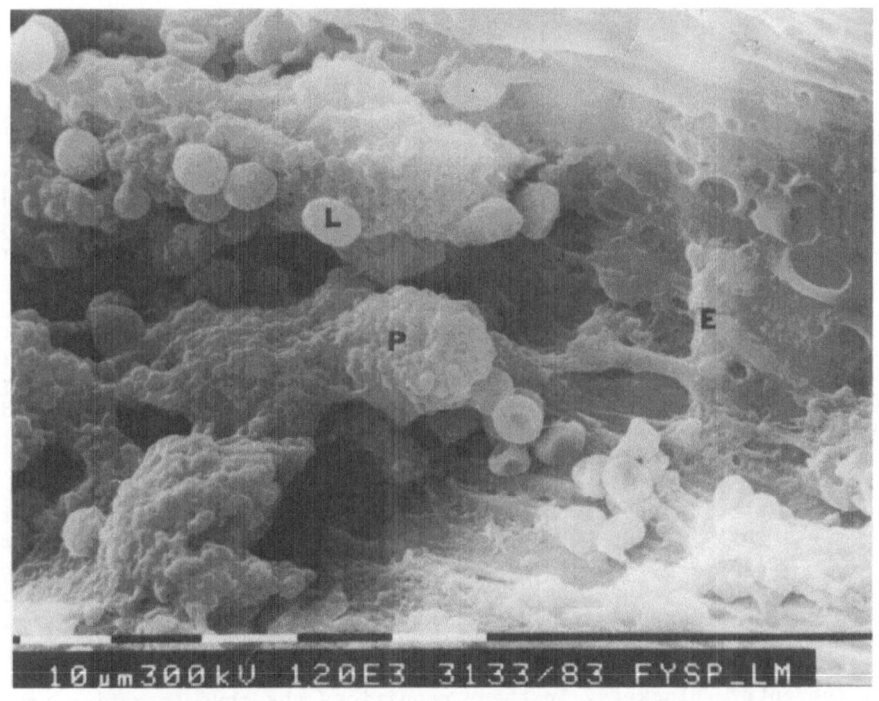

Figure 1. SEM picture of a PAF-acether-induced platelet thrombus (P) in the mesenteric artery of the guinea-pig. Numerous leukocytes (L) are present around the thrombotic mass. Extensive lesions are found in the endothelium (E).

A specific finding in the TEM preparations was the presence of vacuolization and marked bleb formation in the endothelial cells underlying and adjoining the thrombus. Myelin figures were generally present in these blebs (Fig. 2).

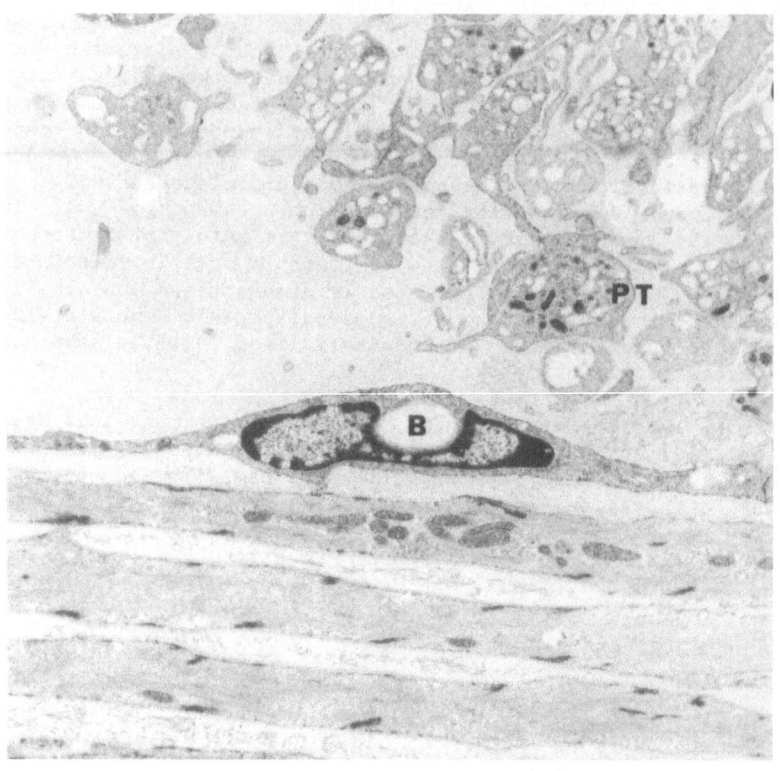

Figure 2. TEM picture of an arterial segment which was submitted to topical superfusion with PAF-acether for 3 minutes. Bleb (B) formation is obvious in the endothelial cells, being particularly visible at the margin of the platelet thrombus (PT) but also at sites underlying the latter.

Some distance downstream from the thrombus, approximately 200-300 μm from the edge of the thrombotic mass, slight, localized retraction of the endothelial cells gave rise to the formation of small aggregates of platelets floating freely in the vascular lumen but remaining firmly anchored to the site of intimal exposure; even here diapedesis of both leukocytes and platelets was still obvious.

In later stages, 20-30 minutes after superfusion with PAF-acether, the thrombotic mass assumed a very condensed appearance. The structure of the platelets became barely recognisable, while more leukocytes became involved and the thrombus increased in size continuously although at a lesser rate. A quite unexpected finding was the behaviour of the thrombotic phenomenon when embolization of the thrombus was initiated by prostacyclin, inhibitors of phospholipase A_2 or calcium-entry blockers. In these conditions thrombus formation recurred immediately. This was not due to the presence of residual, exogenous, PAF-acether, but it could be inhibited by specific antagonists of the mediator (Bourgain et al., 1985).

TEM studies showed that within 10-12 minutes of PAF-acether administration, there was a marked increase of the interstitial tissue separating the smooth muscle cells of the arterial wall, probably due to fluid infiltration and oedema.

DISCUSSION AND CONCLUSIONS

In a previous paper (Bourgain et al., 1985) the effect of PAF-acether applied topically onto a branch of the mesenteric artery in the guinea-pig was described. The present study was undertaken to evaluate thrombus formation over time when investigated by TEM and SEM at well-determined intervals.

In our experimental conditions, it was obvious that the first alteration was the retraction of the endothelial cells in the area underlying the zone of superfusion, followed within a few seconds by adhesion of platelets onto these sites. The platelets involved in this process underwent a marked shape change and demonstrated extensive pseudopod formation. TEM analysis did not reveal any marked alteration within the vessel wall tissue except the denudation of the intimal structures. This effect is quite different from that observed on induction of de-endothelialization by electrical current followed by the topical application of adenosine diphosphate (ADP). Indeed in those conditions, the endothelial cells desquamate rapidly from the underlying structures and few if any platelets adhere to the denuded site which is in sharp contrast to the PAF-acether induced thrombosis. Furthermore, the presence of a thrombus induced by ADP following current application depended entirely on the continuity of the topical superfusion by this mediator; once the superfusion was discontinued, the thrombus disappeared and renewal of thrombus formation occurred only when the topical application of ADP was resumed.

The recruitment of the leukocytes clearly demonstrates the potent chemotactic effect of PAF-acether, as already described (Demopoulos et al., 1979). The platelets of the guinea-pig are extremely sensitive to this mediator and, therefore, rapid platelet activation occurs leading to the generation of platelet thrombi. The ongoing thromboformation and leukocyte recruitment long after the application of the mediator is discontinued, can be explained on the basis of generation of endogenous PAF-acether not only by the activated platelets but also by the surrounding endothelial cells.

Interestingly indeed, Camussi et al. (1983) described the generation and release of PAF-acether by cultured endothelial calls when challenged by thrombin. The extensive vacuolization and bleb formation, associated with the presence in these vacuoles of myelin structures, are lesions observed only in these experimental conditions. They point to the involvement of membrane structures in a continuous process of thrombosis, diapedesis, and leukocyte recruitment, phenomena possibly leading to vascular obstruction. The presence of interstitial oedema clearly indicates that fluid penetrates the vascular wall as the result of the application of the mediator and is highly indicative of the occurrence of plasma extravasation.

The observation that PAF-acether-induced thrombosis can be inhibited by a specific antagonist, BN 52021 (Ipsen Lab), confirmed our previous observations that the exogenous mediator induces the generation of an endogenous PAF-acether-like substance which in a positive-feedback mechanism perpetuates the thrombotic phenomenon and its related changes.

ACKNOWLEDGEMENTS
Investigation supported by grant No. 3. 0067. 83 and 3. 0053. 84 FGWO. We thank Mrs De Backer-De Zutter H. for skilled TEM technical assistance.

REFERENCES

Benveniste, J., Henson, P.M. and Cochrane, C.G. (1972). Leukocyte dependent histamine release from rabbit platelets. The role of IgE, basophils and a platelet-activating factor. J. Exp. Med. 136, 1356

Bjork, J., Lindbom, L., Gerdin, B., Smedegard, G., Arfors, K.-E. and Beneviste, J. (1983). Paf-acether (platelet-activating factor) increases microvascular permeability and affects endothelium-granulocyte interaction in microvascular beds. Acta Physiol. Scand. 119, 305

Bourgain, R.H., Maes, L., Braquet, P., Andries, R., Touqui, L. and Braquet, M. (1985). The effect of 1-0-alkyl-2-acetyl-sn-glycero-3-phosphocholine (Paf-acether) on the arterial wall. Prostaglandins, 30, 185

Bourgain, R.H., Vermarien, H., Andries, R., Vereecke, F., Jacqueloot, J., Rennies, J., Blockeel, E. and Six, F. (1984). A standardized 'in vivo' model for the study of experimental arterial thrombosis. Description of a method. In: Oxygen Transport to Tissue-VI. Eds Bruley, D., Bicher, H.I. and Reneau, D., Plenum Press, New York and London, (Adv. Exp. Med. Biol. 180, 635).

Camussi, G., Aglietta, M., Malayasi, F., Tetta, C., Piacibello, W., Sanavio, F. and Bussolino, F. (1983). The release of Platelet-Activating Factor from human endothelial cells in culture. J. Immunol. 131, 2397

Demopoulos, C.A., Pinckard, R.N. and Hanahan, D.J. (1979). Platelet-activating factor. Evidence for 1-0-alkyl-2-acetyl-sn-glyceryl-3-phosphocholine as the active component (a new class of lipid chemical mediators). J. Biol. Chem. 254, 9355

O'Flaherty, J.T., Miller, C.H., Lewis, J.C., Wykle, R.L., Bass, D.A., McCall, C.E., Waite, M. and De Chatelet, L.R. (1981). Neutrophil responses to platelet-activating factor. Inflammation, 5, 193

Reidy, M.A. and Schwartz, S.M. (1982). A technique to investigate surface morphology and endothelial cell replication of small arteries: a study in acute angiotensin-induced hypertensive rats. Microvasc. Res. 24, 158

Vargaftig, B.B., Chignard, M., Benveniste, J., Lefort, J. and Wal, F. (1981). Background and present status of research on platelet-activating factor (paf-acether). Ann. N.Y. Acad. Sci. 370, 119

ANTI-ANAPHYLACTIC PROPERTIES OF BN 52021: A POTENT PLATELET ACTIVATING FACTOR ANTAGONIST

P. BRAQUET and R.H. BOURGAIN[*]

I.H.B. Research Laboratories, 17 Avenue Descartes, 92350 Le Plessis Robinson, France and [*]Department of Physiology, Medical Faculty, Vrije Universiteit Brussel, Laarbeeklaan 103, B-1090 Brussels, Belgium

INTRODUCTION

PAF-acether is a potential mediator of anaphylaxis and inflammatory reactions (Benveniste, Henson and Cochrane, 1972). It is released by various immune and chemical stimuli from inflammatory cells (basophils, neutrophils, eosinophils, macrophages, platelets) as well as from vascular endothelium; PAF-acether increases vascular permeability and causes sustained hypotension and thrombosis (reviewed by Braquet et al., 1986; Vargaftig and Braquet, 1987). It may also participate in various hypersensitivity reactions including bronchial asthma which cannot be fully accounted for by the known inflammatory mediators, including peptido-leukotrienes (reviewed by Vargaftig et al., 1985).

PAF-acether involvement in airway hypersensitivity is suggested by different lines of evidence:

(i) cyclooxygenase and anti-histamine-independent bronchoconstriction follows human or animal exposure to PAF-acether or to antigen;

(ii) the late reaction of asthma, characterized by bronchial hyper-responsiveness and by the migration of inflammatory cells into the lungs can be duplicated by PAF-acether administration (see below);

(iii) aerosolized PAF-acether causes platelet-independent broncho-constriction in the guinea-pig (Lefort, Rotilio and Vargaftig, 1984). Guinea-pigs which are desensitized to bronchoconstriction following the intra-pulmonary aerosolization of PAF-acether retain their responsiveness to intravenous injections (Lefort et al., 1984), indicating that different cells are involved when the two routes are used: platelets on the one hand, and a component of the alveolar lining, on the other. This component may be the alveolar macrophages which can be activated by PAF-acether (Maridonneau-Parini et al., 1985) and can release lyso-PAF and PAF-acether itself when exposed to specific antigen (Arnoux, Duval and Benveniste, 1980). Parente and Flower (1985) demonstrated that hydrocortisone and its mediator protein, lipocortin, inhibit the formation of lyso-PAF by zymosan-stimulated macrophages and by antigen-stimulated guinea-pig lungs, and it is possible that anti-allergic glucocorticosteroids may act by inhibiting direct or indirect macrophage activation in which PAF-acether is involved;

(iv) PAF-acether is released during systemic anaphylaxis in the rabbit (Kravis and Henson, 1975) and antigen-challenged perfused guinea-pig lungs release its lyso precursor and/or catabolite (Rotilio et al., 1983; Fitzgerald, Moncada and Parente, 1986);

(v) guinea-pigs desensitized to aerosolized antigen maintain their responsiveness to aerosol, whereas guinea-pigs desensitized to PAF-acether do not respond to aerosolized antigen, even though the systemic response persists (Lefort et al., 1984).

Intensive research for specific PAF-acether receptor antagoinists was therefore undertaken in the mid 80's in order to assess the exact role of this mediator in hypersensitivity reactions and with the aim of developing new drugs with potential therapeutic use (reviewed by Braquet and Godfroid, 1986).

In this regard a new compound, BN 52021 (Ginkgolide B; Fig.1), isolated and purified from a crude Ginkgo biloba extract has been shown to display a potent and specific inhibition of the binding of $[^3H]$-PAF-acether to its receptor in rabbit platelets (ki = 1.2 x 10^{-7}M; Braquet, 1984; reviewed by Braquet, Drieu and Etienne, 1986).

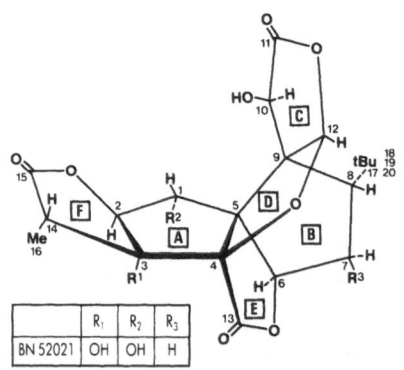

Figure 1. Chemical structure of BN 52021.

BN52021 specifically inhibits in vitro human (Braquet, 1984; Nunez et al., 1986; Kuster, Filep and Frolich, 1986) and rabbit (Braquet et al., 1985a) platelet aggregation and the related events (Lachachi et al., 1985; Baroggi, Etienne and Braquet, 1986; Simon et al., 1986). Similar inhibition is observed in vivo in mesenteric artery of guinea-pigs superfused with PAF-acether (Bourgain et al., 1985). In this model BN 52021 not only prevents thrombus formation but is also able to disaggregate thrombi pre-formed with PAF-acether. BN 52021 antagonizes the renal (Hebert et al., 1986; Plante et al., 1986) cerebral (Plotkine et al., 1986) and cardiovascular (Baranes et al., 1986) impairments due to PAF-acether in various models. BN 52021 also inhibits PAF-acether-induced tissue-type plasminogen activator release (Emeis, 1987) and ion transport impairments (Garay and Braquet, 1986). Finally, BN 52021 blocks specifically rat paw oedema induced by PAF-acether (Cordeiro et al., 1986a,b).

BN 52021 is also effective in complex immune disorders: (i) it inhibits the lethality (Etienne et al., 1985), the haemodynamic changes (Adnot et al., 1986) and the gastric ulcerations (Wallace et al., 1987; Braquet et al., 1987) induced by endotoxin; (ii) it prolongs graft survival in rats (Foegh et al., 1985, 1986); (iii) it antagonizes lethality induced by mycotoxins (Feuerstein et al., 1987); (iv) it inhibits IgG aggregate-induced shock in rats (Sanchez-Crespo et al., 1985); (v) it blocks the haemodynamic changes of experimental cirrhosis in rats (Villamediana et al., 1986).

BN 52021 is also a potent inhibitor of PAF-acether effects in man: it dose-dependently antagonizes both weal and flare (Barnes et al., 1986; Guinot et al., 1986; Chung et al., 1987) and ex vivo platelet aggregation (Barnes et al., 1986) induced by the autacoid.

In this paper the effects of BN 52021 on lung anaphylaxis and airway hyper-responsiveness are reviewed.

INHIBITION OF PAF-ACETHER-INDUCED PULMONARY IMPAIRMENTS BY BN 52021

The I.V. administration of PAF-acether to guinea-pigs induces a dose-dependent fall in systemic blood pressure and heart rate associated with an increase in lung resistance (R_L) and marked decrease in dynamic compliance (C_{Dyn}) which indicates broncho-constriction of the more peripheral airways. This is also accompanied by an increase in non-specific bronchial responsiveness both in animals and in humans, a property not exhibited by other pro-inflammatory mediators (Patterson and Harris, 1983; Cuss, Dixon and Barnes, 1986). The severity of bronchoconstriction also correlates with the modification of the values of blood lactate, pH, PCO_2 and PO_2.

Table 1. Effects of BN 52021 on PAF-acether-induced
bronchoconstriction in guinea-pigs

PAF-acether (dose, route)			BN 52021			Reference
			route	dose	inhibition	
33	ng / kg	i.v.	i.v.	3 mg / kg	− 100 %	*Desquand et al., 1986*
60	ng / kg	i.v.	i.v.	1 mg / kg	− 89 %	*Braquet et al., 1985b*
100	ng / kg	i.v.	i.v.	1 mg / kg	− 67 %	*Touvay et al., 1986a*
45 - 100	ng / kg	i.v.	oral	2 mg / kg	− 59 %	*Touvay et al., 1986c*
100	ng / kg	i.v.	i.v.	1 mg / kg	− 62 %	*Berti et al., 1986*
100	ng / kg	i.v.	i.v.	2 mg / kg	− 94 %	
300 μg/ml for 2 min., aerosol			i.v.	3 mg / kg	− 91 %	*Desquand et al., 1986*

BN 52021 given I.V. (0.5 to 2 mg/kg) or orally (5 to 15 mg/kg) dose-dependently antagonizes both (i) the cardiovascular and pulmonary changes (Fig. 2 and Table 1) and (ii) the increase in bronchial responsiveness induced by PAF-acether (Braquet et al., 1985b; Touvay, Etienne and Braquet, 1985; Berti et al., 1986; Desquand et al., 1986; Touvay et al., 1986a; Vilain et al., 1986). For example BN 52021 administered at a dose of 2 mg/kg I.V. prevents almost completely the lowering of blood pressure and heart rate, and the bronchoconstrictive effect of PAF-acether. It also dose-dependently prevents thrombopenia, leukopenia (Fig. 3) and the increase in circulating thromboxane (at a dose of 2 mg/kg I.V., it keeps this parameter within the normal range). The antagonistic activity of BN 5021 seems to be long lasting since bronchoconstriction is still markedly inhibited after 40 min (Berti et al., 1986). Butoxamine, a β_2-antagonist, moderately antagonizes the protection afforded by BN 52021 (Fig. 2; Desquand et al., 1986; Touvay et al., 1986a).

In an interesting study Arnoux and Gillis (1986) examined the effects of BN 52021 against PAF-acether-induced vasoconstriction and bronchoconstriction in isolated rabbit lung. Given 60 sec before challenge, both Ginkgolide B and SRI 63-072 [another PAF-acether antagonist derived from the mediator framework (Winslow et al., 1985a,b, 1987)] totally block the deleterious effects of PAF-acether. In contrast when given simultaneously with PAF-acether, only BN 52021 prevents the vasoconstriction and the bronchoconstriction of the isolated lung suggesting that its affinity for lung PAF-acether receptors is higher than that of SRI 63-072. Given curatively 40 sec after the challenge, BN 52021 significantly reverses the effects of PAF-acether. However only a slight improvement is observed when the antagonist is given 90 sec after PAF-acether. These data suggest that BN 52021 can reverse the effects of PAF-acether if the subsequent events underlying the cellular effects of the mediator have not yet developed.

Table 2. Effects of BN 52021 on the release of superoxide induced by PAF-acether in human leukocytes

PAF-acether ($\times 10^{-6}$ M)	BN 52021 ($\times 10^{-6}$ M)	Superoxide release (nmol / 10^7 cells / 5 min)
—	—	1.5 ± 1.2
0.01	—	2.3 ± 1.0
	0.5	2.1 ± 1.3
	1.0	1.4 ± 1.3
0.10	—	2.7 ± 1.1
	5	2.1 ± 1.3
	10	1.5 ± 1.1
1.00	—	3.3 ± 1.6
	50	2.5 ± 1.7
	100	1.9 ± 1.5

Aerosolized PAF-acether also induces bronchoconstriction in guinea-pigs (Lefort et al., 1984). Here again BN 52021 blocks this phenomenon provided doses of above 3 mg/kg are used (Table 1; Desquand et al., 1986).

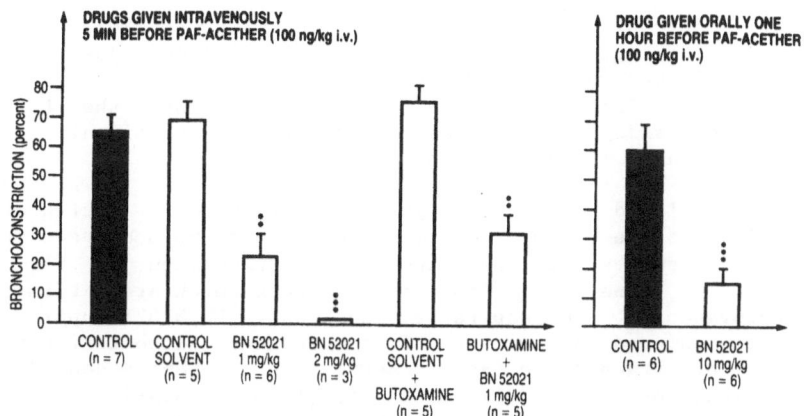

Figure 2. Effect of I.V. (left panel) or oral (right panel) BN 52021
on the bronchoconstriction induced by PAF-acether (100 ng/kg
I.V.) in ethylcarbamate-anaesthetized guinea-pigs. Where
indicated, animals were pretreated with butoxamine (1 mg/kg
I.V.) before receiving BN 52021. Results are expressed as
mean (± S.E.M.) per cent of the maximal resistance obtained
when the tracheal cannula was clamped at the end of the
experiment. **p <0.01, ***p <0.001. (From Desquand et al.,
1986, with permission).

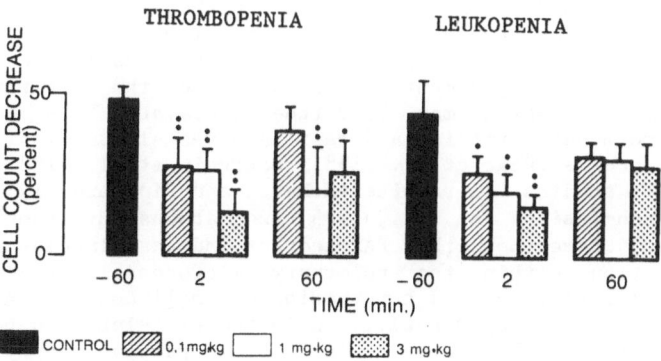

Figure 3. Effect of BN 52021 (I.V.) on thrombopenia and leukopenia
induced by I.V. PAF-acether (33 ng/kg) in guinea-pigs. Cell
count decrease (%) ± S.E.M.; 14 control and 4 treated
animals; *p <0.05, **p <0.01. (Modified from Desquand et
al., 1986, with permission).

BN 52021 also blocks bronchoconstriction and thrombopenia induced by various PAF-acether analogues where the polar head is replaced by isosteric groups e.g. morpholinium, piperidinium etc. (Coeffier et al., 1986).

The antagonistic activity observed with BN 52021 seems to be rather specific against PAF-acether since Ginkgolide B does not affect in a significant way the bronchoconstriction and the lowering of the systemic blood pressure caused by acetylcholine, the tripeptide f-met-leu-phe, serotonin, histamine, collagen, arachidonic acid and LTC$_4$ (Berti et al., 1986; Desquand et al., 1986).

In isolated perfused lungs, PAF-acether activates the arachidonic cascade with a dose-dependent and pronounced generation of thromboxane, suggesting that the lungs stimulated by PAF-acether participate in generating lipidic substances which are detrimental to the cardiovascular and respiratory functions. BN 52021 suppresses the PAF-acether induced generation of thromboxane without affecting the conversion of arachidonic acid (Berti et al., 1986; Desquand et al., 1986). Therefore BN 52021 does not seem to interfere with cyclo-oxygenase but may act at an earlier step involving PAF-acether receptors and phospholipase activation.

Since a variety of effects of PAF-acether in guinea-pigs are platelet-dependent, it is reasonable to presume that the protecting activity of BN 52021 may be primarily ascribed to a prevention of platelet activation and suppression of the generation of thromboxane and other eicosanoids. However a direct activation of bronchial and vascular smooth muscles cannot be ruled out: PAF-acether induces contraction of guinea-pig lung parenchymal strips; BN 52021 inhibits the response to the autacoid in a dose-dependent manner (Fig. 4; Desquand et al., 1986; Touvay et al., 1986a,b). This effect is specific since the contractions of the lung tissue by histamine, LTB$_4$, and LTD$_4$, are not significantly affected by BN 52021 (Touvay et al., 1986c).

BN 52021 also inhibits the PAF-acether-induced contraction of spirally cut large bronchi and of main pulmonary arteries (Berti et al., 1986); this phenomenon should not be underestimated since these PAF-acether-induced impairments contribute to the alteration of the local circulation which may be further aggravated by accumulation of thrombi in pulmonary arterial and capillary vessels and infiltration by inflammatory cells (Page et al., 1985; Bourgain et al., 1985). Indeed, morphological studies in rabbits (McManus and Pinckard, 1985), guinea-pigs (Lellouch-Tubiana et al., 1985) and baboons (B. Arnoux, personal communication) have shown that PAF-acether induces platelet and neutro-phil aggregation within the pulmonary microvessels accompanied by eosinophil infiltration and contraction of small muscular arteries and bronchioles. Most importantly, activated platelets can be seen near the respiratory smooth muscle, as in systemic passive anaphylaxis (Fig. 5B; Lellouch-Tubiana et al., 1985). Bronchoconstriction induced by another cell-stimulating agent, the tripeptide f-met-leu-phe, is not inhibited by BN 52021 but is accompanied by pulmonary leukocyte accumulation. However platelets are only marginally accumulated, as opposed to radio-labelled leukocytes (Boukili et al., 1985). In contrast, BN 52021 antagonizes platelet accumulation induced by PAF-acether, in both mesenteric artery (Fig. 5A) and in pulmonary microvessels (Fig. 5B). This may account for the anti-anaphylactic effect of BN 52021 (see below) since pulmonary recruitment of platelets, particularly to the extra-vascular compartment where they are lodged in apposition to bronchial smooth muscle, was described for both PAF-acether (Lellouch-Tubiana et al., 1985) and antigen (Lellouch-Tubiana et al., 1986). This may also suggest the prophylactic value of the treatment by BN 52021 since platelets may be implicated in smooth muscle hyperplasia by releasing smooth muscle mitogens, such as platelet-derived growth factor (PDGF, Ross and Vogel, 1978). Accordingly, the observation of PAF-acether-induced platelets in close proximity to the bronchial muscle may have important implications for the pathogenesis of bronchial smooth muscle hyperplasia observed in asthma.

Eosinophil infiltration is a major feature of asthma and allergic reactions (De Monchy et al., 1985). Eosinophils are not frequent during the acute phase response, but increase in number, to account for 10 to 80% of the total cell infiltrate during the late phase. Furthermore, released granules are found in the lungs of asthmatic patients, and may participate in bronchial hyper-reactivity (Gleich et al., 1976). These granules contain Major Basic Protein (MBP) which, together with other toxic materials, is found in the sputum of asthmatics. MBP is said to induce destructive changes similar to those found in chronic asthma (Filley et al., 1982).

Since PAF-acether appears to be a potent activator of eosinophil functions, BN 52021 as an antagonist may interfere with the late phase response (Fig. 6). Indeed:

(i) systemic PAF-acether injections induce haemoconcentration and increased total leukocyte counts; the underlying eosinophil and monocyte subsets are decreased by up to a third, which is suppressed by BN 52021 (A. Etienne and P. Braquet, unpublished);

(ii) PAF-acether is one of the most potent chemotactic agents for human eosinophils (Wardlaw and Kay, 1986). This effect is receptor-mediated and BN 52021 dose-dependently inhibits eosinophil locomotion (Wardlaw and Kay, 1986; Braquet and Vargaftig, 1987);

(iii) eosinophils from patients with eosinophilia, including asthmatics, have an increased capacity to release PAF-acether, as compared with those from normal individuals (Lee et al., 1984). Intracutaneous injections of PAF-acether cause a marked eosinophilic infiltration exclusively in allergic patients, a large number of these cells being degranulated. (Henocq and Vargaftig, 1986). In addition, PAF-acether induces LTC_4 formation by human eosinophils; this phenomenon is increased in patients with parasite infestations (Borgeat et al., 1984). BN 52021 antagonizes both LTC_4 and 5-HETE production in PAF-acether and zymosan stimulated eosinophils (Bruynzeel et al., 1986; Braquet and Vargaftig, 1987);

(iv) PAF-acether induces superoxide ($O_2^{-}\cdot$) production and chemoluminescence in various cell systems. BN 52021 selectively inhibits this effect in human eosinophils (Table 2; Bruynzeel et al., 1986), human and rabbit neutrophils (Table 2; Braquet, 1984; Baggiolini and Dewald, 1986; Hoshikawa-Fujimura, Chamone and Jancar, 1986) and guinea-pig macrophages (Desquand et al., 1986). BN 52021 also significantly inhibits aggregation and degranulation of isolated human neutrophils stimulated by PAF-acether (Braquet, 1984). It also inhibits the PAF-acether induced adherence of human leukocytes in various systems (Fink et al., 1986; Vercelotti et al., 1986)

(v)   BN 52021 blocks in a dose-dependent manner the increase in eosinophil cytotoxicity induced by PAF-acether in the presence of <u>Schistosoma mansoni</u> previously coated with C_{3b} and specific antibodies (MacDonald et al., 1986). More interestingly, BN 52021 inhibits IgE-induced cytotoxicity of rat and human eosinophils (M. Capron, personal communication).

(vi)   Finally, preliminary studies using electron microscopy demonstrated that BN 52021 antagonizes accumulation of eosinophils in the lung induced by PAF-acether, in guinea-pigs. This may partially explain the BN 52021-induced inhibition of bronchial hyper-responsiveness following PAF-acether administration.

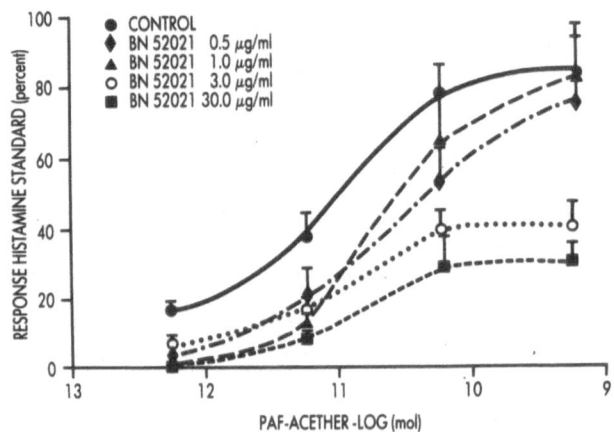

Figure 4.   Effect of BN 52021 on PAF-acether-induced contractions in guinea-pig lung parenchymal strips. Contractions are expressed as per cent of the contractile effect of 2.7 x 10^{-8}M (5ug) histamine. Each point is mean $\pm$ S.E.M. of 6 experiments.

These lines of evidence demonstrate that PAF-acether may participate in eosinophil (and macrophage) recruitment and granule release may account for bronchial hyper-reactivity (Fig. 6), particularly since the removal of the epithelial layer from respiratory smooth muscle <u>in vitro</u> leads to hyper-responsiveness of the smooth muscle to a range of spasmogens in animals (Barnes, Cuss and Palmer, 1985) and in humans (Rubin, Smith and Patterson, 1986). As a PAF-acether antagonist, BN 52021 counteracts these processes and could therefore antagonize airway hyper-responsiveness.

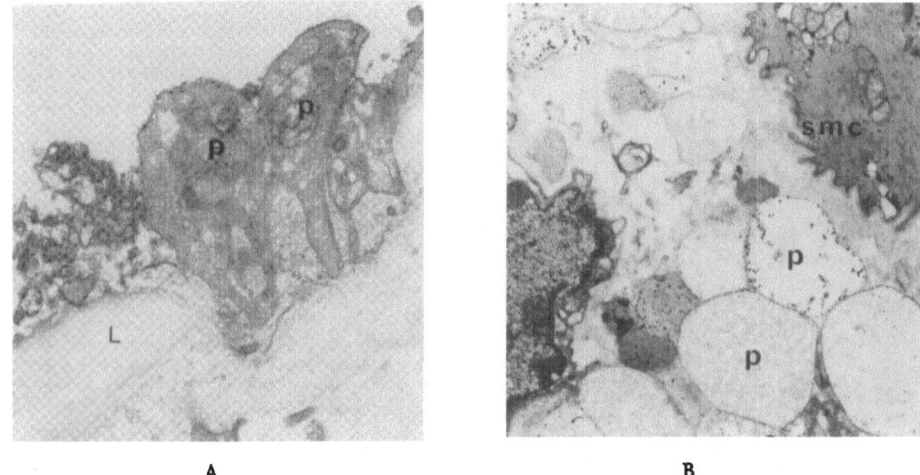

A   B

Figure 5. PAF-acether induces platelet internalization in vascular wall of mesenteric artery (A) and pulmonary microvessels (B). L, lumen; p, platelet; smc, smooth muscle cell.

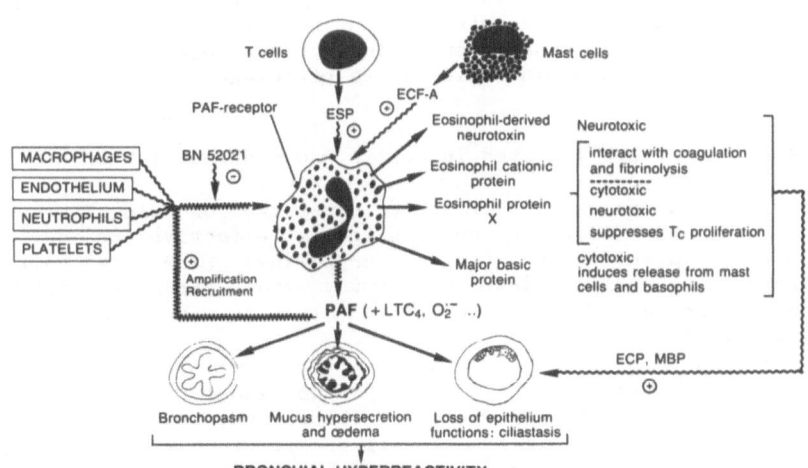

Figure 6. Putative role of eosinophils and PAF-acether in bronchial hyper-reactivity. ESP, eosinophil stimulating promoter; ECF-A, eosinophil chemotactic factor; MBP, major basic protein. (From Braquet and Vargaftig, 1987).

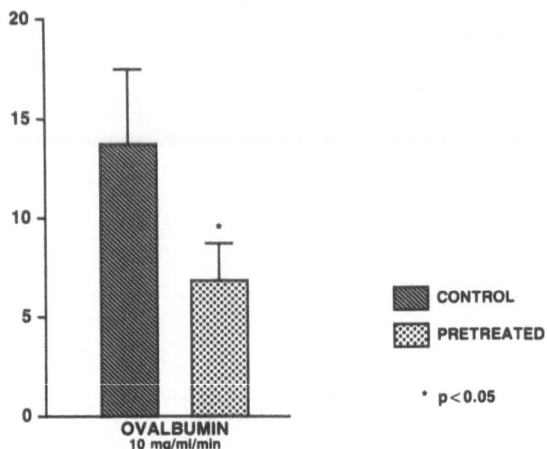

Figure 7.   Effect of BN 52021 (10 mg/kg, I.V.) on antigen (aerosol)-
            induced bronchoconstriction in guinea-pigs (mean $\pm$ S.E.M.).
            Ordinate: Bronchoconstriction   (cmH$_2$0).   (Modified   from
            Cirino et al., 1986; with permission).

INHIBITION OF ANTIGEN-INDUCED PULMONARY IMPAIRMENTS BY BN 52021

BN 52021 given I.V. (0.1 to 2 mg/kg) or orally (10 to 15 mg/kg)
significantly antagonizes both homologous (Vilain et al., 1986; Lagente
et al., 1985, 1987a); and heterologous (Braquet et al., 1985b; Touvay
et al., 1985, 1986a) passive anaphylaxis in guinea-pigs (Table 3). The
PAF-acether antagonist (2 to 4 mg/kg, I.V.) also totally protects the
animals from death and counteracts the sustained bronchoconstriction
observed in actively sensitized guinea-pigs (Berti et al., 1986).
Nevertheless, as shown in Table 3, its efficiency is more variable in
active shock since:

(i)   Larger doses are required in active bronchoconstriction;

(ii)   in certain procedures of active sensitization using low doses of
        antigen, BN 52021 is only slightly effective or ineffective
        whereas H$_1$-blockers (e.g. mepyramine) are very effective
        (Lagente et al., 1985; C. Touvay and P. Braquet unpublished)
        suggesting a more pronounced involvement of histamine in such
        anaphylactic shock.

The protective effect of BN 52021 on immune bronchoconstriction
is associated with the recovery of blood pH, PO$_2$ and TxB$_2$ levels
impaired by antigen challenge (Berti et al., 1986). Interestingly in
both passive and active anaphylaxis, BN 52021 antagonizes the antigen-
induced thrombopenia but the accompanying leukopenia is unaffected
(Berti et al., 1986; Lagente et al., 1987a). These results are in
agreement with those of Adnot et al. (1986) and Beijer et al. (1986)
who respectively showed that BN 52021 in guinea-pigs, (i) antagonizes
the hypotensive response to I.V. <u>Salmonella</u> <u>typhimurium</u> endotoxin
without interfering with leukopenia and (ii) counteracts lung platelet
accumulation induced by endotoxin aerosol.

Table 3. Effects of BN 52021 on anaphylactic bronchoconstriction in guinea-pigs; inh., mean inhibition

Sensitization procedure	Challenge (OA)		BN 52021		Reference
	Route	Dose	Route	Dose and effect	
passive, homologous	i.v.	1 mg/kg	i.v.	1 mg/kg, 93 % inh. (inh. suppressed by propranolol)	*Lagente et al., 1986*
	aerosol	1-10 mg/kg	i.v.	3 mg/kg, 85 % inh. (inh. suppressed by propranolol)	*Cirino et al., 1986*
passive, heterologous	i.v.	1 mg/kg	i.v.	1 mg/kg, 85% inh. 2 mg/kg, 78% inh.	*Braquet et al., 1985b* *Touvay et al., 1986a*
	i.v.	1 mg/kg	oral	20 mg/kg, 55% inh.	*Touvay et al., 1986c*
active, OA 100 mg/ kg i.v. + 100 mg/kg SC	i.v.	5 mg/kg	i.v.	ED_{50} = 1,71 mg/kg	*Berti et al., 1986*
active, OA − 10μg/ animal i.v. + Al (OH)$_3$	i.v.	1 mg/kg	i.v. and oral	inactive	*Lagente et al., 1985* *Touvay et al., 1986a*

BN 52021 also antagonizes immune bronchoconstriction when antigen is given by aerosol in passively sensitized guinea-pigs (Fig. 7); interestingly in the same experiment, another PAF-acether antagonist Ro 19-3704 (Barner et al., 1985) is also effective (although ineffective when antigen is given I.V.; Cirino et al., 1986).

Suppression by BN 52021 of TxB_2, 6-keto-$PGF_{1\alpha}$ and PGE_2 formation by isolated lungs from both passively (Lagente et al., 1987a) or actively (Harczy et al., 1986; Sirois et al., 1986) sensitized guinea-pigs is not accounted for by thromboxane synthetase or cyclooxygenase inhibition since BN 52021 is completely inactive against arachidonic acid under similar conditions (Desquand et al., 1986). In active anaphylaxis of guinea-pig isolated lung, LTB_4 and LTD_4 formation is also antagonized by BN 52021 (Sirois et al., 1986) even though PAF-acether is known not to activate lung mast cells and accordingly fails to release histamine from normal isolated guinea-pig lungs (B.B. Vargaftig, personal communication). The protection exerted by BN 52021 may result either (i) from a specific, PAF-acether-independent effect since it blocks IgE-induced histamine release in rat mast cells (Stanworth, Griffiths and Braquet, 1986) or (ii) a PAF-acether-dependent phenomenon since Lefort et al. (1986) showed that PAF-acether and LTD_4 are converted to histamine-releasing agents when injected into lungs dissected from sensitized animals; formation of PAF-acether by a lung component may thus account for histamine release in shock and accordingly, BN 52021 may interfere as a PAF-acether antagonist.

Of interest, BN 52021 is also effective in cardiac anaphylaxis in both passively (Koltai et al., 1986) and actively (Piper and Stewart, 1986) sensitized isolated guinea-pig hearts and in IgG aggregate-induced hypotension (Sanchez-Crespo et al., 1985), two immune challenges where PAF-acether may also be involved.

Is BN 52021 effective against immune bronchoconstriction strictly because of its anti-PAF-acether properties, or does it display additional anti-allergic effects? It is too early to provide a final statement, but some dissociation may exist since:

(i)   other PAF-acether antagonists like SRI 63-119 (Stenzel, Hummer and Hahn, 1986), 48 740 RP (Stenzel, Sannwal and Hahn, 1986), Ro 19-3704 (Lagente et al., 1987b) do not block antigen-induced anaphylaxis;

(ii)   propanolol removes the in vivo anti-anaphylactic activity of BN 52021 in passive shock (Lagente et al., 1987a) even though the anti-PAF-acether activity was moderately affected by beta adrenergic blockade (Fig.2). As BN 52021 does not interfere with the binding of beta adrenergic ligands to their receptor (Braquet et al., 1985a) this suggests an indirect effect which remains to be elucidated.   Nevertheless it may be argued that propanolol may produce a distortion of the relative importance of mediators of anaphylactic bronchoconstriction, making the shock less PAF-acether-dependent and accordingly less inhibitable by BN 52021;

(iii)   one further complication is the failure of BN 52021 to reduce the in vitro contractions of the parenchymal lung strip and the accompanying release of histamine and thromboxane triggered by antigen (Lagente et al., 1987a), even though BN 52021 blocks the contraction of lung strips induced by PAF-acether (Desquand et al., 1986; Touvay et al., 1986b).   This is also in apparent contradiction to the fact that the release of thromboxane and histamine are significantly reduced when antigen is given intra-arterially to the lungs (see above). In fact, it is difficult to compare results obtained in perfused lungs and isolated strips, since much more PAF-acether is required to induce contraction of the strips than to activate the perfused lungs; in the former case the contractions are long lasting and tachyphylaxis is readily obtained (Desouli, Lefort and Vargaftig, 1985), whereas concentration-dependent activation of perfused lungs is obtained with nanograms of PAF-acether injected intra- arterially at short intervals (Lefort et al., 1984). Finally, BN 52021 reaches the isolated strip by diffusion and the perfused lung by the vascular route to interact with endothelium in the first instance: it is thus likely that the mechanisms accounting for smooth muscle contraction and mediator release may differ, at least partly, in these two preparations of the same tissue.   What ever the explanation for the differences between the two preparations, failure of BN 52021 at effective anti-PAF-acether concentrations, to block passive anaphylaxis triggered in lung parenchymal strips and its effectiveness on isolated perfused lungs, again suggests some degree of dissociation between PAF-acether antagonism and anti-anaphylactic activity;

(iv)   PAF-acether induces a rapid and transient increase in plasma tryptic-like activity which returns to normal levels at about 30 min (Etienne et al., 1987). In comparison, Salmonella enteritidis endotoxin induces a slower but higher and more durable increase in plasma protease activity.   BN 52021 does not exhibit any significant anti-protease effect in vitro.   In contrast, when given I.V. or orally, BN 52021 induces a dose-dependent and very significant decrease in plasma tryptic-like activity triggered in rats by either PAF-acether (Fig.8) or endotoxin (Etienne et al., 1987).   Since protease activity is increased in blood (see for example Neuman, Inbar and Creter, 1984) and bronchial lavage fluid (Pusa and Tchorzewski, 1985) of asthmatic patients the antiprotease activity displayed in vivo by BN 52021 may account for its anti-anaphylactic property, this activity being totally or partially related to its PAF-acether antagonistic behaviour.

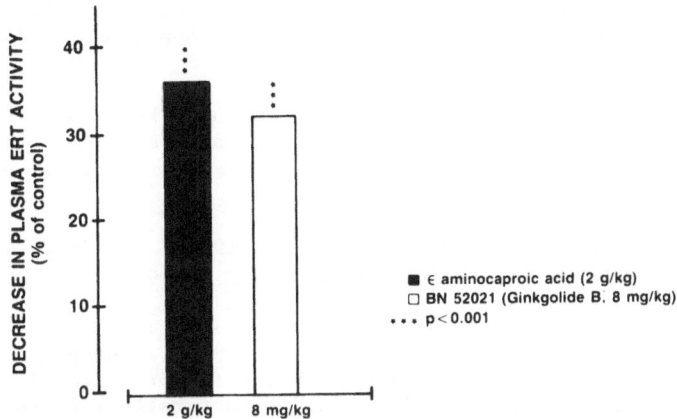

Figure 8.   In vivo antiprotease effect (ERT) of BN 52021 in comparison
with ε-aminocaproic acid. (From Etienne et al., 1987; with
permission).

It is therefore difficult to provide a final statement on the
mechanisms underlying the anti-anaphylactic effect of BN 52021. Nevertheless a recent paper by Cazals-Stenzel (1986) demonstrates that
another PAF-acether antagonist, WEB 2086, a triazolobenzodiazepine, has
an anti-anaphylactic profile very similar to that of BN 52021.
Furthermore using Kadsurenone (Shen et al., 1985) and Alprazolam
(Kornecki, Erhlich and Lenox, 1984), Darius et al. (1986) confirmed our
findings that PAF-acether antagonists can counteract immune bronchoconstriction. These data reinforce the possible involvement of PAF-
acether in immune bronchoconstriction but the mechanisms remain to be
totally elucidated.

REFERENCES

Adnot, S., Lefort, J., Lagente, V., Braquet, P. and Vargaftig, B.B.
(1986). Interference of BN 52021, a PAF-acether antagonist, with
endotoxin-induced hypotension in the guinea-pig. Pharmac. Res. Commun.
18 (Suppl.), 197

Arnoux, B., Duval, D. and Benveniste, J. (1980). Release of platelet-
activating factor (PAF-acether) from alveolar macrophages by the calcium ionophore A23187 and phagocytosis. Eur. J. Clin. Invest. 10, 437

Arnoux, B. and Gillis, C.N. (1986). Role of fatty acids and BN 52021
on vascular and airway actions of platelet-activating factor (PAF) in
isolated rabbit lung in situ. Bull. Eur. Physiopath. Resp. 22 (Suppl.),
51

Baggiolini, M. and Dewald, B. (1986). Stimulus amplification by PAF
and LTB$_4$ in human neutrophils. Pharmac. Res. Commun. 18 (Suppl.), 51

Baranes, J., Hellegouarch, A., Le Hegarat, M., Viossat, I., Auguet, M.,
Chabrier, P.E., Clostre, F. and Braquet, P. (1986). The effects of
PAF-acether on the cardiovascular system and their inhibition by a new
highly specific PAF-acether receptor antagonist BN 52021. Pharmac.
Res. Commun. 18, 737

Barner, R., Hadvary, P., Burri, K., Hirth, G., Cassal, J.M. and Muller, K. (1985). Glycerin derivatives. Eur. Pat. Appl. EP 147,768 12.18.84 [see Chem. Abstracts 104, 88939n, (1986)].

Barnes, P.J., Chung, K.F., Dent, G., Guinot, Ph., MacCusker, M. and Page, C.P. (1986). Effectiveness of a specific antagonist of platelet-activating factor, BN 52063, in man. Brit. J. Pharmac. (Abstr.) (in press).

Barnes, P.J., Cuss, F.M. and Palmer, J.B. (1985). The effect of airway epithelium on smooth muscle contractility in bovine trachea. Br. J. Pharmac. 86, 685

Baroggi, N., Etienne, A. and Braquet, P. (1986). Changes in cytosolic free calcium induced by platelet activating factor in rabbit platelets: specific inhibition by BN 52021 and structurally related compounds. Agents and Actions (in press).

Beijer, L., Botting, J., Oyekan, A.O., Page, C. and Rylander, R. (1986). The involvement of PAF in endotoxin-induced pulmonary recruitment. Brit. J. Pharmac. (Abstr) (in press).

Benveniste, J., Henson, P.M. and Cochrane, C.G. (1972). Leukocyte-dependent histamine release from rabbit platelets. The role of IgE, basophils, and a platelet-activating factor. J. Exp. Med. 136, 1356

Berti, F., Omini, C., Rossoni, G. and Braquet, P. (1986). Protection by two ginkgolides, BN 52020 and BN 52021, against guinea-pig lung anaphylaxis. Pharmac. Res. Commun. 18, 775

Borgeat, P., Fruteau de Laclos, B., Rabinovitch, H., Picard, S., Braquet, P., Hebert,J. and Laviolette, M. (1984). Eosinophil-rich human polymorphonuclear leukocyte preparations characteristically release leukotriene C_4 on ionophore A 23187 challenge. J. Allergy Clin. Immunol. 74, 310

Boukili, M.A., Lagente, V., Lefort, J. and Vargaftig, B.B. (1985). Bronchoconstriction by f-met-leu-phe may involve cyclooxygenase and leukotriene-independent mechanisms. Int. J. Immunopharmac. 7, 382

Bourgain, R.H., Maes, L., Braquet, P. Andries, R., Touqui, L. and Braquet, M. (1985). The effect of 1-0-alkyl-2-acetyl-sn-glycero-3-phosphocholine (PAF-acether) on the arterial wall. Prostaglandins, 30, 185

Braquet, P. (1984). Treatment or prevention of disorders provoked by platelet-activating factor acether, GB Patent, 84/18, 424 (July 19, 1984) Belg. BE 901, 915 [see Chem. Abstracts 103, 189808d, 1985].

Braquet, P., Spinnewyn, B., Braquet, M., Bourgain, R.H., Taylor, J.E., Etienne, A. and Drieu, K. (1985a). BN 52021 and related compounds: a new series of highly specific PAF-acether receptor antagonists isolated from Ginkgo biloba. Blood and Vessels, 16, 559

Braquet, P., Etienne, A., Touvay, C., Bourgain, R.H., Lefort, J. and Vargaftig, B.B. (1985b). Involvement of platelet activating factor in respiratory anaphylaxis, demonstrated by PAF-acether inhibitor BN 52021. Lancet, I, 1501

Braquet, P. and Godfroid, J.J. (1986). PAF-acether specific binding sites : 2. Design of specific antagonists. TIPS 7, 397

Braquet, P., Drieu, K. and Etienne, A. (1986). Le Ginkgolide B (BN 52021) : un puissant inhibiteur du PAF-acether isole du Ginkgo biloba L. Actualites de Chimie Therapeutique (Paris) 13, 237

Braquet, P., Touqui, L., Shen, T.Y. and Vargaftig, B.B. (1986). Perspectives in platelet activating factor research. Pharmac. Rev. (in press).

Braquet, P.. Etienne, A., Mencia-Huerta, J.M. and Clostre, F. (1987). The role of platelet activating factor in gastrointestinal ulcerations. Eur. J. Pharmac. (submitted).

Braquet, P. and Vargaftig, B.B. (1987). Platelet-activating factor today. ISI Atlas of Sciences, (in press).

Bruynzeel, P.L.B., Koenderman, L., Kok, P.T.M., Hameling, M.L. and Verhagen, J. (1986). Platelet-activating factor (PAF-acether)-induced leukotriene C_4 formation and luminol dependent chemiluminescence by human eosinophils. Pharmac. Res. Commun. 18, 61

Casals-Stenzel, J. (1986). Effects of WEB 2086, a novel antagonist of platelet activating factor, in active and passive anaphylaxis. Immunopharmacology (in press).

Chung, K.F., Dent, G., McCusker, M., Guinot, Ph., Page, C.P. and Barnes, P.J. (1987). Effect of a specific antagonist of platelet activating factor (BN 52063) in man. Lancet (in press).

Cirino, M., Lagente, V., Lefort, J. and Vargaftig, B.B. (1986). A study with BN 52021 demonstrates the involvement of PAF-acether in IgE-dependent anaphylactic bronchoconstriction. Prostaglandins, 32, 121

Coeffier, E., Borrel, M.C., Lefort, J., Chignard, M., Broquet, C., Heymans, F., Godfroid, J.J. and Vargaftig, B.B. (1986). Effects of PAF-acether and structural analogues on platelet activation and bronchoconstriction in guinea-pigs. Eur. J. Pharmac. (in press).

Cordeiro, R.S.B., Silva, P.M.R., Martins, M.A. and Vargaftig, B.B. (1986a). Effects of potential antagonists on the rat paw oedema induced by PAF-acether. Prostaglandins (in press).

Cordeiro, R.S.B., Martins, M.A., Silva, P.M.R., Castro Faria Neto, H.C., Castanheira, J.R.C. and Vargaftig, B.B. (1986b). Desensitization to PAF-induced rat paw oedema by repeated intraplantar injections. Life Sci. (in press).

Cuss, F.M., Dixon, C.M. and Barnes, P.J. (1986). Effects of inhaled platelet activating factor on pulmonary function and bronchial responsiveness in man. Lancet, II, 189

Darius, H., Lefer, D.J., Smith, J.B. and Lefer, A.M. (1986). Role of platelet-activating factor-acether in mediating guinea-pig anaphylaxis. Science, 232, 58

De Monchy, J.G., Kauffman, H.F., Vense, P., Koeter, G.H., Jansen, H.M., Sluiter, H.J. and De Vries, K. (1985). Bronchoalveolar eosinophilia during allergen-induced late asthmatic reactions. Am. Rev. Respir. Dis. 135, 373

Desquand, S., Touvay, C., Randon, J., Lagente, V., Vilain, B., Maridonneau-Parini, I., Etienne, A., Lefort, J., Braquet, P. and Vargaftig, B.B. (1986). Interference of BN 52021 (Ginkgolide B) with the bronchopulmonary effects of PAF-acether in the guinea-pig. Eur. J. Pharmac. 127, 83

Detsouli, A., Lefort, J. and Vargaftig, B.B. (1985). Histamine and leukotriene-independent guinea-pig anaphylactic shock unaccounted for by PAF-acether. Br. J. Pharmac. 84, 801

Emeis, J.J. (1987). The role of lipid mediators in blood fibrinolysis. In: Lipid Mediators in Immunology of Burn and Sepsis, Eds Paubert-Braquet, M., Braquet, P., Fletcher, R., Demling, R. and Foegh, F., Plenum Press, New York (NATO ASI Series) (in press).

Etienne, A., Hecquet, F., Guilmard, C., Soulard, C. and Braquet, P. (1987). Inhibition of rat endotoxin-induced lethality by BN 52021 and 52063 compounds with PAF-acether antagonistic effect and protease-inhibitory activity. Tissue Reaction (in press).

Etienne, A., Hecquet, F., Soulard, C., Spinnewyn, B., Clostre, F. and Braquet, P. (1985). In vivo inhibition of plasma protein leakage and salmonella enteritidis-induced mortality in the rat by a specific PAF-acether antagonist : BN 52021. Agents and Actions, 17, 368

Feuerstein, G., Leader, P., Siren, A.L. and Braquet, P. (1987). Protective effect of a PAF-acether antagonist, BN 52021, in trichothecene toxicosis. J. Appl. Pharmac. Toxicol. (in press).

Filley, W.V., Kephart, G.M., Holley, K.E. and Gleich, G.J. (1982). Identification by immunofluorescence of eosinophil granule major basic protein in lung tissues of patients with bronchial asthma. Lancet, II, 11

Fink, A., Afek, A., Shalev, Y., Eliraz, A. and Bentwich, Z. (1986). The sensitivity of peripheral blood leukocytes of ischemic heart patients to PAF-acether and cysteinyl containing leukotrienes: determination by the tube leukocyte adherence inhibition assay. 6th Int. Conference on Prostaglandins and related Compounds, Florence, Italy, June 3-6, Abst. p. 301

Fitzgerald, M.F., Moncada, S. and Parente, L. (1986). The anaphylactic release of platelet-activating factor from perfused guinea-pig lungs. Br. J. Pharmac. 88, 149

Foegh, M., Alijani, M.R., Helfrich, G.B., Khirabadi, B.S. and Ramwell, P. (1985). Prolongation of cardiac allograft survival with the PAF antagonist BN 52021 and with the thromboxane receptor antagonists L640035 and L636499. Advances in Prostaglandin, Thromboxane and Leukotriene Research, 15, 381

Foegh, M.L., Khirabadi, B.S., Rowles, J.R. and Ramwell,P.W. (1986). Prolongation of cardiac allograft survival with BN 52021 a specific antagonist of platelet activating factor. Transplantation, 42, 86

Garay, R. and Braquet, P. (1986). Involvement of K^+ movements in the membrane signal induced by PAF-acether. Biochem. Pharmac. 35, 2811

Gleich, G.J., Loegering, D.A., Mann, K.G. and Maldonado, J.E. (1976). Comparative properties of the Charcot-Leyden crystal protein and the major basic protein from human eosinophils. J. Clin. Invest. 57, 633

Guinot,Ph., Braquet, P., Duchier, J. and Cournot, A. (1986). Inhibition of PAF-acether induced weal and flare reaction in man by a specific PAF-antagonist. Prostaglandins, 32, 160

Harczy, M., Maclouf, J., Pradelles, P., Braquet, P., Borgeat, P. and Sirois, P. (1986). Inhibitory effects of a novel platelet activating factor (PAF) antagonist (BN 52021) on antigen-induced prostaglandin and thromboxane formation by the guinea-pig lung. Pharmac. Res Commun. 18 (Suppl.), 111

Hebert, R.L. Sirois, P., Braquet, P. and Plante, G.E. (1986). Hemodynamic effects of PAF-acether on the dog kidney. Prostagl. Leukot. Med. (in press).

Henocq, E. and Vargaftig, B.B. (1986). Accumulation of eosinophils in response to intracutaneous PAF-acether and allergens in man. Lancet, I, 1378

Hoshikawa-Fujimura, A.Y., Chamone, D.A.F. and Jancar, S. (1986). Effect of BN 52021 on human PMN superoxide anion production. 6th Int. Conference on Prostaglandins and related Compounds, Florence (Italy). June 3-6, Abst. p.388

Koltai, M., Lepran, I., Szekeres, L., Viossat, I., Chabrier, E. and Braquet, P. (1986). Effect of BN 52021, a specific PAF-acether antagonist, on cardiac anaphylaxis in Langendorff hearts isolated from passively sensitized guinea-pigs. Eur. J. Pharmac. 130, 133

Kornecki, E., Ehrlich, Y.H. and Lenox, R.H. (1984). Platelet-activating factor-induced aggregation of human platelets specifically inhibited by triazolobenzodiazepines. Science, 226, 1454

Kravis, T.C. and Henson, P.M. (1975). IgE-induced release of platelet activating factor from rabbit lung. J. Immunol. 115, 1677

Kuster, L.J., Filep, J. and Frolich, J.C. (1986). Mechanisms of PAF-induced platelet aggregation in man. Thrombos. Res., 43, 425

Lachachi, H., Plantavid, M., Simon, M.-F., Chap, H., Braquet, P. and Douste-Blazy, L. (1985). Inhibition of transmembrane movement and metabolism of platelet-activating factor (PAF-acether) by a specific antagonist, BN 52021. Biochem. Biophys. Res. Commun. 132, 460

Lagente, V., Desquand, S., Lefort, J., Cirino, M. and Vargaftig, B.B. (1987b). Interference of the PAF-acether antagonist Ro-19-3704 with PAF-acether and antigen-induced bronchoconstriction in the guinea-pig. Eur. J. Pharmac. (in press).

Lagente, V., Desquand, S., Randon, J., Lefort, J. and Vargaftig, B.B. (1985). Interference of PAF-acether antagonists with the effects of PAF itself and of anaphylactic shock in vitro and in vivo guinea-pig bronchopulmonary preparations. Prostaglandins, 30, 703

Lagente, V., Touvay, C., Randon, J., Desquand, S., Cirino, M., Vilain, B., Lefort, J., Braquet, P. and Vargaftig, B.B. (1987a). Interference of the PAF-acether antagonist BN 52021 with passive anaphylaxis in the guinea-pig. Prostaglandins (in press).

Lee, T.C., Lenihan, D.J., Malone, B., Roddy, L.L. and Wasserman, S.I. (1984). Increased biosynthesis of platelet-activating factor in activated human eosinophils. J. Biol. Chem. 259, 5526

Lefort, J., Malanchere, E., Pretolani, M. and Vargaftig, B.B. (1986). Immunisation induces bronchial hyper-reactivity and increased mediator release from guinea-pig lungs. Brit. J. Pharmac. 89, 768P

Lefort, J., Rotilio, D. and Vargaftig, B.B. (1984). The platelet-independent release of thromboxane A2 by PAF-acether from guinea-pig lungs involves mechanisms distinct from those for leukotriene C4 and bradykinin. Br. J. Pharmac. 82, 525

Lellouch-Tubiana, A., Lefort, J., Pirotzky, E., Vargaftig, B.B. and Pfister, A. (1985). Ultrastructural evidence for extravascular platelet recruitment in the lung upon intravenous injection of platelet-activating factor (PAF-acether) to guinea-pigs. Br. J. Exp. Path. 66, 345

Lellouch-Tubiana, A., Lefort, J., Pfister, A. and Vargaftig, B.B. (1986). Interactions between granulocytes and platelets in the lung in passive anaphylaxis shock. Correlations with PAF-acether-induced lesions. Int. Arch. All. Appl. Immunol. (in press).

MacDonald, A.J., Moqbel, R., Wardlaw, A.J. and Kay, A.B. (1986). Platelet-activating factor (PAF-acether) enhances eosinophil cytotoxicity in vitro. J. Allergy Clin. Immunol. 77, 227

McManus, L.M. and Pinckard, R.N. (1985). Kinetics of acetyl glyceryl ether phosphorylcholine (AGEPC)-induced acute lung alterations in the rabbit. Am. J. Pathol. 121, 55

Maridonneau-Parini, I., Lagente, V., Lefort, J., Randon, J., Russo-Marie, F. and Vargaftig, B.B. (1985). Desensitization to PAF-induced bronchoconstriction and to activation of alveolar macrophages by repeated inhalations of PAF in the guinea-pig. Biochem. Biophys. Res. Commun. 131, 42

Neuman, I., Inbar, O. and Creter, D. (1984). The kinin system in exercise-induced asthma. Ann. Allergy, 53, 351

Nunez, D., Chignard, M., Korth, R., Le Couedic, J.-P., Norel, X., Spinnewyn, B., Braquet, P. and Benveniste, J. (1986). Specific inhibition of PAF-acether-induced platelet activation by BN 52021 and comparison with the PAF-acether inhibitors Kadsurenone and CV-3988. Eur. J. Pharmac. 123, 197

Page, C.P., Paul, W., Basran, G.S. and Morley, J. (1985). Platelet activation in asthma. In: Bronchial Asthma: Mechanisms and Therapeutics, Eds Weiss, E.R., Segal, M.S. and Stein, M., Little, Brown and Co., Boston. pp. 266

Parente, N. and Flower, R.J. (1985). Hydrocortisone and macrocortin inhibit the zymosan-induced release of lyso-PAF from rat peritoneal leucocytes. Life Sci. 36, 1225

Patterson, R. and Harris, K.E. (1983). The activity of aerosolized and intracutaneous synthetic platelet activating factor (AGEPC) in rhesus monkeys with IgE-mediated airway responses and normal monkeys. J. Lab. Clin. Med. 102, 933

Piper, P.J. and Stewart, A.G. (1986). Evidence of a role for platelet-activating factor in antigen-induced coronary vasoconstriction in guinea-pig perfused hearts. Brit. J. Pharmac. 88, 238P

Plante, G.E., Hebert, R.L., Lamoureux, C., Braquet, P. and Sirois, P. (1986). Hemodynamic effects of PAF-acether. Pharmac. Res. Commun. 18 (Suppl.), 173

Plotkine, M., Massad, L., Allix, M., Capdeville, C. and Boulu, R.G. (1986). Cerebral effects of PAF-acether in rats. 6th Conference on Prostaglandins and related Compounds, Florence (Italy) June 3-6 Abstr. p. 300

Pusa, T. and Tchorzewski, H. (1985). Analysis of proteolytic enzymes and their natural inhibitors in serum and bronchial lavage fluid in atopic bronchial asthma. Allerg. Immunol. 31, 169

Robbins, J.C., MaChoy, B.H., Lam, M.J., Ponpipom, M., Rupprecht, K.M. and Shen, T.Y. A synthetic phospholipid inhibitor of platelet-activating factor (PAF) biosynthesis. Fed. Proc. 44, 1269

Ross, R. and Vogel, A. (1978). The platelet derived growth factor. Cell, 14, 203

Rotilio, D., Lefort, J., Detsouli, A. and Vargaftig, B.B. (1983). Absence de contribution du PAF-acether a la reponse anaphylactique pulmonaire in vitro chez le cobaye. J. Pharmac. (Paris) 14 (Suppl. 1), 97

Rubin, A.E., Smith, L.J. and Patterson, R. (1986). Effects of platelet activating factor (PAF) on normal human airways. Am. Rev. Resp. Dis. 133, A91

Sanchez-Crespo, M., Fernandez-Gallardo, S., Nieto, M.L., Baranes, J. and Braquet, P. (1985). Inhibition of the vascular actions of IgG aggregates by BN 52021, a highly specific antagonist of PAF-acether. Immunopharmacology, 10, 69

Shen, T.Y., Hwang, S.B., Chang, M.N., Doebber, T.W., Lam, M.H., Wu, M.S. Wang, X., Han, G.Q. and Li, R.Z. (1985). Characterization of a platelet-activating factor receptor antagonist isolated from haifenteng (Piper futokadsura): specific inhibition of in vitro and in vivo platelet-activating factor-induced effects. Proc. Natl. Acad. Sci. USA 82, 672-676 (see also Patent JP 60 97,972 [85 97, 972] 31 May 1985, US Appl. 541,806 13 Oct. 1983).

Simon, A.F., Chap, H., Braquet, P. and Douste-Blazy, L. (1986). Effect of BN 52021, a specific antagonist of platelet-activating factor (PAF-acether) in Ca^{2+} mobilization and phosphatidic acid production induced by PAF-acether in human platelets. Thrombos. Res. (in press).

Sirois, P., Harczy, M., Robidoux, C., Borgeat, P. and Braquet, P. (1986). Lipid mediators in lung anaphylaxis. In : Lipid Mediators in Immunology of Burn and Sepsis, a NATO ASI, Helsingor (Denmark) July 20-25, Abst. p. 40

Stanworth, D.R., Griffiths, H.R. and Braquet, P. (1986). Effect of PAF-acether antagonists on pre-formed mediator secretion from mast cells In: 2nd World Conference on Inflammation, Monte Carlo, March 19-22, Abst. A.193

Stenzel, H., Hummer, B. and Hahn, H.L. (1986). Effect of the PAF-antagonist SRI 63-441 on the allergic reaction in awake dogs with natural asthma. Platelets, Analgesics and Asthma, IIIrd. Int. Symposium Davos, Switzerland, Sept. 7-9. Abstr. p. 50

Stenzel, H., Sannwal, U. and Hahn, H.L. (1986). Effect of the PAF-antagonist RP 48740 on allergic reactions in awake dogs. 6th Int. Conference on Prostaglandins, Florence (Italy), June 3-6, Abstr. p. 306

Touvay, C., Etienne, A. and Braquet, P. (1985). Inhibition of antigen-induced lung anaphylaxis in the guinea-pig by BN 52021 a new specific PAF-acether receptor antagonist isolated from Ginkgo biloba. Agents and Actions, 17, 371

Touvay, C., Vilain, B., Taylor, J.E., Etienne, A. and Braquet, P. (1986a). Proof of the involvement of platelet-activating factor (PAF-acether) in pulmonary complex immune systems using a specific PAF-acether receptor antagonist : BN 52021. Prog. Lip. Res. (in press).

Touvay, C., Vilain, B., Etienne, A., Sirois, P., Borgeat, P. and Braquet, P. (1986b). Characterization of platelet-activating factor (PAF)-acether-induced contractions of guinea pig lung strips by selected inhibitors of arachidonic acid metabolism and by PAF antagonists. Immunopharmacology (in press).

Touvay, C., Vilain, B., Carre, C., Etienne, A. and Braquet, P. (1986c). Comparative effects of three orally active PAF-receptor antagonists extracted from Ginkgo biloba : on in vivo bronchoconstrictions and in vitro lung strips contractions induced by PAF-acether. 6th Conference on Prostaglandins and related Compounds. Florence, Italy, June 3-6, Abstr. p. 316

Vargaftig, B.B., Borgeat, P., Braquet, P., Braquet, M., Brocklehurst, W.E., Dahlen, S.E., Drazen, J.M., Etienne, A., Fitzpatrick, F.A., Henson, P.M., Holgate, S.T., Israel, E., Mencia-Huerta J.M., Murphy, R.C., Robinson, C. and Shore, S. (1985). Leukotrienes and immediate hypersensitivity. Ann. Inst. Pasteur/Immunol. 136D, 175

Vargaftig, B.B. and Braquet, P. (1987). PAF-acether today : relevance for acute experimental anaphylaxis. Brit. Med. Bull. (in press).

Vercelotti, G.M., Huh, P.W., Yin, H.Q., Nelson, R.D. and Jacob, H.S. (1986). Enhancement of PMN-mediated endothelial damage by platelet activating factor (PAF): PAF primes PMN responses to activating stimuli. 28th. Annual Meeting of the American Society of Hematology (December) Abstr. p. 204

Vilain, B., Lagente, V., Touvay, C., Desquand, S., Randon, J., Lefort, J., Braquet, P. and Vargaftig, B.B. (1986). Pharmacological control of the in vivo passive anaphylactic shock by the PAF-acether antagonist compound BN 52021. Pharmac. Res. Commun. 18 (Suppl.), 119

Villamediana, L.M., Sanz, E., Fernandez-Gallardo, S., Caramelo, C., Sanchez-Crespo, M., Braquet, P. and Lopes-Novoa, J.M. (1986). Effects of the platelet-activating factor antagonist BN 52021 on the hemo-dynamics of rats with experimental cirrhosis of the livers. Life Sci., 39, 201

Wallace, J.L., Stell, G., Whittle, B.J.R., Lagente, V. and Vargaftig, B.B. (1987). Evidence for platelet activating factor (PAF) as a mediator of endotoxin-induced gastrointestinal damage in the rat : effects of three PAF antagonists. Gastroenterology, (in press).

Wardlaw, A.J. and Kay, A.B. (1986). PAF-acether is a potent chemo-tactic factor for human eosinophils. J. Allergy Clin. Immunol. 77, 236

Winslow, C.M., Anderson, R.C., D'Aries, F.J., Frish, G.E., DeLillo, A.K., Lee, M.L. and Saunders, R.N. (1987). Toward understanding the mechanism of action of PAF receptor antagonists. In: New Horizons in Platelet Activating Factor Research. Eds Lee, M.L. and Winslow, C.M., John Wiley, London (in press).

Winslow, C.M., Vallespir, S.R., Frisch, G.E., D'Aries, F.J., DeLillo, A.K., Houlihan, W.J., Parrino, V., Schmitt, G. and Saunders, R.N. (1985a). A novel platelet activating factor receptor antagonist. Prostaglandins, 30, 697

Winslow, C.M., Frisch, G.E., D'Aries, F.J., Handely, D.A., Melden, M.K., Deacon, R.W., Houlihan, W.J., Parrino, V., Schmitt, G. and Saunders, R.N. (1985b). A platelet-activating factor (PAF) receptor antagonist which influences the primary physiological responses of rodents to PAF. Prostaglandins, 30, 698

ROLE OF HYPOXIA AND ACETYLCHOLINE IN THE REGULATION OF CEREBRAL BLOOD FLOW

E. DORA and A.G.B. KOVACH

Experimental Research Department and Second Institute of Physiology, Semmelweis University Medical School, Budapest, Hungary

INTRODUCTION

It is well-known that the cerebral cortex is heterogeneously supplied with oxygen. According to microelectrode measurements, PO_2 at some microregions of the brain cortex is close to zero mmHg (Silver, 1966; Metzger and Heuber, 1977). Supporting these data, Rosenthal et al. (1976) showed that the oxido-reduction state of cytochrome a,a_3 in the brain cortex is shifted toward a more oxidized state if cerebral oxygen supply is increased. On the basis of this finding, Rosenthal et al. (1976), in accordance with Davies and Bronk (1957), suggested that under physiological conditions the brain cortex is on the border of slight hypoxia. This assumption seemed to be supported by the data of Kontos et al. (1978) and Morii, Winn and Berne, (1983). They showed that the autoregulatory dilatation of pial arteries can be counteracted by superfusion of the brain cortex with oxygen-saturated mock cerebrospinal fluid (CSF) or fluorocarbon (Kontos et al., 1978), and the resting cerebral blood flow (CBF) can be decreased by systemic theophylline treatment (Morii et al., 1983). However, other investigators (Leniger-Follert, 1985; Rubin and Bohlen, 1985) found no change in cortical PO_2 during autoregulation, and resting CBF was not altered by systemic theophylline treatment in Emerson and Raymond's (1981) experiments. This controversy in the literature led us to investigate the effects of an excess of oxygen on the steady NAD/NADH redox state and autoregulation of the brain cortex. In addition, we studied the effect of topical adenosine deaminase and theophylline treatment of the brain cortex in order to see whether adenosine (ADO) contributes to the maintenance of resting CBF.

In experiments in vitro we investigated the effects of acetylcholine (ACh) on isometric tension of ring segments of the middle cerebral artery (MCA). This study was initiated by the findings of Gibson, Duffy and Plum (1980), which showed that cerebral ACh synthesis is decreased by even the slightest arterial hypoxia, and according to in vitro experiments, ACh stimulates vascular endothelial cells to produce potent dilator and constrictor substances (Furchgott and Zawadski, 1980; Furchgott, 1983). The few available studies show that in the middle cerebral artery, as in other vascular beds, ACh causes dilatation in the concentration range of approximately 10^{-8} to

10^{-6} mol/l and constriction at concentrations higher than 10^{-6} mol/l (Edvinsson and Owman, 1977; Lee, 1981). The ACh-induced dilatation is attributed to the action of some endothelium-derived dilator substances (EDRF), and the constriction, to the direct effect on vascular smooth muscle (Furchgott and Zawadski, 1980; Furchgott, 1983). Since from the physiological point of view the vasoactive effect of low ACh concentrations ($< 10^{-8}$ mol/l) can be of interest, we were particularly concerned to see the effects of ACh in the concentration range of 10^{-17} to 10^{-8} mol/l.

EXPERIMENTAL PROCEDURES AND ANALYSIS OF THE DATA

The experiments were carried out partly _in vivo_ on the intact brain cortex, partly _in vitro_ on ring segments of the middle cerebral artery. In the studies _in vivo_, the cats were anaesthetized with chloralose, immobilized with flaxedil, and ventilated artificially. For the measurements of cortical NADH fluorescence and microcirculation a cranial window was fixed into the left parietal bone (Dora, 1984). NADH fluorescence and microcirculation were measured through the cranial window with a microscope fluororeflectometer and a Zeiss camera (Eke, Hutiray and Kovach, 1979; Dora and Kovach, 1983; Dora, 1984, 1985).

In three sets of experiments we studied the effect of ACh on cortical microcirculation and NAD/NADH redox state; the effect of theophylline (10^{-4} mol/l, 1 ml/min superfusion) and adenosine deaminase (0.025 units/ml, 1 ml/min superfusion) on resting CBF; and the effect of superfusion of the brain cortex with oxygen-saturated mock CSF (PO_2 approximately 500 mmHg) on CBF autoregulation evoked by moderate arterial hypotension (mean arterial blood pressure was decreased by bleeding from the control level to 75-85 mmHg).

In the studies _in vitro_, the cats were anaesthetized with ether and exsanguinated. The main trunk of the middle cerebral artery was gently dissected under a Zeiss operating microscope. The vessels were cut with a razor blade into 1-3 mm-long ring segments and stored in Krebs solution at 4°C. The composition of the solution (Hogestatt, Andersson and Edvinsson, 1983) was: NaCl 119, KCl 4.6, $CaCl_2$ 2.5, $MgCl_2$ 1.2, $NaHCO_3$ 15.0, NaH_2PO_4 1.2, glucose 6.0 (all concentrations in mmol/l). The Krebs solution was thermostated at 37°C and equilibrated with a gas mixture containing 5% CO_2 in 95% O_2. Its pH was 7.4. The ring segments were mounted on two L-shaped specimen holders (stainless steel wires, diameter 0.1 mm) and their isometric tension was measured with Grass FT-03 transducers (Hogestatt et al., 1983). The double-walled chamber containing the specimen and the Krebs solution had a volume of 2.5 ml. After mounting, the vessel strips were stretched with a force of 400 mg (4 mN) and were incubated for 60 min. During the incubation, the Krebs solution in the chamber was changed every 15 min. The effects of ACh on isometric tension was tested on vessels precontracted with norepinephine (NE, 10^{-6} mol/l) or prostaglandin $F_{2\alpha}$ ($PGF_{2\alpha}$; 5 x 10^{-6} mol/l). Adrenergic beta receptors were inhibited with 5 x 10^{-7} mol/l propranolol. The various substances were administered with Hamilton microsyringes (10-25 µl) into the chamber. NE and $PGF_{2\alpha}$ were obtained from Chinoin, Budapest, Hungary. ACh and propranolol from Sigma, St. Louis, U.S.A. Concentrations shown in the figures represent final concentrations in the chamber.

The results are expressed as mean $\pm$ standard error. Statistical analysis of the data was carried out with Student's t test.

RESULTS AND DISCUSSION

ACh applied to the brain cortex by superfusion dilated pial arteries concentration-dependently (Fig.1). The diameter of pial arteries was increased by approximately 50% at an ACh concentration of 10^{-5} mol/l. Pial veins showed a slight constriction in response to ACh.

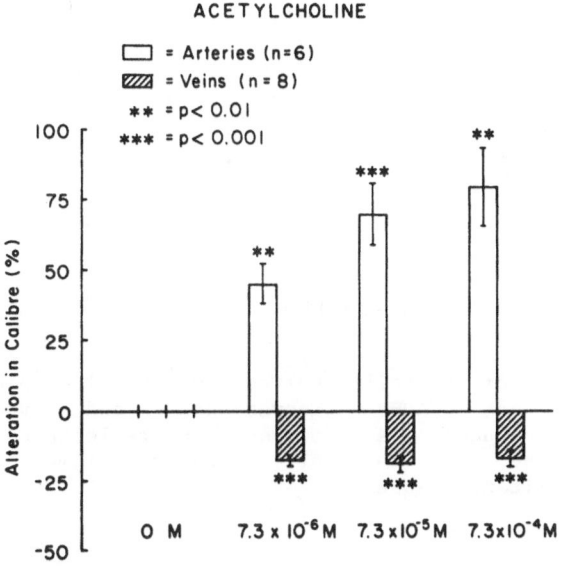

Figure 1. Effect of topically applied acetylcholine on the diameter of pial arteries and veins. Asterisks show significant alterations from control (0 M).

As indicated by the decrease of cortical reflectance, ACh markedly increased cortical vascular volume (Fig. 2). However, although ACh vastly increased cortical oxygen supply, it did not shift the steady NAD/NADH redox state toward oxidation. This finding strongly questions the assumption of Davies and Bronk (1957) and Rosenthal et al. (1976), according to which microregions of the brain cortex with a PO_2 lower than 4 mmHg are biochemically hypoxic.

We could not confirm the findings of Morii et al. (1983) concerning the role of such hypoxia-related metabolites as adenosine in maintaining resting CBF. As Figures 3 and 4 demonstrate, both adenosine deaminase and theophylline considerably inhibit the CBF-increasing effect of topically applied adenosine, but they were unable to alter resting CBF. Since in the case of topical application the systemic effects of the drugs are avoided, it seems very unlikely, on the basis of our results, that adenosine contributes to the maintenance of resting cerebrovascular tone.

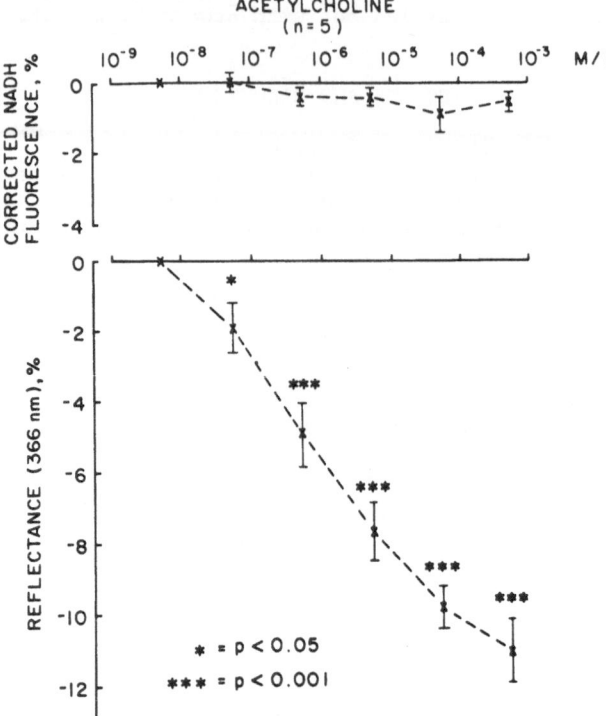

Figure 2. Effect of topically applied acetylcholine on cerebrocortical corrected NADH fluorescence and vascular volume (reflectance). A decrease in reflectance indicates an increase in cortical vascular volume. Asterisks show significant changes.

When mean arterial blood pressure is decreased arterial vessels of the brain dilate. The exact mechanism of this so-called autoregulatory dilatation is still obscure. Among many theories, the 'metabolic theory' suggests that the autoregulatory dilatation is due to hypoxia and the accumulation of such anaerobic tissue metabolites as adenosine, H^+, etc. (Kontos et al., 1978; Winn, Rubio and Berne, 1981). If this were the case, we would expect CBF to decrease in the brain cortex superfused with oxygen-saturated mock CSF during moderate arterial hypotension. However, as Figure 5 shows, autoregulation was functioning well even in this case. Our finding corresponds with the data of Leniger-Follert (1985) and Rubin and Bohlen (1985) who revealed no decrease in cortical PO_2 during moderate arterial hypotension, but disagrees with the data of Kontos et al. (1978). The latter authors found pial artery constriction in response to topical application of oxygen during arterial hypotension. Our disparate finding can most probably be attributed to a better maintained autoregulation. However, since we measured blood flow in the outer 0.3 - 0.4 mm of the cortex, but Kontos et al. (1978) recorded only the diameter of individual pial arteries, our results do not exclude the possibility that individual pial arteries may react with constriction to topically applied oxygen in a hypotensive animal.

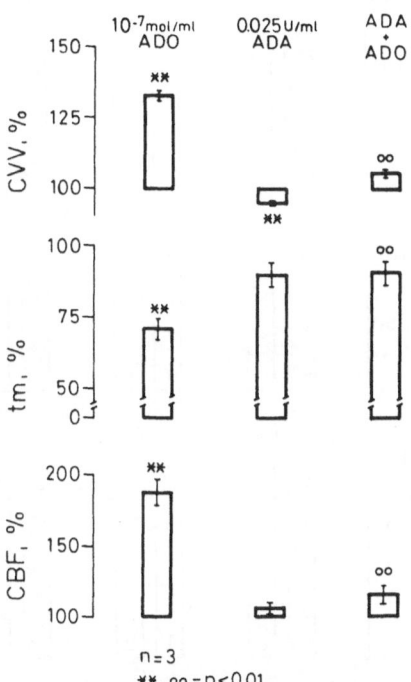

Figure 3. Effect of topically applied adenosine deaminase (ADA) on the cortical microcirculation (middle column) and on the adenosine (ADO)-induced changes in microcirculation (the top and bottom row of columns). Asterisks show the significant alterations as compared with the control values; oo, shows the significant changes after ADA treatment. CVV, cortical vascular volume; t_m, vascular mean transit time; CBF, cortical blood flow. The control values were regarded as 100%.

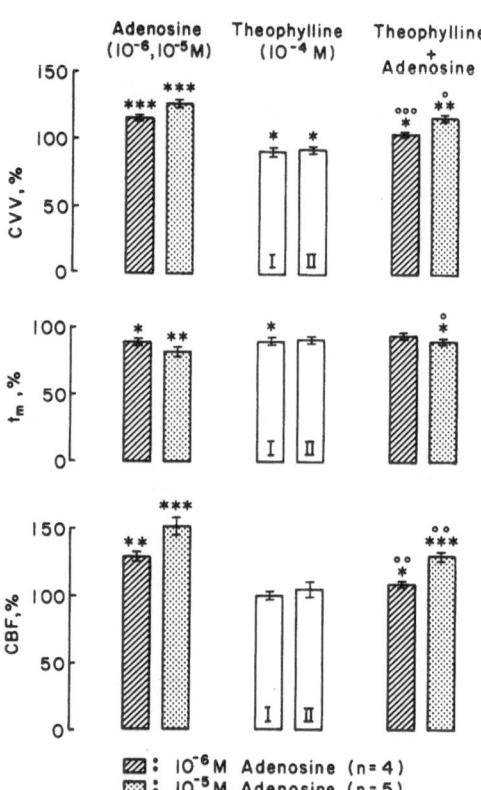

Figure 4. Effect of topically applied theophylline on cortical micro-circulation (middle columns) and on the adenosine-induced changes in microcirculation (top and bottom row of columns). Abbreviations are as in Fig. 3; xxx, ooo, = p< 0.001; xx, oo = p< 0.01; x, o, = p< 0.05.

Unexpectedly, we did not obtain uniform responses to ACh in our $\underline{in}$ $\underline{vitro}$ experiments. The accepted reaction, i.e. no response at concentrations below about 10^{-8} mol/1, with dilatation between approximately 10^{-8} and 10^{-6} mol/1, occurred in only 30% of the

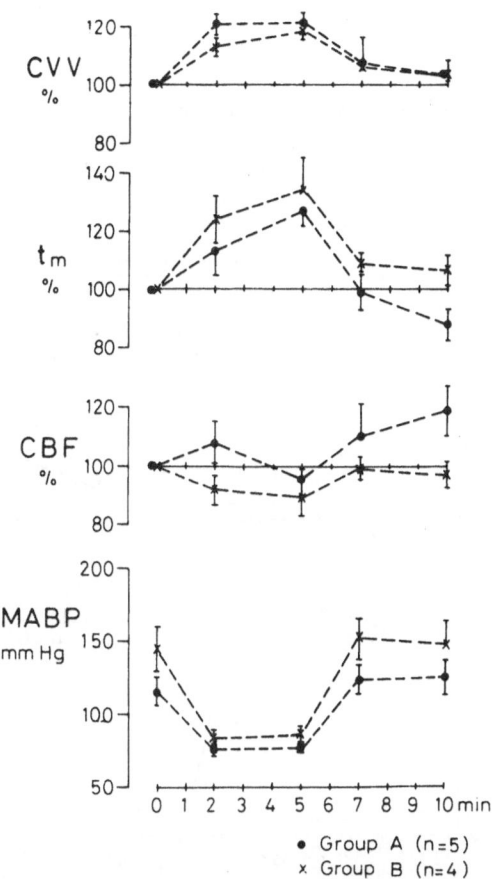

Figure 5. Effect of superfusion of the brain cortex with oxygen-saturated mock CSF on the autoregulatory responses of cortical microcirculation evoked by moderate arterial hypotension. Group A, superfused; Group B, non-superfused; MABP, mean arterial blood pressure. Other abbreviations are as in Fig. 3.

experiments (top panel, Fig. 6). In 40% of the cases the dilatatory effect of ACh was already evident in the range 10^{-17} to 10^{-15} mol/1 (middle panel, Fig. 6) and in the remaining 30% of the experiments the extremely low ACh concentrations produced constriction (bottom panel, Fig. 6). Typical examples of the diverse ACh responses are shown in Figures 7, 8 and 9.

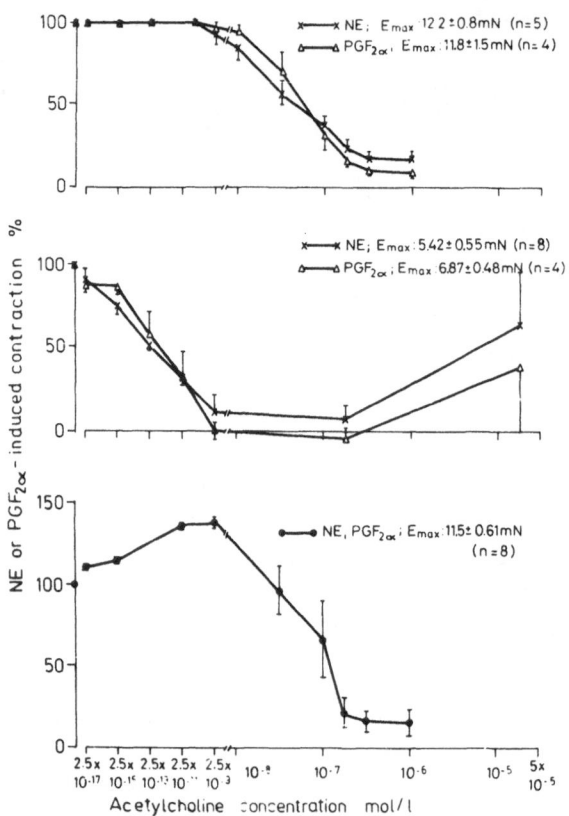

Figure 6. Diversity of acetylcholine response in ring segments of the middle cerebral artery precontracted with 5×10^{-6} mol/l $PGF_{2\alpha}$ or norepinephrine (NE). Abscissa shows acetylcholine concentrations applied to the precontracted vessels. E_{max}, maximum contraction in mN.

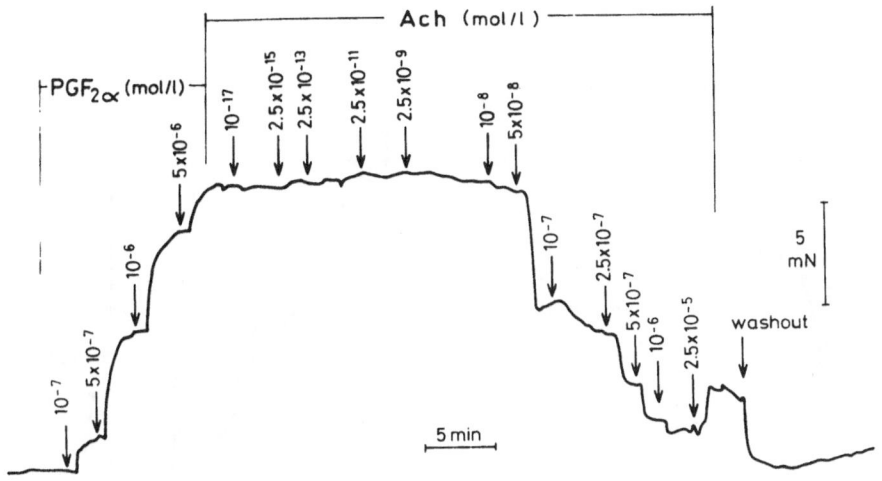

Figure 7. A typical dose response curve for the first type of acetyl
choline (ACh) response.

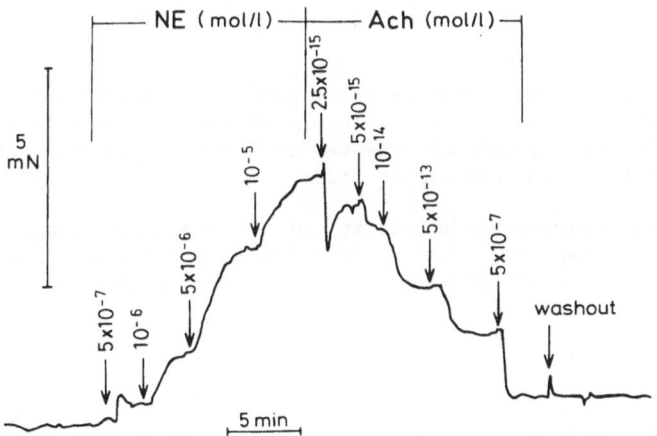

Figure 8. A typical dose response curve for the second type of ACh
response.

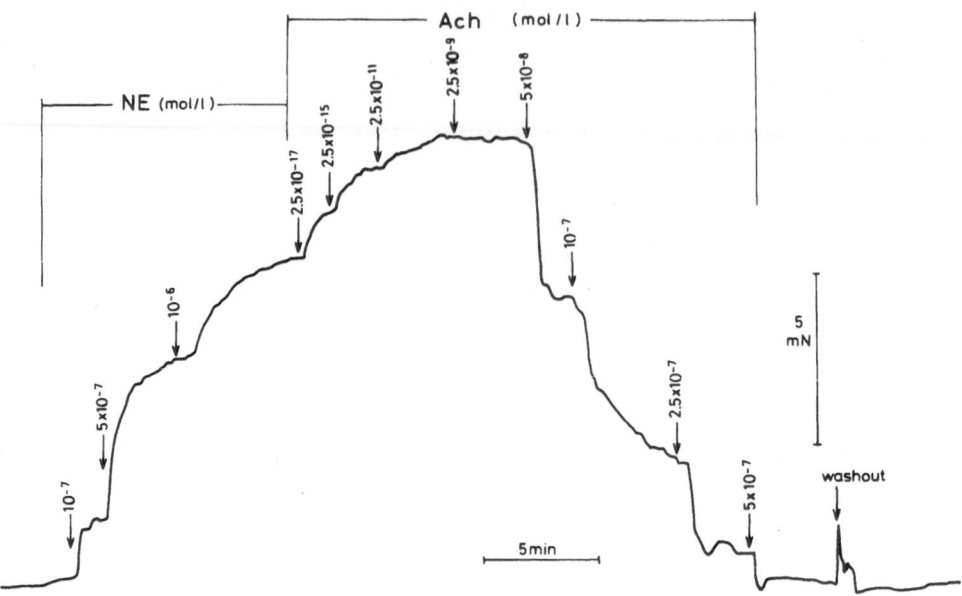

Figure 9. A typical dose response curve for the third type of ACh response.

CONCLUSIONS

1). Since the enhancement of cortical oxygen supply by topically applied acetylcholine did not shift the steady NAD/NADH redox state toward oxidation, it seems very unlikely that the microregions of the brain cortex with very low PO_2 are biochemically hypoxic.

2). Since theophylline and adenosine deaminase failed to alter blood flow in the resting brain cortex, it seems very unlikely that adenosine contributes to the maintenance of the tone of the cerebro-cortical arteries.

3). Since superfusion of the brain cortex with mock CSF saturated with oxygen did not alter autoregulation, it is very unlikely that hypoxia and the associated tissue metabolites play a significant role in the autoregulatory process of brain circulation.

4). The extreme sensitivity of the middle cerebral artery to acetylcholine suggests that acetylcholine and the endothelium-derived substances may have great impact on the regulation of cerebral circulation.

REFERENCES

Davies, P.W. and Bronk, D.W. (1957). Oxygen tension in mammalian brain. Fed. Proc. 16, 689

Dora, E. (1984). A simple cranial window technique for optical monitoring of cerebrocortical microcirculation and NAD/NADH redox state. Effect of mitochondrial electron transport inhibitors and anoxic anoxia. J. Neurochem. 42, 101

Dora, E. (1985). Further studies on the reflectometric monitoring of cerebrocortical microcirculation. Importance of lactate anions in coupling between cerebral blood flow and metabolism. Acta Physiol. Hung. 66, 199

Dora, E. and Kovach, A.G.B. (1983). Effect of topically administered epinephrine, norepinephrine, and acetylcholine on cerebrocortical circulation and the NAD/NADH redox state. J. Cereb. Blood Flow Metabol. 3, 161

Edvinsson, L. and Owman, C. (1977). Pharmacological characterization of postsynaptic vasomotor receptors in brain vessels. In: Neurogenic Control of Brain Circulation. Eds Owman, C. and Edvinsson, L., Pergamon Press, New York, p. 167

Eke, A., Hutiray, Gy. and Kovach, A.G.B. (1979). Induced hemodilution detected by reflectometry for measuring microregional blood flow and blood volume in cat brain cortex. Am. J. Physiol. 236, H759

Emerson, T.E. and Raymond, R.M. (1981). Involvement of adenosine in cerebral hypoxic hyperemia in dog. Am.J.Physiol. 241, H134

Furchgott, R.F. (1983). Role of endothelium in responses of vascular smooth muscle. Circ. Res. 53, 557

Furchgott, R.F. and Zawadski, J.V. (1980). The obligatory role of endothelial cells in the relaxation of arterial smooth muscle by acetylcholine. Nature, 288, 273

Gibson, G.E., Duffy, T.E. and Plum, F. (1980). Acetylcholine synthesis and cerebral blood flow during hypoxaemia. In: Cerebral Circulation and Neurotransmitters. Eds Bes, A. and Geraud, G., Excerpta Medica, Amsterdam, p. 199

Hogestatt, E.D., Andersson, K.-E. and Edvinsson, L. (1983). Mechanical properties of rat cerebral arteries as studied by a sensitive device for recording of mechanical activity in isolated small blood vessels. Acta Physiol. Scand. 117, 49

Kontos, H.A., Wei, E.P., Raper, A.J., Rosenblum, W.I., Navari, R.M. and Patterson, J.L. (1978). Role of tissue hypoxia in local regulation of cerebral microcirculation. Am. J.Physiol. 234, H582

Lee, T.J.F. (1981). Is acetylcholine the dilator transmitter in cerebral blood vessels? A critical examination. J. Cereb. Blood Flow Metabol. 1 (Suppl), S305

Leniger-Follert, E. (1985). Oxygen supply and microcirculation of the brain cortex. In: Oxygen Transport to Tissue-VII. Eds Kreuzer, F., Cain, S.M., Turek, Z. and Goldstick, T.K., Plenum Press, New York and London, (Adv. Exp. Med. Biol. 191, 3).

Metzger, H. and Heuber, S. (1977). Local oxygen tension and spike activity of the cerebral grey matter of the rat and its response to short intervals of O_2 deficiency or CO_2 excess. Pflugers Arch. 370, 201

Morii, S., Winn, H.R. and Berne, R.M. (1983). Effect of theophylline, an adenosine receptor blocker, on cerebral blood flow (CBF) during rest and transient hypoxia. J. Cereb. Blood Flow Metabol. 3 (Suppl), S480

Rosenthal, M., LaManna, J.C., Jobsis, F.F., Levasseur, J.E., Kontos, H.A. and Patterson, J.L. (1976). Effects of respiratory gases on cytochrome a in intact cerebral cortex: Is there a critical PO_2? Brain Res. 108, 143

Rubin, M.J. and Bohlen, H.G. (1985). Cerebral vascular autoregulation of blood flow and tissue PO_2 in diabetic rats. Am. J. Physiol. 249, H540

Silver, I.A. (1966). The measurement of oxygen tension in tissues. In: Oxygen Measurements in Blood and Tissues and their Significance. Eds Payne, J.P. and Hill, D.W., Churchill, London, p. 135

Winn, R.H., Rubio, G.R. and Berne, R.M. (1981). The role of adenosine in the regulation of cerebral blood flow. J. Cereb. Blood Flow Metabol. 1, 239

ADAPTATION TO HYPOXIA

I.S. LONGMUIR

Department of Biochemistry, North Carolina State University, Raleigh, NC 27695-7622, U.S.A.

SUMMARY

Many biochemical changes occur during acclimation to altitude. Some appear to be deleterious, but the increase in cytochrome-P 450 and the changes in the endoplasmic reticulum seem to be beneficial. The latter changes could reduce the capillary to oxidase gradient to compensate for the lower capillary oxygen tensions. The possibility of accelerating acclimation to hypoxia by using drugs which produce similar changes is being explored.

INTRODUCTION

An abrupt fall in arterial oxygen tension can occur physiologically on reduction of the inspired partial pressure of oxygen, for example by ascending to a high altitude. Pathologically the same change can be produced by a variety of intrathoracic events which impair the oxygenation of arterial blood. In either case if the fall is severe enough, unconsciousness or death will occur (Robin, 1980). However, if the abrupt fall in PaO_2 is preceded by a less severe hypoxia, as in the slow ascent of a mountain, changes occur which permit survival (Hurtado, 1964). Such acclimation is not associated with a reduced oxygen consumption (Moore, 1956). It is, however, associated with a number of biochemical changes some of which, perhaps not all, must be of such a nature as to enable the organism to take up oxygen more efficiently as a result of this acclimation.

CHANGES IN THE BLOOD

A number of changes in blood chemistry occur during acclimation which have been interpreted as being able to increase the efficiency of oxygen transport in the blood. The first change seen after a brief period of exposure, about a day in man, is an increase in 2:3 diphosphoglycerate (2:3-DPG; Torrance et al., 1971). This has the effect of shifting the oxyhaemoglobin dissociation curve to the right. Such a shift will certainly reduce the percentage saturation of arterial blood. However, it may also reduce the saturation of the venous blood by a greater amount to give a net gain in oxygen delivery at moderate altitudes. But at more extreme altitudes there is a net loss (Fig. 1) and it has been shown theoretically that a shift of the curve to the left is needed for acclimation to extreme altitudes, such as the summit of Mount Everest (Turek, Kreuzer and Hoofd, 1973). This was confirmed during a scientific expedition to Mount Everest (West, 1982). Indeed only those who showed no increase in 2:3-DPG were able to reach the summit without supplemental oxygen (West, 1982). Thus the increase in 2:3-DPG is not beneficial in acclimation to severe hypoxia.

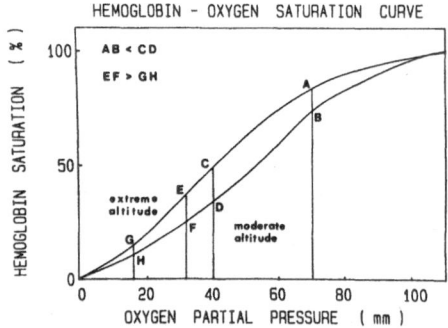

Figure 1. Effect of the shift of the oxyhaemoglobin curve on oxygen delivery at moderate and high altitudes.

The second change which occurs over a longer period is an increase in haemoglobin concentration. However, not all acclimated subjects show this increase and it has been shown that the optimal haemoglobin concentration for exercise at altitude is very near that at sea level (Cerretelli, 1976). An increase in haemoglobin should increase oxygen capacity, but this is cancelled by a greater increase in blood viscosity (Fig. 2.). The oxygen capacity rises linearly with haematocrit but the viscosity rises exponentially so that tissue perfusion falls faster than the increased capacity resulting in a reduced 'tissue oxygen offering'.

The blood changes probably represent part of the organism's response to anaemia rather than hypoxia.

TISSUE CHANGES

There are in general two possible tissue changes which might enable an organism to acclimate to hypoxia. The concentration of respiratory enzymes might rise or their affinity for oxygen might increase. Alternatively the transport of oxygen from blood to the site of utilization might become more efficient so that the reduction in arterial oxygen is compensated for by a smaller tissue fall.

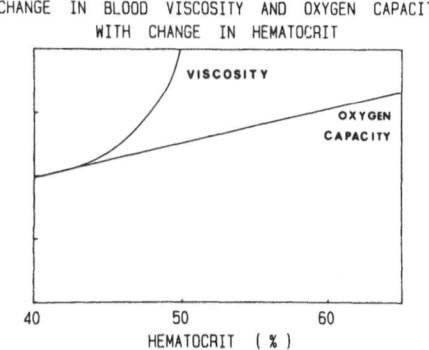

CHANGE IN BLOOD VISCOSITY AND OXYGEN CAPACITY
WITH CHANGE IN HEMATOCRIT

Figure 2. Relation between haematocrit and oxygen capacity and viscosity.

Figure 3 shows the general relationship between the respiration rates of three groups of oxidases and ambient oxygen tension. The major oxidase, cytochrome a, a_3, has a very high affinity for oxygen (Longmuir, 1954). Thus it continues to respire at its maximal rate until virtually all the oxygen has gone. This being the case, an increase in amount or affinity of this enzyme could have only a trivial benefit. The next group of enzymes, the cytochrome-P 450s. have a lower affinity, a K_m of about 1 mmHg. Now there might be some acclimation advantages to an increase in activity or affinity for oxygen. When mice are exposed to a PIO_2 of 100 mmHg for four hours the level of this enzyme doubles, but there is no change in affinity to be found (Longmuir and Pashko, 1977). The final group of oxidases, mainly the flavoproteins, have very low affinities for oxygen. However, they are present in great excess and their activity seems to be controlled by the availability of the second substrate. Thus hypoxia may not limit their activity. It appears that enzymic changes alone cannot account for acclimation to hypoxia.

If simple passive diffusion is the sole mechanism of tissue oxygen transport as proposed by Krogh (1919), it is difficult to see how acclimation changes could occur in the transport of oxygen from the capillary to the site of utilization on the cell except by a reduction of the diffusion coefficient. Some evidence of this was found by Longmuir and Bourke (1965). However, further examination of the problem suggested a different interpretation of the data (Longmuir and McCabe, 1964, 1965; McCabe, 1964): the kinetics of oxygen transport

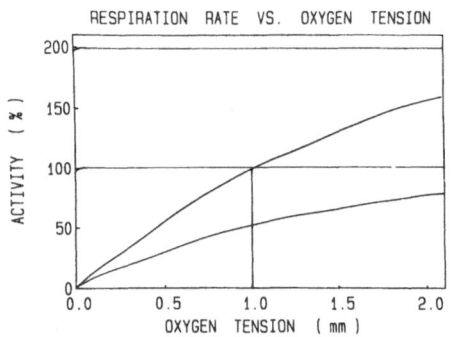

Figure 3. Activity of oxidases as functions of oxygen tension.

cannot be adequately explained by simple passive diffusion alone. A search for other mechanisms (Longmuir, 1977; Longmuir, Young and Mailman, 1978) led to the observation that all the kinetic data could be explained by channels in cells along which oxygen diffuses faster than in water (M. Mochizuki, personal communication). Vanderkooi and Callis (1974) have shown that oxygen moves along membranes about six times as fast as through water. Young (1977) has shown by electron microscopy that the endoplasmic reticulum (e.r.) is continuous with the plasma membrane and the outer mitochondrial membrane. Since the e.r. amounts to about 4% of the volume of liver cells on average, and about twice as much as this in the centrilobular cells which are the most hypoxic, this might be the channel for oxygen. Preliminary observations on the livers of mice exposed to hypoxia (PIO_2 = 100 mmHg) showed striking changes (Betts and Longmuir, 1986). Subsequent work, published in this volume (Betts and Longmuir, 1987), has shown that hypoxia halves the length of liver e.r. without affecting the number of elements. Thus if a significant fraction of intracellular oxygen transport occurs along the e.r. a halving of its length would restore normal oxygen delivery following a halving of capillary oxygen tension.

REFERENCES

Betts, W. and Longmuir, I.S. (1986). Effect of hypoxia on endoplasmic reticulum in mouse liver. Oxygen Transport to Tissue-VIII. Ed. Longmuir, I.S., Plenum Press, New York and London, (Adv. Exp. Med. Biol. 200, 425-427).

Betts, W. and Longmuir, I.S. (1987). Changes in the endoplasmic reticulum following exposure to hypoxia. This volume.

Cerretelli, P. (1976). Limiting factors to oxygen transport on Mount Everest. J. Appl. Physiol. 40, 658-667.

Hurtado, A. (1964). Animals in high altitudes: resident man. In: Handbook of Physiology, Section 4, Adaptation to the Environment. Eds Dill, D.B., Adolph, E.F. and Wilber, C.G., American Physiological Society, Washington, D.C., pp. 843-859.

Krogh, A.S. (1919). The rate of diffusion of oxygen through animal tissues, with some remarks on the coefficient of invasion. J. Physiol. 52, 391-408.

Longmuir, I.S. (1954). The respiration rate of bacteria as a function of oxygen concentration. Biochem. J. 57, 81-87.

Longmuir, I.S. (1977). Search for alternative oxygen carriers. In: Oxygen and Physiological Functions, Ed. Jobsis, F. Professional Information Library, Dallas, Texas, U.S.A.

Longmuir, I.S., and Bourke, A. (1960). The measurement of the diffusion of oxygen through respiring tissue. Biochem. J. 76, 225-229.

Longmuir, I.S. and McCabe, M.G.P. (1964). Evidence for an oxygen carrier in tissue. J. Polarograph. Soc. x, 45-48.

Longmuir, I.S. and McCabe, M.G.P. (1965). Tissue adaptation to oxygen lack. J. Theoret. Biol. 8, 124-129.

Longmuir, I.S. and Pashko, L. (1977). The role of facilitated diffusion of oxygen in tissue hypoxia. Int.J. Biometeor. 27, 179-187.

Longmuir, I.S., Young, A. and Mailman, R. (1978). Induction by hypoxia of a new haemoglobin-like pigment. In: Oxygen Transport to Tissue-III. Eds Silver, I.A., Erecinska, M. and Bicher, H.I., Plenum Press, New York and London, (Adv. Exp. Med. Biol. 94, 297-300).

McCabe, M.G.P. (1964). Tissue respiration and anaesthetics. Ph.D. Thesis, University of London, U.K.

Moore, R.E. (1956). Response of new-born kittens to hypoxia. J. Physiol. 133, 60-70.

Robin, E.D. (1980). Amberson Lecture. Amer. Rev. Resp. Dis. 121. 774.

Torrance, J.D., Lenfant, C., Cruz, J. and Marticorena, E. (1971). Oxygen transport mechanisms in residents at high altitude. Respir. Physiol. 11, 1-15.

Turek, Z., Kreuzer, F. and Hoofd, L.J.C. (1973). Advantage or disadvantage of a decrease of blood oxygen affinity for tissue oxygen supply at hypoxia: a theoretical study comparing man and rat. Pflugers Arch. 342, 185-197.

Vanderkooi, J. and Callis, J.B. (1974). Pyrene. A probe of lateral diffusion in the hydrophobic region of membranes. Biochemistry, 13, 4000-4006.

West, J.B. (1982). American Medical Research Expedition to Everest. Physiologist, 25, 36-38.

Young, A.J. (1977). Facilitated diffusion of oxygen in tissue. Ph.D. Thesis, North Carolina State University, U.S.A.

CHANGES IN THE ENDOPLASMIC RETICULUM FOLLOWING EXPOSURE TO HYPOXIA

W.F. BETTS and I.S. LONGMUIR

Department of Biochemistry, North Carolina State University, Raleigh, NC 26795-7622, U.S.A.

INTRODUCTION

The effects of severe hypoxia on mammals are well documented. If hypoxia occurs abruptly death may follow. However, if the arterial oxygen tension falls gradually the organism may survive. It has become clear that many changes in the tissue occur during such hypoxia. Tissue changes which take place under these conditions may perform an acclimatizing role to assist in the organism's survival. The purpose of this paper is to report a study of changes in hypoxic tissue at the subcellular level that may constitute the mechanism of acclimation.

The nature of oxygen transport has been investigated by many workers. Longmuir and Bourke (1959) have shown that the uptake of oxygen by the liver does not conform to the Warburg Model. To resolve this dilemma, Longmuir and Sun (1970) hypothesized the existence of a tissue oxygen carrier perhaps located in the endoplasmic reticulum (e.r.). In work done by Gold (1969) it was noted that the carrier was more likely to be a fixed site carrier rather than a mobile one. Both these ideas follow a reasonable assumption that the e.r. is continuous with the plasma and mitochondrial membranes and that the fixed site carrier could be the cytochrome-P 450 on the e.r. (Longmuir and Sun, 1970).

Since previous studies have shown that hypoxic liver tissue shows two distinct sets of changes, an increase in cytochrome-P 450 (Longmuir and Pashko, 1976) and various alterations in the e.r. (Betts and Longmuir, 1985), the question arises as to how the changes may be measured accurately and how they may benefit a hypoxic animal. It is the purpose of this paper to explore and evaluate these changes.

MATERIALS AND METHODS

Hypoxic exposure

ICR Dublin mice aged 7-9 weeks were placed in a warm desiccator evacuated to give a PIO_2 of 100 mmHg. A leak was introduced into the chamber in order to maintain the pressure constant and an air flow of at least one $l.min^{-1}$ to remove carbon dioxide. The temperature was checked several times during the exposure to insure that the animal did not become hypothermic. The animals were maintained in this hypoxic environment for four hours.

Tissue preparation and cytochrome-P 450 assay

Control and hypoxic mice were killed by cervical dislocation. The livers were exposed by bilateral abdominal incisions and were perfused with warm normal saline via the portal vein. The livers were then excised and divided, one part being used for electron microscopic (EM) preparation and the other for cytochrome-P 450 determination.

Specimens for EM were placed in 3% glutaraldehyde (GTA) at $4^{\circ}C$ for one hour. After initial fixation the tissue was rinsed several times with buffer and post-fixed with 1% osmium tetroxide. Standard dehydration was carried out with ethanol at $4^{\circ}C$ and the tissue was then infiltrated with 1:1 propylene oxide:Medcast epoxy resin solution. Embedding and polymerization were carried out utilizing Luft's procedure (1961) in the epoxy resin mentioned above. After embedding, the tissue was thin sectioned, placed on grids, and stained. Staining was accomplished by a double staining method utilizing uranyl acetate in 70% ethanol and lead hydroxide chelated with citrate. After staining, the grids were rinsed with distilled water, dried, and allowed to stand overnight before examination.

It was noted in this study that the centrilobular and perilobular cells appeared to have different amounts of e.r. Therefore, a comparison was carried out of each of these areas separately, between normoxic and hypoxic liver tissue.

Analyses of EM micrographs were performed using Weibel's method of curved line stereology (Weibel and Elias, 1967) for studying the rough e.r. (r.e.r.) to remove uncertainty of identification. The number of intersections between evenly spaced grid lines and r.e.r. were counted and used as an indicator of e.r. length in each sample. The total number of e.r. elements was also counted.

A spectrophotometric determination of the concentration of cytochrome-P 450 as described by Pashko (1977) was made on the remaining tissue. The liver tissue separated for this determination was blotted dry, weighed, and homogenized with a Potter-Elvehjem homogenizer. The homogenate was treated with carbon monoxide (CO) to bring about reduction of haemoglobin and cytochrome a, a_3, and a sample was further treated with sodium dithionite to reduce the cytochrome-P 450. The concentration of P 450 was determined by taking the difference spectra between this reference sample in which the reduced P 450 was able to react with CO, and a dithionite-free sample, in which no reaction occurred. The absorbance difference between the two samples was measured at 450 and 490 nm, and by using a millimolar extinction coefficient for cytochrome-P 450 of 91, the concentration was calculated.

RESULTS

Elevation of the concentration of cytochrome-P 450 followed hypoxia in accordance with the findings of Longmuir and Pashko (1976).

Table 1 shows the means of the ratios between the number of intersections to the number of e.r. elements. The figures indicate a difference between perilobular and centrilobular areas in the cell.

These numbers are proportional to the length of the e.r. Although in most samples the centrilobular area of the cell had more elements and more intersections than the perilobular area, these ratios were the same. Most importantly, shown in Table 1 is a comparison of these regions under normoxic and hypoxic conditions. The perilobular normoxic and hypoxic ratios were found to be different ($p < 0.01$) utilizing Student's t-test. Analogous ratios for the centrilobular regions were also found to be significantly different ($p < 0.05$). In both cases the ratio had decreased under hypoxic conditions.

Table 1. Mean ratios of intersections to e.r. elements
(i.e. proportional to the length of each element)

	Hypoxic tissue	Normoxic tissue
Perilobular	** 1.1989 (8)	1.8577 (8)
Centrilobular	* 1.3169 (8)	1.8669 (8)

The differences between hypoxic and normoxic tissue were highly significant (**$p < 0.01$) for perilobular tissue and significant (*$p < 0.05$) for the centrilobular samples. Figures in parentheses denote numbers.

In an additional analysis, the number of polyribosomes per 10 mm length (in the EM photograph) of e.r. element were counted. The data concerning these mice appear in Table 2 which compares the mean values of these measurements. Again a difference ($p < 0.05$) was seen between normoxic and hypoxic tissue, with the hypoxic tissue showing a greater number of polyribosomes.

Table 2. Mean of the number of polyribosomes per 10 mm length of e.r.

	Hypoxic tissue	Normoxic tissue
Set (1)	*10.633 (30)	7.666 (12)
Set (2)	*10.407 (30)	7.333 (12)

The mean number of polyribosomes per 10 mm (photographic length equal to 0.54 micron of cell) length of e.r. element. These numbers were found to be significantly different (*$p < 0.05$) between hypoxic and normoxic samples using Student's t-test.

DISCUSSION

To date there has been no complete three dimensional reconstruction of a high magnification image of the whole cell. Each e.r. picture consists at most of about 0.5% of a cell in our study. Thus it is only possible to arrive at a picture of the whole cell by induction from a series of e.r. pictures. Those that we have examined in this study are consistent with a model of the e.r. double membrane continuous with both the plasma and mitochondrial membranes.

Since the e.r. constitutes about 4% of the volume of the cell, and oxygen diffuses about six times as fast along membranes as through water (Vanderkooi and Callis, 1976) and may diffuse through cytosol about half as fast as through water (Ho, Ju and Ho, 1986), a significant proportion of oxygen will travel along the e.r. This proportion may be increased further by the anatomical relations described earlier.

Our results show that acclimation to hypoxia reduces the length of each e.r. element (i.e. the ratio of intersections to total number). Thus the length of a major oxygen pathway is reduced by about the same amount as the capillary PO_2. Thus oxygen transport by this route will be restored towards its normoxic level.

REFERENCES

Betts, W.J. and Longmuir, I.S. (1986). Effect of hypoxia on the endoplasmic reticulum of mouse liver. In: Oxygen Transport to Tissue-VIII. Ed. Longmuir, I.S., Plenum Press, New York and London, (Adv. Exp. Med. Biol. 200, 425-427).

Gold, H. (1969). Kinetics of facilitated diffusion of oxygen in tissue slices. J. Theoret. Biol. 23, 455-462.

Ho, C.S., Ju, L. and Ho, C.T. (1986). Measuring oxygen diffusion coefficients with polarographic oxygen electrodes. II. Fermentation media. Biotech. and Bioeng. 28, 1086-1092.

Longmuir, I.S. and Bourke, A. (1959). Application of Warburg's equation to tissue slices. Nature, 184, 634-635.

Longmuir, I.S. and Pashko, L. (1976). The induction of cytochrome P-450 by hypoxia. In: Oxygen Transport to Tissue-II. Eds Grote, J., Reneau, D. and Thews, G., Plenum Press, New York and London, (Adv. Exp. Med. Biol. 75, 171-175).

Longmuir, I.S. and Sun, S. (1970). A hypothetical tissue oxygen carrier. Microvasc. Res. 2, 287-289.

Luft, H.J. (1961). Improvements in epoxy resin embedding methods. J. Biophys. Biochem. Cytol. 9, 409-414.

Pashko, L. (1977). Thesis presented to the North Carolina State University Faculty, Raleigh, NC, U.S.A.

Vanderkooi, J.M. and Callis, J.B. (1976). Pyrene. A probe of lateral diffusion in the hydrophobic region of membranes. Biochemistry, 13, 4000-4004.

Weibel, E.R. and Elias, H. (1967). Introduction to stereologic principles. In: Quantitative Methods in Morphology. Eds Weibel, E.R. and Elias, H., Springer-Verlag, Berlin and New York, pp. 89-98.

CEREBRAL MICROCIRCULATORY CHANGES DURING EXPOSURE TO HYPOXIA

P. WEINBRECHT, I. LONGMUIR, J. KNOPP and M. MILLS

Department of Biochemistry, North Carolina State University, Raleigh, NC 27695-7622, U.S.A.

SUMMARY

Changes in the red blood cell content of the cat cerebral cortical vessels were monitored using reflecting light during a decrease in PIO_2. The following observations were made:
1) the capillary bed red blood cell content increased during hypoxia prior to the arteriole response at an arterial oxygen tension of 52.0 mmHg;
2) at arterial oxygen tensions below 52 mmHg the red blood cell content increased in all vessels simultaneously;
3) as the PIO_2 decreased, the increase in red blood cell content occurred earlier in all vessels.

Capillaries, therefore, can respond to hypoxia independently of the arteriole in cat cerebal cortical tissue.

MATERIALS AND METHODS

Female cats were anaesthetized with Nembutal I.P. (30 mg/kg), and anaesthesia was maintained throughout the experiment with I.P. or I.V. injections of Nembutal (50 mg/ml). The depth of anaesthesia was determined by pupil reflex to light, and respiration rates. The aim was to keep the animal lightly but adequately anaesthetized. A skull window was surgically implanted over the sensory cortex for observations of light reflection from the cortex during the inspiration of hypoxic gases (Longmuir, Knopp and Pittman, 1984). Catheterization was performed for drug administration I.V. and for arterial blood gas analysis. The arterial catheter was inserted up to the thoracic aorta. The animal was mechanically ventilated after paralysis was induced by administration of flaxedil.

The brain surface was illuminated with light from a mercury arc filtered to obtain incident light isosbestic to oxy- and deoxyhaemoglobin. The reflectance signal was detected with a Silicon Intensified Target vidicon. The data collection and analysis was controlled by an Apple II microprocessor (Weinbrecht et al., 1985). The cerebral microvasculature was identified on the basis of oxygenation as indicated by the colour difference when exposed to unfiltered white light, and the pulsatile nature of some arterioles. A 3.3X objective was used giving a 10 micron resolution limit. Thus any areas showing no vessels contained only capillaries and cortical tissue (Benson, Knopp and Longmuir, 1981; Weinbrecht et al., 1985). The reflectance light penetrated to a depth of 600 microns, and the tissue was in focus to a depth of 3000 microns, therefore, all illuminated tissue was in focus (Longmuir, Knopp and Weinbrecht, 1985). The reflectance analysis was performed on the video recording of the experiment. The arteriole and venule reflectance scans were performed on a volume of tissue that included the vessel and some surrounding areas. The capillary scan contained no discernible vessels. The time course of change in the reflectance signal during hypoxia was the mean reflectance of the scanned tissue calculated repetitively over time. The time of initial increase in RBC content was determined by different analytical methods performed on the raw data and the data after binomial smoothing (Weinbrecht et al., 1985).

Hypoxia was induced after a stable reflectance reading was obtained on air, and blood gas analysis confirmed a normoxic and normocapnic animal. The hypoxic gas was made from mixing 100% oxygen and 100% nitrogen. This gas was diluted with 100% nitrogen to create gases for deeper hypoxia. Each induced hypoxic gas cycle involved the inspiration of air, then the hypoxic gas, followed by 100% oxygen, and finally air, with a full recovery of the reflectance signal before the next hypoxic gas cycle. Blood gas analysis during the hypoxic gas confirmed the degree of arterial hypoxia. It was possible to cycle the animals through a hypoxic gas and back to normoxic reflectance levels only when the length of hypoxic gas exposure did not exceed five minutes. Three minutes exposure to a hypoxic gas seems to be sufficient for a qualitative measurement such as the time course of the increase in red blood cell content in the microcirculation.

RESULTS

The time of initial decrease of reflectance in the cerebral microvasculature and the induced change in the arterial blood gases for each of the three hypoxic gas cycles are shown in Table 1. The diameter of the vessels and the dimensions of the reflectance scane are also listed. In all three hypoxic gases, the arterial oxygen tension was more than 40% below normoxic levels and it is doubtful if the third hypoxic gas was compatible with life. On inspiration of the first hypoxic gas, the increase in RBC content occurred in the capillaries before that in the other vessels investigated. Inspiration of a more hypoxic gas promoted simultaneous increases of RBC content in all types of vessel. It was also found that as arterial oxygen tension decreased the latency of the increase in RBC content shortened in all vessel types.

Figures 1 to 3 show the reflectance profiles as a function of time during exposure to the hypoxic gas (arterial oxygen tension of 52.0 mmHg) for the arteriole, capillary and venule respectively. The time of blood gas sampling is indicated along with the range in the time at which the initial reflectance decrease occurred. It was also observed that under greater hypoxia not only did the increase in RBC content occur earlier and simultaneously in all vessels, but the reflectance profile changed its character towards a more sharply falling signal.

Table 1. Time of reduced reflectance after change in PIO_2

VESSEL (diameter) [scan volume]	HYPOXIC GAS 52.0 mmHg* secs**	HYPOXIC GAS 44.6 mmHg* secs**	HYPOXIC GAS 26.7 mmHg* secs**
ARTERIOLE (50 – 60 μ) [20 x 350 x 600 μ³]	53.8 ± 5.6	38.3 ± 4.3	28.4 ± 3.9
SMALL ARTERIOLE (< 20 μ) [100 x 250 x 600 μ³]	54.4 ± 4.3	35.8 ± 2.6	30.4 ± 3.5
CAPILLARY BED [200 x 200 x 600 μ³]	42.9 ± 2.9	38.0 ± 2.4	27.0 ± 1.5
SMALL VENULE (< 20 μ) [100 x 100 x 600 μ³]	47.0 ± 4.1	37.0 ± 1.4	29.7 ± 0.7
VENULE (40 – 50 μ) [100 x 270 x 600 μ³]	53.1 ± 5.2	37.2 ± 2.0	29.5 ± 3.5

*Arterial oxygen tension
**Time represents the mean ± the range from various techniques used to determine the time of $\overline{RBC}$ content increase.

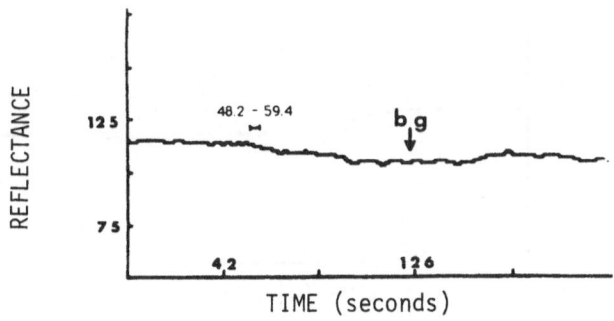

Figure 1. ARTERIOLE. Time course of the reflectance signal change from a cerebral arteriole. The time of arterial blood gas sampling is denoted by bg; ⊢ represents the range in the time at which the initial decrease in reflectance signal occurred.

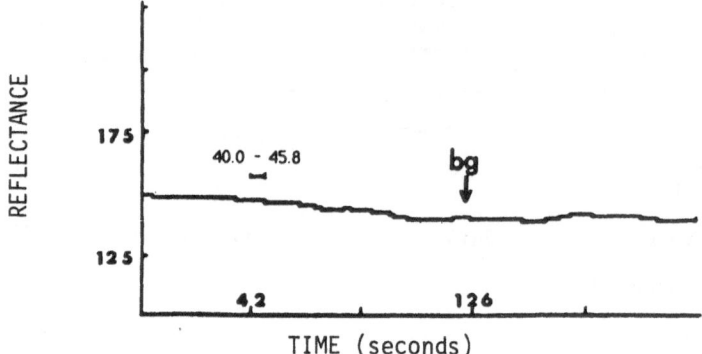

Figure 2. CAPILLARY. Time course of the reflectance signal change from a cerebral capillary. The time of arterial blood gas sampling is denoted by bg; ⊢ represents the range in the time at which the initial decrease in reflectance signal occurred.

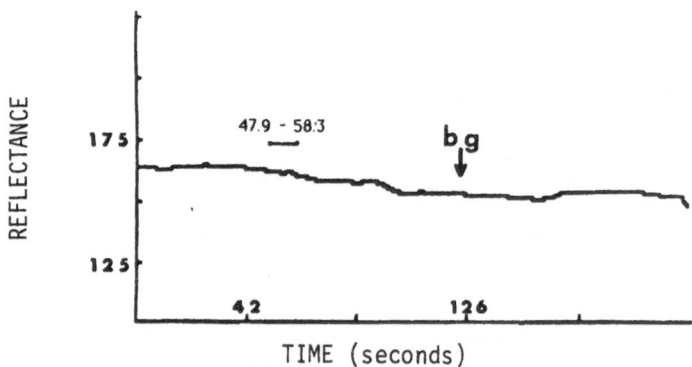

Figure 3. VENULE. Time course of the reflectance signal change from a cerebral venule. The time of arterial blood gas sampling is denoted by bg; ⊢ represents the range in the time at which the initial decrease in reflectance signal occurred.

262

DISCUSSION

These results show that, in the cat brain, a change in the capillary bed perfusion is the initial response to local hypoxia. The results also show that at arterial oxygen tensions below 52.0 mmHg the capillary bed RBC increase no longer precedes the changes in the other vessels. The observation of deeper hypoxia causing an earlier vascular response and the simultaneous increase in RBC content of the capillary bed, the arterioles, and the venules indicates that all the autoregulatory mechanisms can detect hypoxia but come into play at different levels and that a threshold exists above which the increase in capillary RBC content can precede an arteriole response.

It should be emphasized that the observed vascular responses may be modified by the drugs used during the experiment. Drug metabolism for each experimental animal is dependent on enzyme levels that are themselves sensitive to parameters such as age, sex, and environment, and this variation can easily create differences in a response sensitive to drug levels. Vascular tone and its effect on autoregulatory capacity is such a response.

Future experiments are to be performed under milder hypoxia with cycling of the animal to deeper levels of hypoxia and back again, to prove the repeatability of the response. Drugs are to be administered between repetitive cycles to determine their effect on the response.

ACKNOWLEDGEMENT

This work was supported in part by NIH grant HL 16828

REFERENCES

Benson, D.M., Knopp, J.A. and Longmuir, I.S. (1981). Effect of optical configuration and tissue absorbance on the depth of light penetration in light microscopy. Fed. Proc. 40, 587.

Longmuir, I.S., Knopp, J.A. and Pittman, J.L. (1984). Changes in cerebral oxygen tension and red cell content on sensory stimulation. In: Oxygen Transport to Tissue-VI. Eds Bruley, D., Bicher, H.I. and Reneau, D., Plenum Press, New York and London, (Adv. Exp. Med. Biol. 180, 185-190).

Longmuir, I.S.,Knopp, J.A. and Weinbrecht, P.T. (1985). Capillary red cell residence as a measure of tissue oxygen delivery. In: Oxygen Transport to Tissue-VII. Eds Kreuzer,F., Cain, S.M., Turek, Z. and Goldstick, T.K., Plenum Press, New York and London, (Adv. Exp. Med. Biol. 191, 917-921).

Weinbrecht, P.T., Johnson, L., Longmuir, I.S. and Knopp, J.A. (1986). Influence of PaO_2 on cerebral macro- and microcirculation as observed by light reflection: time course of changes. In: Oxygen Transport to Tissue-VIII. Ed. Longmuir, I.S., Plenum Press, New York and London, (Adv. Exp. Med. Biol. 200, 131-136).

METABOLIC, IONIC AND ELECTRICAL ACTIVITIES DURING AND AFTER INCOMPLETE OR COMPLETE CEREBRAL ISCHAEMIA IN THE MONGOLIAN GERBIL

A. MAYEVSKY and N. ZARCHIN

Department of Life Sciences, Bar-Ilan University Ramat-Gan 52 100, Israel

INTRODUCTION

During the last 10 years the Mongolian gerbil (<u>Meriones unguic-ulatus</u>) has been adopted as an animal model for cerebral ischaemia because of the incompleteness of its Circle of Willis (Levy and Brierley, 1974; Levy, Brierley and Plum, 1975; Kobayashi, Lust and Passonneau, 1977; Levy and Duffy, 1977). The posterior connection between the vertebro-basilar and the carotid arteries is missing in all animals hence complete cerebral ischaemia develops after bilateral carotid artery occlusion. The degree of connection between the two anterior cerebral arteries varies in different individuals, so with unilateral carotid artery occlusion, the level of ischaemia developing in the ipsilateral hemisphere can vary between 0 and 100%. Using fibre optic surface fluorometry we monitored the intramitochondrial NADH redox state under partial or complete ischaemia in the gerbil brain (Mayevsky, Lebourdais and Chance, 1980; Mayevsky, Zarchin and Friedli, 1982; Mayevsky and Chance, 1983). The purpose of the present study was: 1) to correlate the degree of ischaemia developed under unilateral and bilateral carotid artery occlusion with the impairment of ionic homeostasis developed under these conditions; 2) to test the recovery rate of the extracellular K^+ level (K^+_e) under the same conditions.

In order to evaluate the functional state of the brain a multiprobe assembly was used (Friedli, Sclarsky and Mayevsky, 1982). Energy metabolism was evaluated by monitoring the intramitochondrial NADH redox state simultaneously with tissue surface PO_2. The reflected light (366 nm) measured simultaneously with the fluorescence signal (450 nm) served as an indirect measure of blood volume changes in the observed area. The ionic activity was monitored by measuring the extracellular K^+ levels using the valinomycine type of electrode. Electrical activities were evaluated by bipolar electrocorticography (ECoG) and the DC steady potential.

METHODS

The multiprobe assembly (MPA) used in the present study was described in detail in our previous publication (Friedli et al., 1982) hence only a few major points will be described. The MPA has the following main features: 1) all components have the same type of non-invasive surface contact with the cortex; 2) it can very easily be removed from the skull at the end of the experiment without any damage, so that repetitive experiments can be performed in a short period of time; 3) all signals monitored by the MPA have a very low sensitivity to movements of the animal, even while exposed to hyperbaric pressure.

Figure 1 shows schematically the connection between the MPA and the cerebral cortex of the gerbil brain. The probes are mounted in a lucite cannula (c) which can be cemented directly to the skull. The holes accommodating the specific sensing electrodes (K^+, PO_2) converge at the implantation site to occupy less space in the brain, and diverge at the top of the cannula to leave maximum room for mounting and replacing the electrodes. Each sensor chamber is pierced

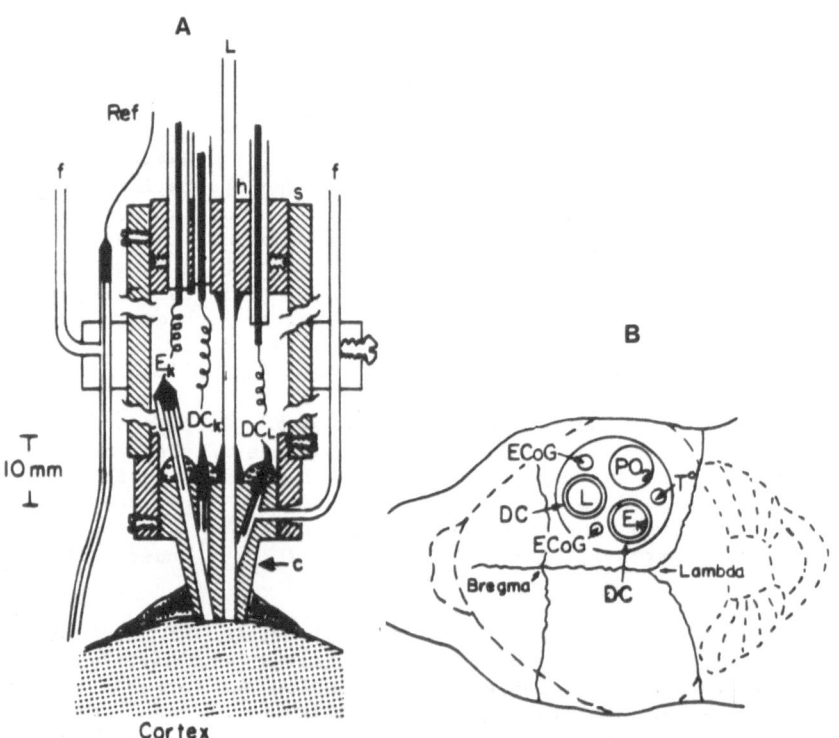

Figure 1.   Schematic presentation of the multiprobe assembly used in the present study.   Part A shows a longitudinal section of the cannula and assembly implanted over the gerbil cerebral cortex.   B shows a surface view of the gerbil skull. Ref, reference Ag/AgCl electrode; f, refill tube for Ref or DC electrode; c, Lucite cannula; s, Plexiglass sleeve; L, light guide; h, cable holder; E_k, K^+ electrode; DC_k, DC_1, Ag/AgCl electrode; PO_2, O_2 electrode; ECoG, electrocorticogram electrode (from Mayevsky, 1983).

approximately half way down by a hole drilled obliquely from the top which houses an Ag/AgCl wire to record DC potential around the sensing electrode. One more hole drilled horizontally into the DC wire chamber allows for filling with saline and for cleaning ECF and blood from around the electrodes at the end of an experiment. A lucite sleeve (s) placed over the entire assembly is held in place with set screws. This sleeve holds the assembly together as one unit, protects the electrodes and wiring, and also provides shielding for the high impedance K^+ electrode. The sensing electrodes, Ag/AgCl wires for DC measurement, stainless steel wires for ECoG measurement, and a polyethylene-sleeved temperature probe are held in and sealed to the cannula with 5-minute epoxy, or by melting wax around them. Another block of lucite attached to the outer shield holds the reference assembly and the tubes (f) for filling the DC chambers.

Surface fluorometry

The intramitochondrial NADH redox state was monitored by a light guide DC fluorometer/reflectometer (Mayevsky and Chance, 1983). The source for the 366 nm light was a 100 watt mercury arc cooled by water. The emitted light was split in a 90:10 ratio for the measurement of the NADH fluorescence and reflectance, respectively. The common part of the light guide was cemented inside the multiprobe assembly in the same way as the electrodes. A branch of the excitation and emission fibre bundle was connected to the common part.

Animal preparation

The gerbil was anaesthetized by I.P. injection of Equithesin (0.3 ml/100 g body weight). (Each ml contains: pentobarbital, 9.72 mg; chloral hydrate, 42.51 mg; magnesium sulphate, 21.25 mg; propylene glycol, 44.34% w/v; alcohol, 11.5%, and water). The skull was exposed through a midline incision and a hole of 6.0 mm diameter was drilled in the right parietal bone. Four holes (1.5 mm diameter) were drilled in various points of the skull and stainless steel screws were used to hold the cement to the skull. The dura mater was removed carefully, only from the area where the K^+ electrode was placed. The MPA was held in a micromanipulator and located over the cortex without creating any extra pressure on the brain. The ground and reference electrodes were placed below the skin and were cemented to the skull, together with the MPA and the screws, by dental acrylic cement.

RESULTS

Figure 2 shows the effects of complete ischaemia induced by bilateral carotid artery occlusion ($L+R_{occl}$), on the various parameters monitored from the gerbil brain. Upon occlusion, tissue oxygen level decreased to the anoxic level and stayed low until reperfusion started ($L+R_{open}$). As oxygen became limited, the NADH level (CF) increased and reached its maximum soon after the occlusion. During the occlusion a large increase in the reflectance signal (R) was recorded, which led to an almost similar change in the uncorrected fluorescence (F). The NADH level recovered to its pre-ischaemic value within 2 min after the reperfusion started. Due to the secondary reflectance increase (SRI), the NADH trace showed an apparent oxidation although the PO_2 at this time was at the anoxic level. The recovery of the reflectance signal exhibits two-step kinetics: immediately after reopening the blood vessels a fast decrease was recorded, followed by a further slow decrease, which recovered afterwards to the pre-ischaemic level. The effect of the occlusion on the extracellular K^+ level (E_{K^+}) shows clear two-step kinetics. Soon after the occlusion a slow gradual increase in E_{K^+} was recorded followed by a rapid large increase.   During   the   first   phase, E_K+ changed from approximately 3 mM to 6 mM, whereas during the second phase it reached the level of 40–50 mM.

The   rapid   increase   in E_K+ was   concomitant   with   the   negative shift   in   the   DC   steady   potential   measured   around   the   potassium electrode (DC_K+) as well as around the light guide (DC_F). Both the E_K+ and the DC steady potential recovered to pre-ischaemic levels within 10 min after reopening the two carotid arteries. The ECoG trace shows that the EEG disappeared within 30 sec after bilateral carotid occlusion.   The recovery of the EEG (not shown in the figure) was much slower than any other parameter measured and took about 30 min in this particular ischaemic episode.

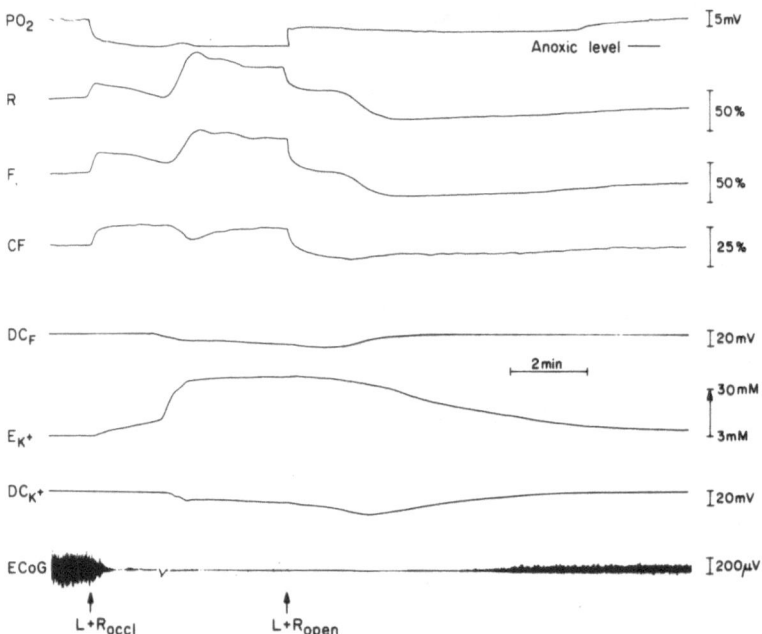

Figure 2.   Effects of bilateral carotid artery occlusion ($L+R_{occl}$) on metabolic, ionic and electrical activities measured from the right hemisphere of the gerbil brain. PO_2, tissue oxygen partial   pressure;   R,   reflectance;   F,   fluorescence;   CF, corrected   fluorescence; E_K+,   extracellular K^+ level; DC_K+ and DC_F, DC steady potential measured around the K^+ electrode and the light guide respectively; ECoG, elec-trocorticogram.   The PO_2 anoxic level was determined by exposing the gerbil to 100% N_2 for 1 min.

Figure 3 shows the response of a gerbil brain to 5 min unilateral carotid   artery   ligation.   The   main   difference   between   the   responses   in this ischaemic episode and those shown in Figure 2 is due to the lower level of ischaemic insult.   The PO_2 trace shows an initial decrease followed by a slow recovery due to compensatory flow through the con-tralateral anterior cerebral artery.   The same pattern of events is seen in the NADH redox state. As soon as the large and rapid increase in E_K+ occurred, the SRI was recorded in the reflectance trace simultaneously with a sharp decrease in PO_2.

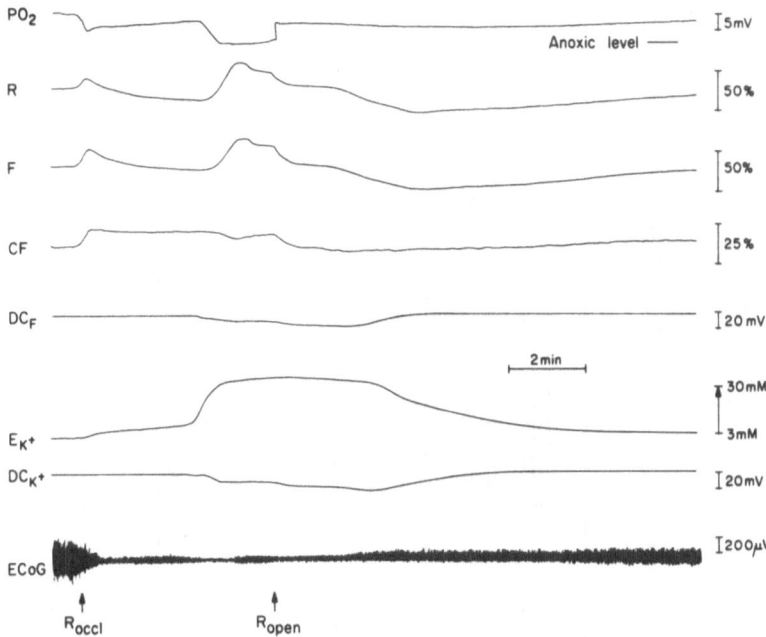

Figure 3. Responses of the various parameters, measured in the same gerbil as in Fig. 2, to unilateral carotid artery occlusion (R_{occl}).

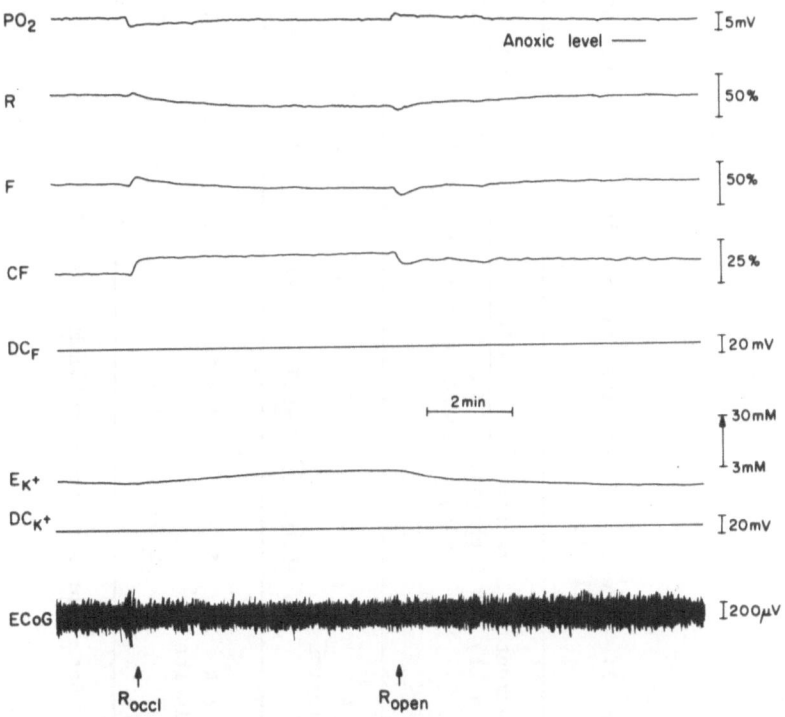

Figure 4. The effects of unilateral carotid artery occlusion (R_{occl}) on the metabolic, ionic and electrical activities measured in a different gerbil brain from that in Figs 2 and 3.

Table 1. Effects of partial or complete ischaemia on NADH redox state and extracellular K^+ levels during occlusion

Treatment (n = 11)	Duration of ischaemia (min)	NADH		K^+_e during Phase I			K^+_e during Phase II		
		% increase	% ischaemia	Base line (mM)	mM increase	Duration (min)	Max increase (mM)	Time to max (min)	Time to max/2 (min)
R occlusion	5.1 ±0.04	8 ±0.7	66 ±6	3.1 ±0.3	4.5 ±0.4	2.1 ±0.3	41 ±2.6	2.8 ±0.3	0.62 ±0.05
L + R occlusion	5.1 ±0.06	13*** ±0.5	100	3.3 ±0.4	6.1** ±0.7	1.4* ±0.1	49* ±3.8	3.6 ±0.2	0.61 ±0.05

Data represent Means ± S.E.M.   * p< 0.05;   ** p< 0.01;   *** p< 0.001

Leakage of K^+ from the cells in Phase I of the insult was slower and smaller compared with that produced by bilateral occlusion. The effect of partial ischaemia on the ECoG activity was also smaller and recovery was faster. Figure 4 shows a different type of response to unilateral occlusion found in 40-50% of the gerbils used. The level of ischaemia achieved was smaller and, as a result, probably the changes in E_K^+ were more gradual and did not reach the second phase of large K^+ leakage. The ECoG trace shows only a very small change, if any. The analog signals recorded from 11 gerbils included in this study were analysed for quantitative changes in NADH redox state and E_K^+ and are presented in Tables 1 and 2. As can be seen in Table 1, the level of ischaemia produced by unilateral artery occlusion was significantly lower than that produced by bilateral occlusion. During Phase I of potassium leakage, two significant differences were found between partial and complete ischaemia as well as differences in the maximum level of E_K^+.

Table 2 presents the recovery of the reflectance trace and the E_K^+ after ischaemic episodes. The time to reach the second reflectance decrease was longer, but was not significant. All parameters representing the recovery of E_K^+ show it was slower in the completely ischaemic episodes but only the time to reach the E_K^+ minimum level was significantly longer. The new E_K^+ level after recovery was not different from the pre-ischaemic values with either insult.

Table 2. Recovery of extracellular K^+ levels after 5 min partial or complete ischaemia

Treatment	Reflectance	K^+_e, Time (min) to:			K^+_e New
(n = 11)	Time to second decrease (min)	Initial decrease	Minimum	Minimum 2	baseline level (mM)
R Reopening	2.8	1.58	13.2	3.9	2.95
	+0.7	+0.44	+1.7	+0.6	+0.34
L + R Reopening	3.8	2.8	21.6**	6.5	3.06
	+0.8	+0.9	+2.8	+1.3	+0.38

Data represent Means $\pm$ S.E.M.   ** p< 0.01

DISCUSSION

The Mongolian gerbil provides a useful model for studying brain responses to partial or complete ischaemia (see Introduction for references). In our previous study (Mayevsky, Friedli and Reivich, 1985) we compared the effects of 5 min and 30 min complete cerebral ischaemia induced by bilateral carotid artery occlusion. In the present experiments we compared the events which occurred in the brain under partial or complete cerebral ischaemia in the same animal. The effects of bilateral occlusion were as expected from previous work but with unilateral carotid artery occlusion two very different responses were recorded. The reason for the variation between animals is not only that the level of ischaemia induced differs in different animals, but also involves other factors, such as membrane permeability to K^+. The type of response is very much dependent on the level of ischaemia induced. Due to the ischaemic depolarization that developed in about 50% of the gerbils exposed to unilateral carotid artery occlusion, the functional state of the brain will be completely different as can be seen in Figures 3 and 4. The results found can be summarized as follows: 1) bilateral occlusion led to the development of complete ischaemia, while after unilateral carotid occlusion the level of ischaemia (as evaluated by the NADH redox state) was only 50-60%; 2) extracellular K^+ level under bilateral occlusion showed a two-step increase reaching its maximal level (40-50 mM) within 3.5 min; 3) under partial ischaemia, two types of response in K^+_e were noted, the first was similar to that recorded under complete ischaemia but in the second, only a small monotonic increase in K^+_e was seen; 4) during the recovery phase, the kinetics of K^+_e toward the base line level were significantly faster after partial ischaemia as compared with complete ischaemia. We concluded that after incomplete ischaemia, the recovery processes were faster and this was borne out by the recovery of the ECoG traces.

ACKNOWLEDGEMENTS

Supported by a grant from the Chief Scientist's Office, Ministry of Health, Israel; by the Health Sciences Research Center, Department of Life Sciences, Bar-Ilan University, and by the NIH Grant NS-22881, RR-02305; NIH SBIR Grant NS-22309 and Advanced Technology Center of South-Eastern Pennsylvania.

REFERENCES

Friedli, C.M., Sclarsky, D.L. and Mayevsky, A. (1982). A new multiprobe assembly for surface monitoring of ionic, metabolic, and electrical activities in the awake brain. Am. J. Physiol. **243**, R462-R469.

Kobayashi, M., Lust, W.D. and Passonneau, J.V. (1977). Concentrations of energy metabolites and cyclic nucleotides during and after bilateral ischemia in the gerbil cerebral cortex. J. Neurochem. **29**, 53-59.

Levy, D.E. and Brierley, J.B. (1974). Communications between vertebro-basilar and carotid arterial circulations in the gerbil. Exp. Neurol. **45**, 503-508.

Levy, D.E., Brierley, J.B. and Plum, F. (1975). Ischaemic brain damage in the gerbil in the absence of 'no-reflow'. J. Neurol. Neurosurg. Psychiat. **38**, 1197-1205.

Levy, D.E. and Duffy, T.E. (1977). Cerebral energy metabolism during transient ischemia and recovery in the gerbil. J. Neurochem. **28**, 63-70.

Mayevsky, A. (1983). Multiparameter monitoring of the awake brain under hyperbaric oxygenation. J. Appl. Physiol. **54**, 740-748.

Mayevsky, A. and Chance, B. (1983). Multisite measurements of the NADH redox state from cerebral cortex of the awake animal. In: Oxygen Transport to Tissue-IV. Eds Bicher, H.I. and Bruley, D.F., Plenum Press, New York and London, (Adv. Exp. Med. Biol. 159, 143-155).

Mayevsky, A., Friedli, C.M. and Reivich, M. (1985). Metabolic, ionic and electrical responses of the gerbil brain to ischemia. Am. J. Physiol. 248, R99-R107.

Mayevsky, A., Lebourdais, S. and Chance, B. (1980). The interrelation between brain pO_2 and NADH oxidation-reduction state in the gerbil. J. Neurosci. Res. 5, 173-182.

Mayevsky, A., Zarchin, N. and Friedli, C.M. (1982). Factors affecting the oxygen balance in the awake cerebral cortex exposed to spreading depression. Brain Res. 236, 93-105.

INSIDE A QUANTITATIVE ANALYSIS OF IN VIVO NEAR INFRARED MONITORING

F. CARTA, M. FERRARI* and I. GIANNINI

Eniricerche, Monterotondo, Rome, and *Laboratorio di Fisiopatologia di Organo di Sistema, Istituto Superiore di Sanita, Rome, Italy

INTRODUCTION

Measuring human brain function is one of the most formidable scientific/engineering endeavours. Recently spectacular methods such as CT, NMR and PET have eclipsed the importance of relatively simple continuous non-invasive techniques based on the analysis of electromagnetic signals and/or optical features of brain tissue.

Brain spectrophotometric and fluororeflectometric observations have received great attention from physiologists and clinicians over the past twenty years. In particular spectroscopy using near infrared (NIRS) wavelengths (w.l.) (750-900 nm) has been used to view the cerebrum non-invasively since the skull is relatively transparent at these wavelengths. This technique was introduced for animals and humans by Jobsis and co-workers (Jobsis, 1977) and independently developed by ourselves (Giannini et al., 1982).

Haemoglobin (Hb) and cytochrome-c-oxidase (EC 1.9.3.1.) (cyt a, a_3) can be used as markers for cerebral Hb content (HbV) and oxygenation (HbS), as well as intracellular oxygen availability related to cyt a,a_3 redox level (R).

The experimental and pilot clinical studies thus far performed indicate that the use of more accurate models in the analysis of the spectroscopic signals, the development of more refined instrumentation and a further experimental validation of the sensitivity and specificity of this methodology are necessary steps towards NIRS use in the clinical field.

We present here some recent data from work performed along these lines.

BIOCHEMICAL AND PHYSIOLOGICAL BACKGROUND

The basis of NIRS for HbV, HbS and R measurements relies on the difference in optical absorption spectra of a) deoxygenated Hb, b) oxygenated Hb (HbO_2), c) oxidized cyt a,a_3, and d) reduced cyt a,a_3, as well as brain tissue scattering effects.

275

Hb content of the human brain can be estimated to be 0.5 g/100 g (30 uM/100 g) on the basis of: a) a cerebral blood volume (CBV) of 4.8 ml/100 g (Sakai et al., 1985); b) a mean cerebral haematocrit of 31.3 (Sakai et al., 1985) and c) a mean Hb concentration of 34 g/100 ml of red blood cells. However, it must be remembered that CBV values are affected by changes in blood PCO_2 and/or PO_2; figures ranging from -10% to + 20% are frequently observed.

In the IR spectral region, the Hb isosbestic point is 810 nm with a molar absorption coefficient of 220 μmol per cm (Takatani and Graham, 1979). Under normal physiological conditions about 98% of the O_2 present in blood is combined with Hb. The amount of O_2 bound to Hb in the arterial blood is usually expressed in terms of percent saturation (SaO_2). Arterial PO_2 and SaO_2, both of which are good indicators of blood oxygenation, have a different physiological meaning; obviously direct measurements of SaO_2 are more accurate than values calculated from PO_2.

HbS does not correspond to SaO_2 because it is a function of blood volume distribution between the arterial and the venous compartment and of regional blood flow (rCBF) both of which affect cerebral arteriovenous oxygen difference. CBV and rCBF are affected by changes in arterial carbon dioxide tension due to CO_2 inhalation or hyperventilation (Greenberg et al., 1978). The primary response to cerebral hypoxia is an increase of CBF due to arteriolar dilatation and an increase in the number of perfused capillaries (capillary recruitment) which results in a CBV increase (Francois-Dainville, Buchweitz and Weiss, 1986).

Cyt a,a_3 is the terminal oxidase of the electron-transport system of all higher organisms and represents the most complicated member of the respiratory chain. Four known redox centres have been identified, namely haem a associated with Cu_A and haem a.3 with Cu_B. In addition, the constant presence of extra copper (Cu_X), zinc and magnesium has recently been described (Reddy, Hendler and Bunow, 1986). Cyt a,a_3 catalyses the electron transfer from cytochrome c to dioxygen, which is reduced to water. The energy made available by this reaction is used for ATP synthesis. This interaction between electron transfer and energy transduction has for many years stimulated intensive studies involving different techniques. The smallest functional unit contains two haem groups and two copper atoms. The copper Cu_A is also known as the 'visible' copper of cyt a,a_3 and Cu_B as the IR 'invisible' one. This is based on optical and electron paramagnetic resonance investigations (EPR), which led to the assigning of the 830-850 nm spectrum to the Cu_A^{2+} cupric ion atom. A possible small contribution (about 10%) from Cu_B^{2+} is controversial (Beinert et al., 1980). According to the previously published spectra (Brunori, Antonini and Wilson, 1981) the molar absorption coefficient of the oxidized chromophores of the enzyme should be 1800 μmol per cm at 830 nm.

All these studies were usually performed on the purified enzyme extracted from heart of different animals. Preparation of cyt a,a_3 as well as of brain mitochondria is a difficult problem due to heterogeneity of the central nervous system and the presence of large amounts of lipid and membrane material. Up to now at least two nonsynaptic and two synaptic mitochondrial populations have been isolated from rat brain (Lai and Clark, 1979).

According to studies performed in experimental animals or carried out on post-mortem human specimens, the amount of brain mitochondrial cyt a,a₃ is regulated by the long-term functional activity of anatomical systems. It is difficult to extrapolate these values for intact brain and reliable data on the mitochondrial proteins/tissue weight ratio are lacking. Recent studies on the distribution of rat brain cyt a,a₃ activity have not revealed any significant difference, but in general, the activity in grey matter was three to five times greater than in white matter, without relevant variations among grey matter structures (Darriet, Der and Collins, 1986). Unfortunately similar data are not available for humans and, in addition, these findings do not fit with known copper concentrations in human grey and white matter (Greiner, Chan and Nicolson, 1975) which suggests that in man copper-enzymes are not homogeneously distributed within the brain regions.

SIMULATED IN VIVO NEAR INFRARED SPECTRA

According to the literature it is possible to calculate the concentrations in brain tissue and relative spectral contributions in the near infrared region of the four chromophores described above.

We have performed a series of computer simulations in order to investigate the most probable spectral variations due to changes in cerebral cyt a,a₃ and Hb under different physiological conditions.

Table 1. Estimated percentage changes of cerebral haemodynamic parameters and concentrations of the brain chromophores during different respiratory conditions corresponding to mild and to more consistent assumed haemodynamic changes.

Conditions	$\triangle$CBV%	$\triangle$CBF%	Hb%	HbO$_2$%	Cyt a,a$_3$ Ox%	Cyt a,a$_3$ Red%
Normoxia	0	0	32	68	55	45
A. Mild haemodynamic changes						
Hyperoxia (100% O$_2$)	0	0	27	73	55	45
Moderate Hypoxia (10% O$_2$-90%N$_2$)	8	10	35	65	50	50
Hypercapnia (5% CO$_2$-95% O$_2$)	13	60	16	84	70	30
B. Consistent haemodynamic changes						
Hyperoxia (100% O$_2$)	0	−15	35	65	55	45
Moderate Hypoxia (10% O$_2$-90%N$_2$)	10	35	28	72	48	52
Hypercapnia (5% CO$_2$-95%O$_2$)	20	70	13	87	75	25

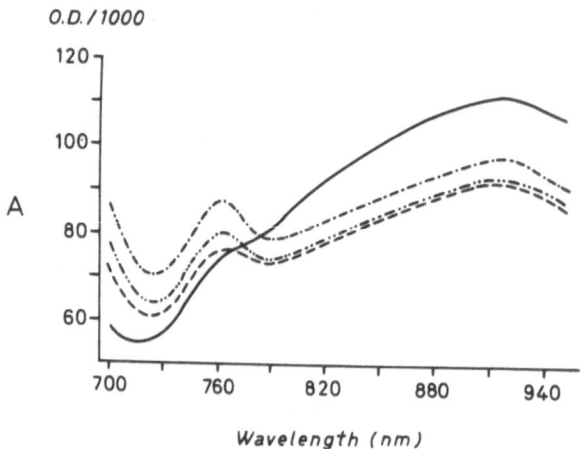

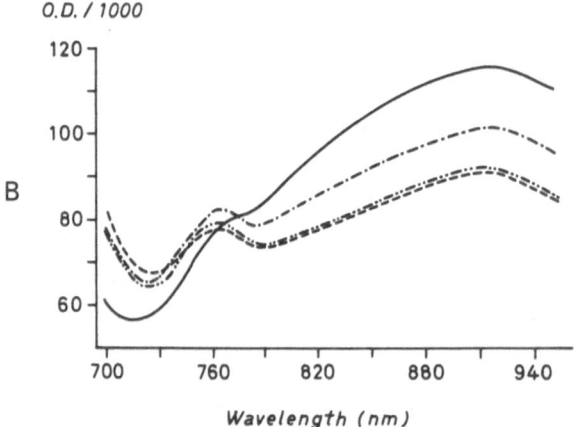

Figure 1.   Calculated   cerebral   absolute   spectra   during   different
              respiratory   conditions.   HB   and   cyt   a,a3   relative   concen-
              trations   and   CBV/CBF   changes   are   reported   in   Table   1.
              Curve   a   (-·-)   normoxic   normocarbia.   Curve b (------)
              hyperoxia   (100% O_2).   Curve   c   (-·-·)   moderate   hypoxia
              (10% O_2-90%N_2).   Curve   d   (_____)   hyperoxic   hyper-
              capnia   (5% CO_2-95% O_2).   The   two   sets   of   spectra   are
              referable   to two steady state conditions in which mild (A)
              and   consistent   (B)   haemodynamic   changes   occur,   assuming   a
              constant   tissue   oxygen   consumption.

The absolute spectra in Figures 1A and 1B have been calculated by assuming different degrees of oxygen and carbon dioxide supply as well as different values of CBV and CBF (see also Table 1). Absorption values during normoxic normocarbia (curve a) were calculated using cerebral Hb and cyt a,a_3 concentrations of 30 and 9 $\mu M/100$ g respectively. The ratio between Hb and cyt a,a_3 concentrations is compatible with our previous studies on perfluorocarbon exchange-transfused rats. Other curves are obtained by changing Hb concentrations and saturation as well as R according to the well known physiological responses to moderate hypoxia (10% O_2, curve c), hyperoxia (100% O_2, curve b) and hyperoxic hypercapnia (5% CO_2-95% O_2, curve d). From an overall evaluation of the four spectra of Figures 1A and 1B it appears that <u>in vivo</u> HbV changes can be easily determined by taking into account the fact that Hb is the most abundant chromophore of this spectral region and that the Hb isosbestic point is not affected by variation of Hb oxygen saturation. Conversely the 750-770 nm wavelength region is extremely sensitive to Hb oxygen saturation. It is significant that the spectral data of Figures 1A and 1B correspond to the measurements carried out on human volunteers and patients.

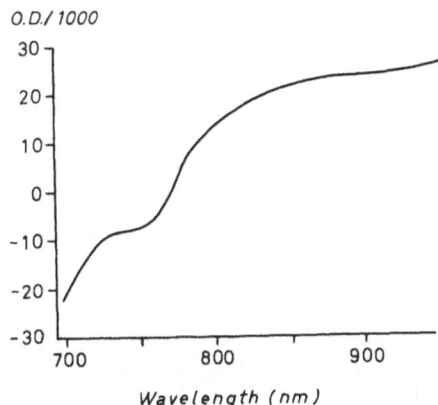

Figure 2. Differential spectrum between Curve d and b of Fig. 1B. The main contribution is the different cyt a,a_3 oxidation level between hyperoxic hypercapnia and hyperoxia.

More complex is a reliable evaluation of cyt a,a_3 and scattering contributions in the same spectral region. While brain cyt a,a_3 concentration can be considered constant, R can change over a relatively wide range (Sylvia, Piantadosi and Jobsis-VanderVliet, 1985). Scattering is certainly the most relevant process affecting light propagation in the tissue but only small changes in apparent OD caused by diffusion have been observed in animal experiments and they have been evaluated by a parametric model.

Cyt   a,a$_3$ R changes are clearly shown in Figure 2 which
exhibits the differential spectrum between curve d and b of Figure 1B
in which the percent oxidation of cyt a,a$_3$ is 75% and 55% respectively.   In addition to the impact of the vascular term upon the
estimated values, the main contribution to the spectral change is
attributable to the different cyt a,a$_3$ oxidation levels. Unfortunately P$_{50}$ is approximately 38 torr (5 kPa) for rats and 27 (3.5 kPa)
for man, and consequently, the previously reported in vivo rat brain
differential spectra should not be equated directly with those of the
human brain.   It is questionable whether spectra obtained in man under
different respiratory conditions can provide a better evaluation of the
spectral contributions: in our experience, cyt a,a$_3$ redox fluctuations are not easily triggered by small O$_2$ supply variations.

IN VIVO MEASUREMENTS

Figure 3 shows the two instruments presently working in our
laboratory for NIRS measurements (on the right the 4 w.l. photometer
employed on adults and newborn babies and on the left an LPDA spectrophotometer).

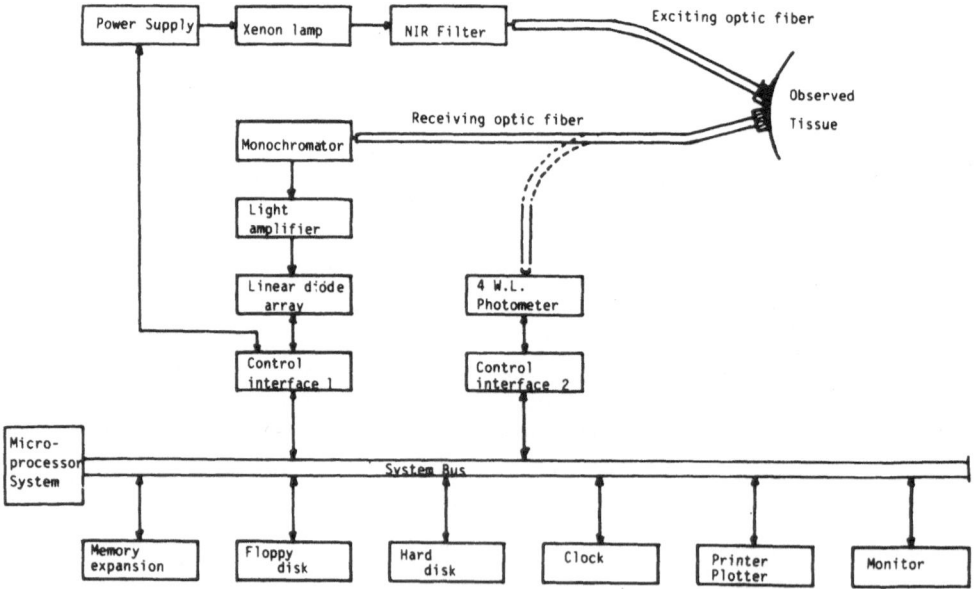

Figure 3.   Block diagram of an LDPA (left) and a four-wavelength
             spectrophotometer (right).

The photometer has been used since 1984 on volunteers, adult
patients with cerebrovascular disease and, in the last year, on the
newborn.   Figure 4 summarizes the dispersion of the apparent optical
densities in a set of clinical measurements carried out on symptom-free
newborn babies.

In a typical IR tracing of a preterm infant, during fluctuations of respiratory activity, HbS and transcutaneous oxygen tension ($TcPO_2$) recordings, measured on the upper part of the thorax, had the same trend. Conversely the redox trace was generally stable. All these findings suggest the reliability of cortical oximetry, which, obviously, is more clinically significant than peripheral measurements.

Recent developments in solid state image detectors have yielded small integrated photodiode arrays with accurately defined geometries and good infrared responsiveness which incorporate integral circuits to generate serial video output signals. This new detector system might provide a logical alternative to photomultipliers (Jones, 1985). It allows the collection of all wavelengths in the near infrared region simultaneously, in a few milliseconds, thus increasing significantly the available information. The hardware and software features of this spectrophotometer are similar to other computer-assisted scanning spectrophotometers employed in in vivo visible reflectance spectroscopy (Pirasky et al., 1985).

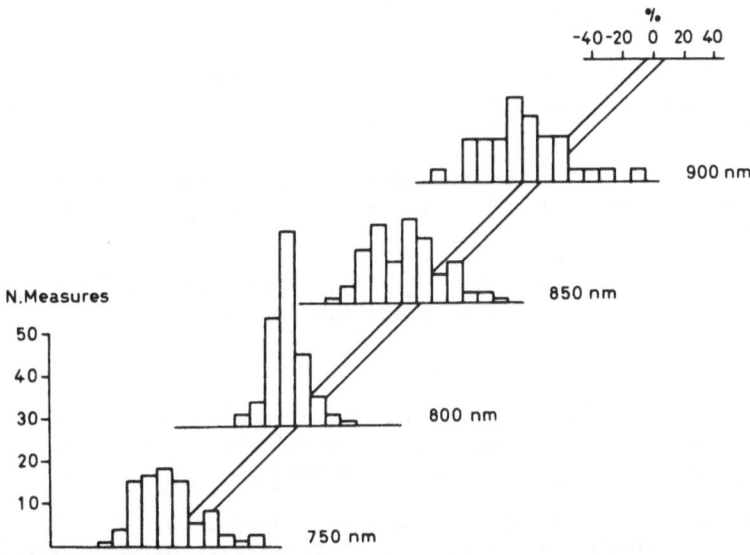

Figure 4.   Dispersion of the apparent optical densities in a set of clinical measurements at the four monitored wavelengths.

Spectra generated while 750, 800, 850 nm interference filters were placed in the light path demonstrated good overall instrumental resolution. Further in vitro-in vivo calibrations/measurements are in progress.

REFERENCES

Beinert, R.W., Shaw, R.S., Hansen, R.E. and Hartzell, C.R. (1980). Studies on the origin of the near-infrared (800-900 nm) absorption of cytochrome c oxidase. Biochim. Biophys. Acta, 591, 458-470.

Brunori, M., Antonini, E. and Wilson , M.T. (1981). Copper proteins. In: Metal Ions in Biological Systems, Vol. 13. Ed. Seigel, H., Marcell Dekker, New York and Basel, pp. 187–228.

Darriet, D., Der, T. and Collins, R.C. (1986). Distribution of cytochrome oxidase in rat brain. J. Cereb. Blood Flow Metabol. 6, 8–14.

Francois-Dainville, E., Buchweitz, E. and Weiss, H.R. (1986). Effect of hypoxia on percent of arteriolar and capillary beds perfused in the rat brain. J. Appl. Physiol. 60, 280–288.

Giannini, I., Ferrari, M., Carpi, A. and Fasella, P. (1982). Rat brain monitoring by near infrared spectroscopy: an assessment of possible clinical significance. Physiol. Chem. Phys. 14, 295–305.

Greenberg, J., Alavi, A., Reivich, M., Kuhl, D. and Uzzell, B. (1978). Local cerebral blood volume response to carbon dioxide in man. Circ. Res. 43, 324–331.

Greiner, A.C., Chan, S.C. and Nicolson, G.A. (1975). Human brain contents of calcium, copper, magnesium and zinc in some neurological pathologies. Clin. Chim. Acta, 64, 211–213.

Jobsis, F.F. (1977). Non-invasive, infrared monitoring of cerebral and myocardial oxygen sufficiency and circulatory parameters. Science, 198, 1264–1267.

Jones, D.G. (1985). Photodiode array detectors in U.V.-VIS spectroscopy, Part I & II. Anal. Chem. 57, 1057A–1073A, and 1207A–1214A.

Lai, J.L.K. and Clark, J.B. (1979). Preparation of synaptic and nonsynaptic mitochondria from mammalian brain. Methods in Enzymology, 55, 51–60.

Pirasky, S.M., LaManna, J.C., Sick, T.J. and Rosenthal, M. (1985). A computer-assisted rapid scanning spectrophotometer with application to tissues in vitro and in vivo. Comp. Biomed. Res. 18, 408–421.

Reddy, K.V. Subba, Hendler, R.W. and Bunow, B. (1986). Complete analysis of cytochrome components of beef heart mitochondria in terms of spectra and redox properties. Cytochromes a,a$_3$. Biophys. J. 49, 705–715.

Sakai, F., Nakazawa, K., Tazaki, Y., Ishii, K., Hino, H., Igarashi, H. and Kanda, T. (1985). Regional cerebral blood volume and hematocrit measured in normal human volunteers by single photon emission computed tomography. J. Cereb. Blood Flow Metabol. 5, 207–213.

Sylvia, A.L., Piantadosi, C.A. and Jobsis-VanderVliet, F.F. (1985). Energy metabolism and in vivo cytochrome c oxidase redox relationships in hypoxic rat brain. Neurol. Res. 7, 81–88.

Takatani, S. and Graham, M.D. (1979). Theoretical analysis of diffuse reflectance from a two-layers tissue model. IEEE Trans. Biomed. Eng. BME-26, 656–664.

NEAR-INFRARED SPECTROPHOTOMETRIC MONITORING OF HAEMOGLOBIN AND

CYTOCHROME a,a3 IN SITU

O. HAZEKI, A. SEIYAMA and M. TAMURA

Biophysics Division, Research Institute of Applied Electricity, Hokkaido University, Sapporo 060, Japan

INTRODUCTION

In the near-infrared region, two chromophores of biological interest, haemoglobin and cytochrome a,a3, have specific absorption bands, which respond to the oxygenation state of living tissues. Several authors have applied near-infrared spectroscopy to their analysis in situ (e.g. Jobsis, 1977; Gianni et al., 1982; Kariman and Burkhart, 1985a), but quantitative analyses in tissue have encountered difficulties from the overlap of changes in the amount and oxygenation state of haemoglobin and the redox state of cytochrome a,a3.

We have developed a method to measure changes in the content of oxy- and total (oxy- plus deoxy-) haemoglobin in rat head (Hazeki and Tamura, unpublished) and skeletal muscle (Seiyama, Hazeki and Tamura, 1987). In these studies we illuminated rat tissues with near-infrared light and analysed the light transmitted through them. The quantification was made on the basis of the linear relationship between the changes of optical absorbance and haemoglobin content. In this paper we present a method to monitor the redox changes of cytochrome a,a3 based on our previous work.

METHODS

Optical techniques

The centre of the rat head was illuminated by light from an air-cooled tungsten-iodine lamp through a light-guide of 4 mm diameter. The light transmitted through skin, cranial bone, mandible and brain tissue was collected by a three-branched light-guide with optical filters of 700, 730 and 830 nm, and led to photomultipliers. The light intensities at each wavelength were recorded and the absorbance changes calculated.

Perfused rat head preparation

Male Wistar rats weighing 180-280 g were anaesthetized with urethane (0.8 g/kg I.P.). The left and right external carotid arteries were ligated at their origins. The common carotid arteries were cannulated with polyethylene tubes and Krebs Ringer bicarbonate buffer was infused slowly. The ascending aorta was occluded quickly at its origin and again at the point where it turned downward. The perfusate was then changed to saline containing 10 mM $Na_2S_2O_4$ or 1 mM NaCN, and the infusion rate raised to 3 ml/min.

RESULTS AND DISCUSSION

Figure 1A shows the difference absorption spectrum of isolated haemoglobin solution. An isosbestic point for oxy and deoxy species was found at 805 nm. Figure 1B shows the difference absorption spectrum of rat head induced by anoxia. The spectrum in vivo was similar to that of haemoglobin solution, but 805 nm was no longer the isosbestic point. This could be attributed either to the decrease in amount of haemoglobin or to the decrease in absorption of other tissue component(s). In order to estimate the contribution of haemoglobin to in vivo spectra, we measured the absorption spectrum of haemoglobin-free rat head (Fig. 1C), in which blood was replaced with Fluosol-43 emulsion.

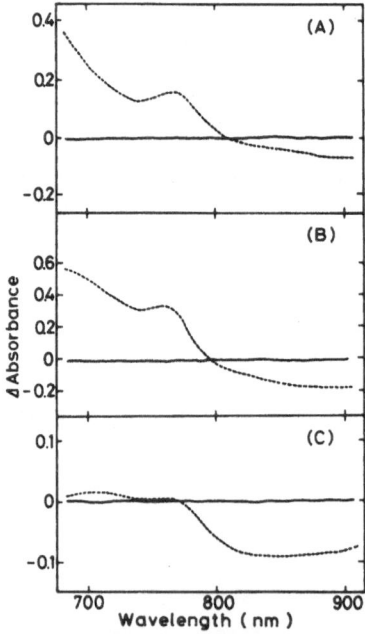

Figure 1. Difference absorption spectra of haemoglobin solution (A), normal rat head (B), and haemoglobin-free rat head (C). The spectra at oxygenated states were used as baselines (solid lines), and the anoxic spectra (broken lines) were recorded.

In this rat head the absorbance change in the region of 700–780 nm disappeared, showing that the change in this region of normal rat was mostly due to the change in haemoglobin. In contrast, the absorbance above 780 nm was decreased by anoxia even in the haemoglobin-free rat. The change was attributable to the reduction of cytochrome a,a₃ in mitochondria (Griffith and Wharton, 1961; Chance, 1966 and Ferrari et al., 1983).

From these results we can express the absorbance changes at 700, 730 and 830 nm as follows:

$$\Delta A_{700} = k_{700}\Delta[HbO_2]d + k'_{700}\Delta[Hb]d \tag{1}$$

$$\Delta A_{730} = k_{730}\Delta[HbO_2]d + k'_{730}\Delta[Hb]d \tag{2}$$

$$\Delta A_{830} = k_{830}\Delta[HbO_2]d + k'_{830}\Delta[Hb]d + k''_{830}\Delta[cyt\ a,a_3]d \tag{3}$$

where A_i is the absorbance change at wavelength i nm, d is the light path length, k_i and k'_i indicate the absorption constants <u>in situ</u> for oxy- and deoxyhaemoglobin, respectively; k" is the difference in absorbtion coefficients between the oxidized and reduced forms of cytochrome oxidase. From Equations (1) and (2) we can get

$$\Delta[HbO_2]d \alpha \Delta A_{700} - (k'_{700}/k'_{730})\Delta A_{730} \tag{4}$$

In order to determine the optical constant in this expression, we perfused the rat head with a blood cell suspension which had been deoxygenated by washing three times with saline containing 10 mM $Na_2S_2O_4$. Under this condition the following equations are obtained, because $\Delta[HbO_2]$ and $\Delta[cyt\ a,a_3]$ are expected to be zero:

$$\Delta A_{700} = (k'_{700}/k'_{730})\Delta A_{730} \tag{5}$$

$$\Delta A_{730} = (k'_{730}/k'_{830})\Delta A_{830} \tag{6}$$

Thus, there should be linear relationships between ΔA_{700}, ΔA_{730} and ΔA_{830}. The results are shown in Figure 2.

There were good correlations between these absorbance changes. From the slopes of the straight lines, we obtained values of 1.20 and 1.40 for k'_{700}/k'_{730} and k'_{730}/k'_{830}, respectively. Thus, from Expression (4) the change of oxyhaemoglobin content in rat head can be monitored by:

$$\Delta A_{700} - 1.20\Delta A_{730} \tag{4)'}$$

From Equations (1) - (3), we can also get the following expression:

$$\Delta[cyt\ a,a_3]d \alpha (\Delta A_{730} - 1.40\Delta A_{830}) - K(\Delta A_{700} - 1.20\Delta A_{730}) \tag{7}$$

where K is a constant, expressed as:

$$\frac{k'_{730}(k_{730}k'_{830} - k'_{730}k_{830})}{k'_{830}(k_{700}k'_{730} - k'_{700}k_{730})}$$

Under conditions where the redox state of cytochrome a,a_3 does not change ($\Delta[cyt\ a,a_3] = 0$), the following equation is expected:

$$(\Delta A_{730} - 1.40\Delta A_{830}) = K(\Delta A_{700} - 1.20\Delta A_{730}) \tag{8}$$

To confirm this equation, we measured the absorbance changes of rat head perfused with an oxygenated blood cell suspension containing 1 mM NaCN. As shown in Figure 3 there was a linear relationship between $\Delta A_{730} - 1.40\Delta A_{830}$ and $\Delta A_{700} - 1.20\Delta A_{730}$ for haematocrit values ranging from 5 to 63%.

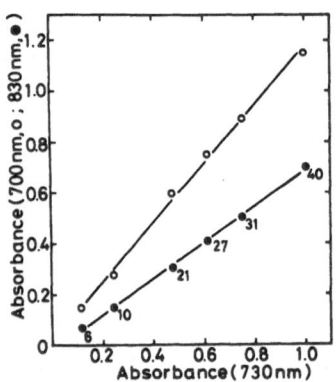

Figure 2. Determination of (k'_{700}/k'_{730}) and (k'_{730}/k'_{830}). The rat head was perfused with saline containing 10 mM $Na_2S_2O_4$ and intensity of transmitted light recorded. Then, deoxygenated blood cell suspensions (haematocrit values are shown on the figure) were infused and the absorbance changes measured at 700, 730 and 830 nm.

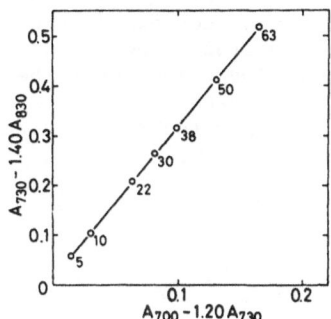

Figure 3. Determination of K. Rat head was perfused with saline containing 1 mM NaCN and the intensity of transmitted light was recorded. Then blood cell suspensions (haematocrit values are shown on the figure) were infused and the absorbance changes measured.

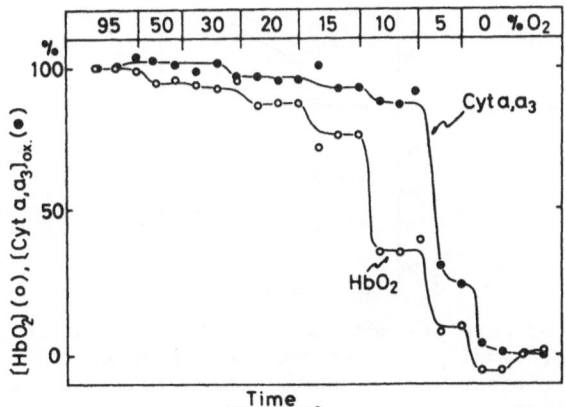

Figure 4. The effect of oxygen concentration in inspired gas on the content of oxyhaemoglobin (o) and the redox state of cytochrome a,a_3 (●) in rat head.

The value of K was thus estimated to be 3.05 from the slope of the straight line and Expression (7) is now written as:

$$\Delta[cyt\ a,a_3] \propto (\Delta A_{730} - 1.40\Delta A_{830}) - 3.05(\Delta A_{700} - 1.20\Delta A_{730})$$

$$= -3.05\Delta A_{700} + 2.66\Delta A_{730} - 1.40\Delta A_{830} \tag{7)'}$$

The Expressions (4) and (7) were applied to the analysis of data obtained from rats in situ. Oxyhaemoglobin and the oxidized form of cytochrome a,a_3 in rat head decreased when the oxygen concentration in the inspired gas was lowered stepwise under hypercapnic conditions (5% CO_2, Fig.4). The ordinate indicates the contents as per cent of those at 95% O_2; a 50% decrease in haemoglobin was seen at oxygen concentrations of 15-10% and in cytochrome a,a_3 at 10-5%.

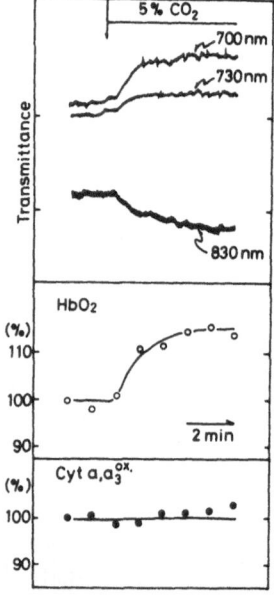

Figure 5. The effect of addition of CO_2 to inspired gas (oxygen) on the intensity of transmitted light at 700, 730 and 830 nm, the content of oxyhaemoglobin (o) and the redox state of cytochrome a,a_3 (●) in rat head. Abscissa, time (min).

Addition of 5% CO_2 to inspired oxygen increased the intensity of transmitted light at 700 and 730 nm and decreased it at 830 nm. These changes are expressed as the changes in haemoglobin and cytochrome a,a_3 (Fig. 5.). The oxyhaemoglobin content increased by 15%.

CO_2 is known to dilate cerebral vessels, lower resistance and increase blood flow. The resultant oxygenation of venous blood may explain this observation. In contrast, the redox state of cytochrome a,a_3 did not change. Recently using 3 wavelength reflectance spectro -photometry around 605 nm, based on a principle similar to ours, Kariman and Burkhart (1985b) showed that cytochrome a,a_3 was oxidized significantly in the presence of CO_2. This discrepancy might arise from the difference in the kinetics of cytochrome a,a_3 at 605 nm and 830 nm (Chance, 1966; Kariman and Burkhart, 1985a). This possibility is under investigation using fluorocarbon-perfused rat organs (the intensity of transmitted light around 605 nm is too weak to be measured precisely in normal rat).

REFERENCES

Chance, B. (1966). Spectrophotometric observations of absorbance changes in the infrared region in suspensions of mitochondria and in submitochondrial particles. In: The Biochemistry of Copper. Eds Peisach, G., Aisen, P. and Blumberg, W.E., Academic Press, New York, pp. 293-301.

Ferrari, M., Gianni, I., Carpi, A. and Fasella, P. (1983). Noninvasive near-infrared spectroscopy of brain in fluorocarbon exchange transfused rats. Physiol. Chem. Phys. Med. NMR, 15, 107-113.

Gianni, L., Ferrari, M., Carpi, A. and Fasella, P. (1982). Rat brain monitoring by near-infrared spectroscopy: an assessment of possible clinical significance. Physiol. Chem. Phys. 14, 295-305.

Griffith, D.E. and Wharton, D.C. (1961). Studies of the electron transport system. XXXV. Purification and properties of cytochrome oxidase. J. Biol. Chem. 236, 1850-1856.

Jobsis, F.F. (1977). Noninvasive, infrared monitoring of cerebral and myocardial oxygen sufficiency and circulatory parameters. Science, 198, 1264-1267.

Kariman, K. and Burkhart, D.S. (1985a). Heme-copper relationship of cytochrome oxidase in rat brain in situ. Biochem. Biophys. Res. Commun. 126, 1022-1028.

Kariman, K. and Burkhart, D.S. (1985b). Non-invasive in vivo spectrophotometric monitoring of brain cytochrome aa_3 revisited. Brain Res. 360, 203-213.

Seiyama, A., Hazeki, O. and Tamura, M. (1987). Simultaneous measurement of haemoglobin oxygenation of brain and skeletal muscle of rat in vivo by near-infrared spectrophotometry. This volume.

SIMULTANEOUS MEASUREMENT OF HAEMOGLOBIN OXYGENATION OF BRAIN AND SKELETAL MUSCLE OF RAT <u>IN VIVO</u> BY NEAR-INFRARED SPECTROPHOTOMETRY

A. SEIYAMA, O. HAZEKI and M. TAMURA

Research Institute of Applied Electricity, Hokkaido University, 060 Sapporo, Japan

INTRODUCTION

Analysing transmitted light at 700 and 730 nm, we found a linear relationship between the blood concentration and the absorbance change in perfused hindlimb of rat when physiological haematocrits (Hc) in the range of 15 to 50% were used for the inflowing perfusate. Several optical parameters, which are required for calculation of changes in oxyhaemoglobin, deoxyhaemoglobin and blood volume, were determined. In the present paper, we report the application of a two-wavelength method to the simultaneous measurement of haemoglobin oxygenation of rat brain and skeletal muscle at steady state <u>in vivo</u>.

EQUATIONS

Basic equations for the one component system

For calculation of changes of oxyhaemoglobin ($[HbO_2]$), deoxy-haemoglobin ($[Hb]$), and blood volume ($[{|}Hb]$), equations can be derived as follows according to Lambert-Beer's Law:

$$\Delta A_{700} = k_1 \Delta[HbO_2] + k'_1 \Delta[Hb] \tag{1}$$

$$\Delta A_{730} = k_2 \Delta[HbO_2] + k'_2 \Delta[Hb] \tag{2}$$

$$\Delta A_{805} = k_3 \Delta[HbO_2] + k_3 \Delta[Hb] \tag{3}$$

where $k_1(k'_1)$, $k_2(k'_2)$, and k_3 are constants depending on wavelengths $\lambda_1 = 700$ nm, $\lambda_2 = 730$ nm, and $\lambda_3 = 805$ nm (isosbestic point), respectively. Then Equation (3) can be shown as follows:

$$\Delta A_{805} = k_3(\Delta[HbO_2] + \Delta[Hb])$$

$$= k_3 \Delta[{|}Hb] \tag{4}$$

Combining Equations (1) and (2), we can derive:

$$\Delta[HbO_2] = (k'_2 \Delta A_{700} - k'_1 \Delta A_{730})/(k_1 k'_2 - k_2 k'_1) \tag{5}$$

$$\Delta[Hb] = (k_2 \Delta A_{700} - k_1 \Delta A_{730})/(k_2 k'_1 - k_1 k'_2) \qquad (6)$$

$$\Delta[\dot{H}b] = \frac{[(k'_2 - k_2)\Delta A_{700} - (k'_1 - k_1)\Delta A_{730}]}{k_1 k'_2 - k_2 k'_1)} \qquad (7)$$

Rearranging Equations (5), (6) and (7) by using k_3, we get the following equations:

$$\Delta[HbO_2]k_3 = \frac{(k'_2/k_3)\Delta A_{700} - (k'_1/k_3)\Delta A_{730}}{(k_1/k_3)(k'_2/k_3) - (k_2/k_3)(k'_1/k_3)} \qquad (5')$$

$$\Delta[Hb]k_3 = \frac{(k_2/k_3)\Delta A_{700} - (k_1/k_3)\Delta A_{730}}{(k_2/k_3)(k'_1/k_3) - (k_1/k_3)(k'_2/k_3)} \qquad (6')$$

$$\Delta[\dot{H}b]k_3 = \frac{(k'_2/k_3-k_2/k_3)\Delta A_{700}-(k'_1/k_3-k_1/k_3)\Delta A_{730}}{(k_1/k_3)(k'_2/k_3) - (k_2/k_3)(k'_1/k_3)} \qquad (7')$$

Equations (5'), (6') and (7') are fundamental equations in this study. To use these equations, it is necessary to find values of k_1/k_3, k'_1/k_3, k_2/k_3 and k'_2/k_3. To find these values, the _in situ_ experiments were performed in perfused hindlimb of rat. These values are shown in Table 1.

Table 1. Optical parameters required for calculations of $[HbO_2]$, $[Hb]$ and $[\dot{H}b]$, determined _in situ_

Parameter	Oxy	Deoxy
$\Delta A_{700}/\Delta A_{805}$	0.80 $(=k_1/k_3)^w$	1.50 $(=k'_1/k_3)^y$
$\Delta A_{730}/\Delta A_{805}$	0.85 $(=k_2/k_3)^x$	1.23 $(=k'_2/k_3)^z$

w. $n = 3$; Hc = 3.5 ~ 51%; SO_2 of blood samples, higher than 98.5%

x. $n = 3$; Hc = 7.0 ~ 61%; SO_2 of blood samples, higher than 99.0%

y. $n = 3$; Hc = 3.0 ~ 55%; blood samples reduced with $Na_2S_2O_4$

z. $n = 3$; Hc = 4.0 ~ 44%; blood samples reduced with $Na_2S_2O_4$

Development of basic equations

For the calculation of oxyhaemoglobin saturation (SO_2) substitutions of optical parameters in Equations (5'), (6') and (7') yield:

$$\Delta[HbO_2]k_3 = -4.23\Delta A_{700} + 5.15\Delta A_{730} \qquad (5'')$$

$$\Delta[Hb]k_3 = -2.75\Delta A_{730} + 2.92\Delta A_{700} \qquad (6'')$$

$$\Delta[\dot{H}b]k_3 = -1.31\Delta A_{700} + 2.41\Delta A_{730} \qquad (7'')$$

Let state 'a' be the hyperoxic condition where the rat inspires 95% O_2 + 5% CO_2 (in our experiments under these conditions the values of oxyhaemoglobin saturation for aorta and vena cava were greater than 99.7% and 90% respectively) and let state 'b' be the anoxic condition where the oxyhaemoglobin saturation of artery and vein may be taken as nearly zero. Then the total oxyhaemoglobin concentration ($[HbO_2]_n$) in a state 'n' may be expressed as follows:

$$[HbO_2]_n k_3 = (\Delta[HbO_2]_n - \Delta[HbO_2]_b)k_3$$

$$= 4.23 \log I_n^{700}/I_b^{700} - 5.15 \log I_n^{730}/I_b^{730} \tag{8}$$

Similarly, we may obtain the total concentration of deoxyhaemoglobin ($[Hb]_n$) and the blood volume ($[\dot{|}Hb]_n$). Assuming SO_2 in state 'a' to be 100% we get the following equations (9) and (10):

$$[Hb]_n k_3 = (\Delta[Hb]_n - \Delta[Hb]_a)k_3$$

$$= 2.92 \log I_a^{700}/I_n^{700} - 2.75 \log I_a^{730}/I_n^{730} \tag{9}$$

$$[\dot{|}Hb]_n k_3 = [HbO_2]_n k_3 + [Hb]_n k_3 \tag{10}$$

Determination of the value of k_3 is impossible, therefore values obtained for total concentrations of oxyhaemoglobin and deoxyhaemoglobin, and blood volume are expressed as the product of the constant value of k_3. Dividing Equation (8) by Equation (10), we get oxyhaemoglobin saturation in a state 'n' as:

$$SO_2 = [HbO_2]_n k_3/[\dot{|}Hb]_n k_3, \quad (0 \leq SO_2 \leq 1).$$

METHODS

Spectrophotometric measurement

A computer-controlled wavelength scanning spectrophotometer and 2-channel wavelength spectrophotometer operated in the photon counting mode were used. The light (700 - 1000 nm) from the air-cooled tungsten-iodine source illuminated the head or the thigh muscle through the light guide of 4 mm diameter. The transmitted light was guided into the photometers through the other light guide.

Determination of optical parameters

Male Wistar rats (180 to 230 g body weight) fed on a commercial diet were used. Rats were anaesthetized with an intraperitoneal injection of 1.2 g/kg urethane. The preparation of the isolated rat hindlimb was carried out by a slight modification of the method of Ruderman, Houghton and Hems (1971). The perfusion apparatus was essentially the same as that used in liver perfusion (Sugano et al., 1978). The isolated hindlimb was preperfused for 15 min in a flow-through mode at 32°C with the flow rate at 16 ml·min^{-1} per leg in a 180 g rat. After preperfusion, blood samples were injected to measure absorption changes. SO_2 values of the collected blood samples were measured by using a Gas Analyzer, ABL-30.

Surgical preparation for in vivo measurement

Animals were anaesthetized with urethane (0.8 g/kg, I.P.) and respired artificially, through a tracheal cannula, with a mixture of O_2 and N_2 containing 5% CO_2 from a respiratory pump (70 strokes/min and 140 ml/min). Arterial blood pressure was measured at the femoral artery.

RESULTS AND DISCUSSION

Figure 1 shows the relationship of the absorbance change between 700 nm and 805 nm in perfused hindlimb of rat. A similar relationship was obtained between 730 nm and 805 nm. There were linear relationships between the blood concentration and the absorbance change in the oxy- and deoxy- form of the blood samples, when physiological haematocrits of 15-50% were used.

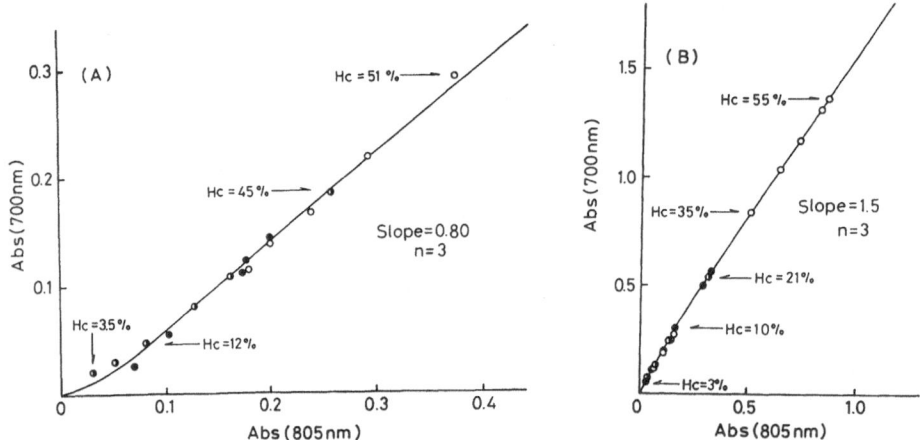

Figure 1. Relationship of the absorbance change between 700 nm and 805 nm in the perfused hindlimb of the rat. The relationship in a fully oxygenated condition is shown in (A), and that in a fully reduced condition in (B).

The slope ($\Delta A \lambda_1 / \Delta A \lambda_3$) gives the optical parameter, used in this study, as shown in Table 1). Values of these parameters in the thigh muscle were in good agreement with those in the head. Optical determination of oxyhaemoglobin saturation was carried out using Equations (8), (9) and (10). In this wavelength region (700-730 nm), the influence exerted by cytochrome a, a_3 may be neglected and that exerted by myoglobin in the thigh muscles was less than 10% of the absorbance change, estimated from absorbance changes in the head and the thigh muscle of rat exchange-transfused with Fluosol-43 (unpublished data). The changes in total oxyhaemoglobin and deoxyhaemoglobin and in blood volume and oxyhaemoglobin saturation are shown in Figure 2. The values of SO_2 were between the SO_2 values of aorta and vena cava measured by gas analysis (Fig. 2 bottom) probably reflecting the distribution of the blood between artery and vein (Smith and Kampine, 1980). Simultaneous measurement of haemoglobin oxygenation in the head and the thigh muscles of rat was also performed. SO_2 in the head was lower than that in the muscle of the hindlimb.

In summary, the quantitative analysis employed here is satisfactory for further applications.

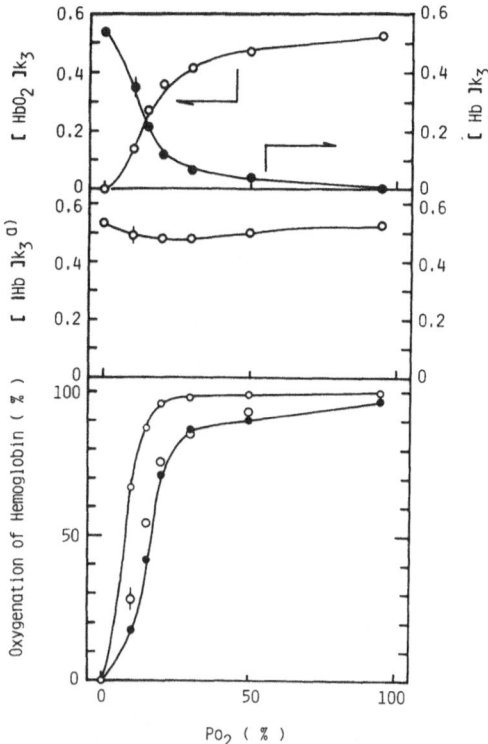

Figure 2. Optical assessment of changes in the rat head of [HbO2]
and [Hb] top trace; [¦Hb], middle trace; and SO2, bottom
trace. The calculated SO2 values (O) were compared with
SO2 values in the aorta (o) and vena cava (•) (bottom
traces). a; blood samples were taken from the aorta and vena
cava at each PO2 to measure SO2.

REFERENCES

Ruderman, N.B., Houghton, C.R.S. and Hems, R. (1971). Evaluation of
the isolated perfused rat hindquarter for the study of muscle
metabolism. Biochem. J. 124, 639-651.

Smith, J.J. and Kampine, J.P. (1980). Circulatory Physiology - The
Essentials. The Williams & Wilkins Company, Baltimore, MD.

Sugano, T., Suda, K., Shimada, M. and Oshino, N. (1978). Biochemical
and ultrastructural evaluation of isolated rat liver systems perfused
with hemoglobin-free medium. J. Biochem. (Tokyo), 83, 995-1007.

SPECTROSCOPIC CHARACTERISTICS OF RAT SKELETAL AND CARDIAC TISSUES IN THE VISIBLE AND NEAR-INFRARED REGION

M. TAMURA, A. SEIYAMA and O. HAZEKI

Biophysics Division, Research Institute of Applied Electricity, Hokkaido University, Sapporo 060, Japan

INTRODUCTION

Cardiac and skeletal muscle cells contain myoglobin and mito-chondrial cytochromes, both of which respond to oxygen concentration. In this paper, we will present the optical characteristics of rat cardiac and skeletal muscles in the range of 500 to 900 nm.

EXPERIMENT

Langendorff preparations of isolated rat heart were perfused with Krebs-Ringer bicarbonate buffer solution (Araki, Tamura and Yamazaki, 1983). For measurements on hindlimb muscle, rats were used in which the blood was replaced with Fluosol-43.

RESULTS

Figure 1 shows the spectral changes of perfused rat heart induced by the transition from the aerobic to the anaerobic state. The absorption minima at 580 and 540 nm are those of oxymyoglobin. The absorption maximum at 650 nm is due to the reduction of cytochrome a, a_3.

Figure 2 shows the relationship between percent deoxygenation of myoglobin and percent reduction of copper in cytochrome a, a_3 in the perfused heart. The redox changes of the haem moeity (605-620 nm) and copper (830-940 nm) are also given in the figure. Both plots show a deviation from a straight-line relationship.

Figure 3 shows the relation between the oxygen consumption rate and the redox state of copper and haem in cytochrome a, a_3 and of cytochrome c. Oxygen consumption rate was varied by changing the perfusion pressure (Araki et al., 1983). The redox state of copper in cytochrome a, a_3 shifted linearly to the reduced state with increasing oxygen consumption. The haem moiety in cytochrome a, a_3 behaved differently from copper.

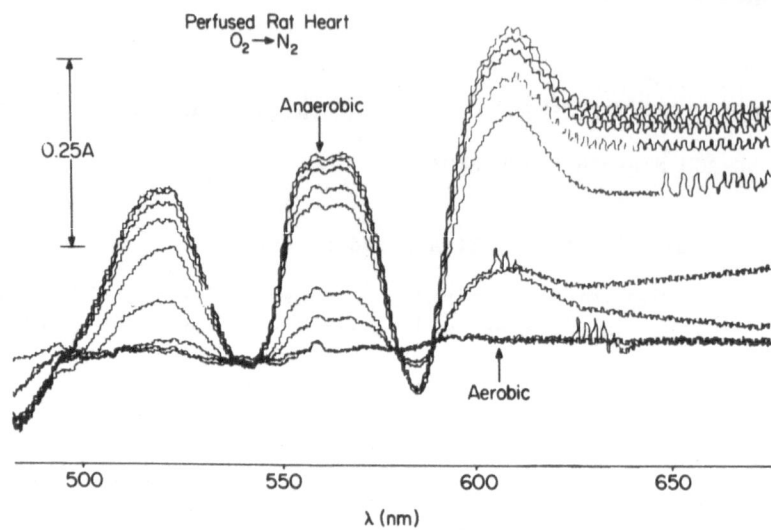

Figure 1. Spectral changes of perfused rat heart induced by aerobic-anaerobic transition.

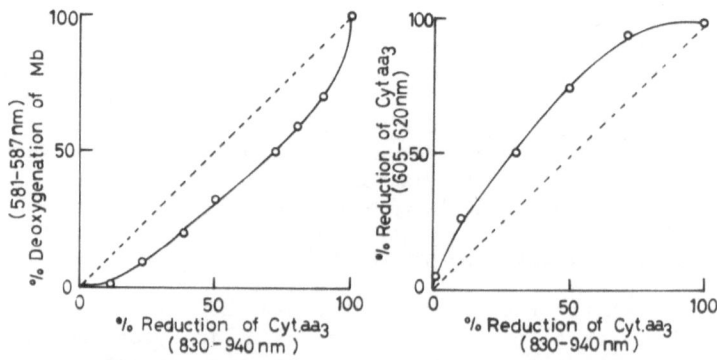

Figure 2. Relationship between per cent deoxygenation of myoglobin (Mb) and per cent reduction of haem and copper in cytochrome a,a₃ in the perfused rat heart.

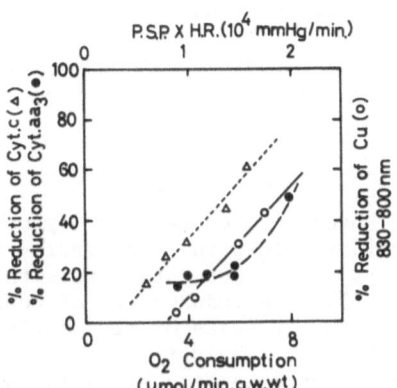

Figure 3.   Relationship between the redox state of cytochrome a,a₃
(605 – 620 nm), copper and oxygen consumption in the
perfused rat heart.   The redox state of cytochrome c (550 –
530 nm) is also shown.   P.S.P.   shows the product of peak
systolic pressure and heart rate.

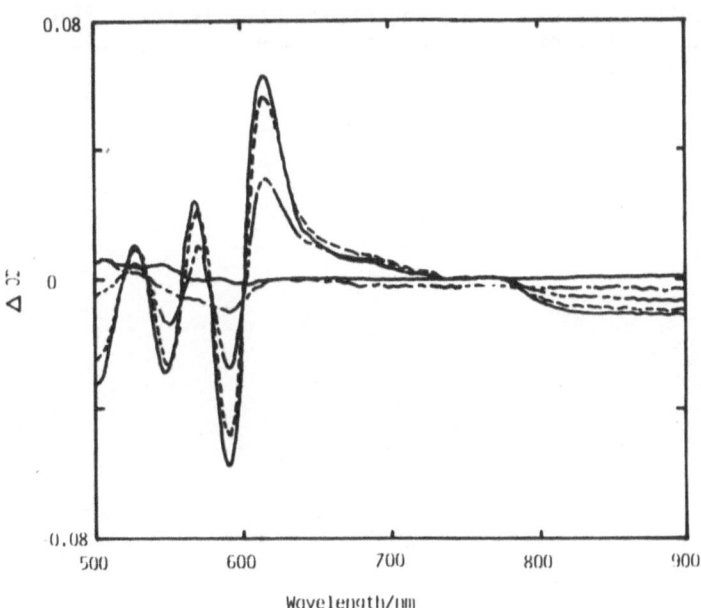

Figure 4.   Spectral   changes   of   Fluosol–43–substituted   rat   hindlimb
induced by a change from the aerobic to the anaerobic state.

Figure 4 shows the spectral changes in rat hindlimb induced by aerobic-anaerobic transition. In this study, in which rats were exchange-transfused with Fluosol-43, aerobic conditions were obtained by artificial respiration with 95% O_2 + 5% CO_2, and anaerobic conditions with 95% N_2 + 5% CO_2. The flat base-line is the aerobic state. The absorption minima at 540 and 580 nm are those of oxymyoglobin, and absorption maximum at 605 nm is more clearly seen in hindlimb than in heart (Fig. 1). The absorption minimum around 830 nm is due to the reduction of copper in cytochrome a,a_3. The molar ratio of myoglobin to cytochrome a,a_3 is about 3:1.

DISCUSSION

The near-infrared band of copper in cytochrome a,a_3 has been used as an oxygen indicator in situ. In cardiac tissue, myoglobin is very abundant, but this does not interfere with measurements of the redox state (Figs 2 and 3). The redox state of copper can also be used as an indicator of the energy state of tissue (Fig. 3.). In skeletal muscle of rat hindlimb, the absorption peak of cytochrome a,a_3 is more clearly seen than in cardiac tissue. This suggests that skeletal muscle is well suited for the optical measurement of cytochromes and myoglobin as well as haemoglobin.

REFERENCE

Araki, R., Tamura, M. and Yamazaki, I. (1983). The effect of intracellular oxygen concentration on lactate release, pyridine nucleotide reduction and respiration rate in rat cardiac tissue. Circ. Res. 53, 448-455.

NON-INVASIVE NEAR-INFRARED MEASUREMENTS OF HUMAN ARM TISSUES IN VIVO

M. TAKADA, T. TAMURA and M. TAMURA[*]

R & D Engineering-Spectrophotometric Instrument, Analytical Instrument Division, Shimadzu Corporation, Nakagyo-ku, Kyoto 604, and [*]Biophysics Division, Research Institute of Applied Electricity, Hokkaido University, Sapporo 060, Japan

INTRODUCTION

Near-infrared spectroscopy has been used for the measurement of tissue oxygenation in animals (Jobsis, 1977; Gianni et al., 1982; Kariman and Burkhart, 1985). To extend this technique for clinical use, we have tried to measure the haemoglobin oxygenation state in human muscle tissues under various conditions.

EXPERIMENTS

Human arm was measured with the system shown in Figure 1.

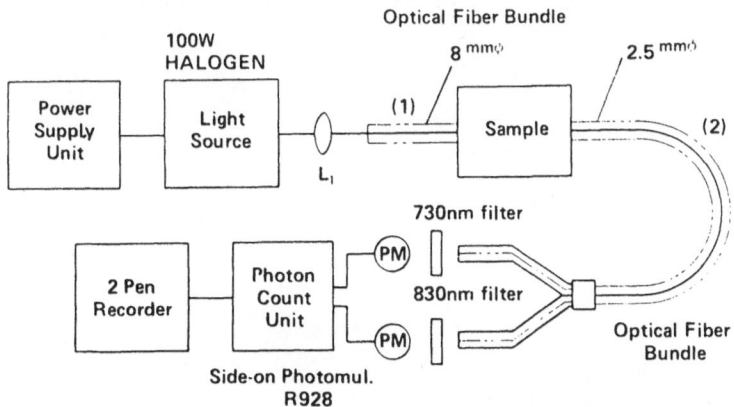

Figure 1. Schematic diagram of measuring system.

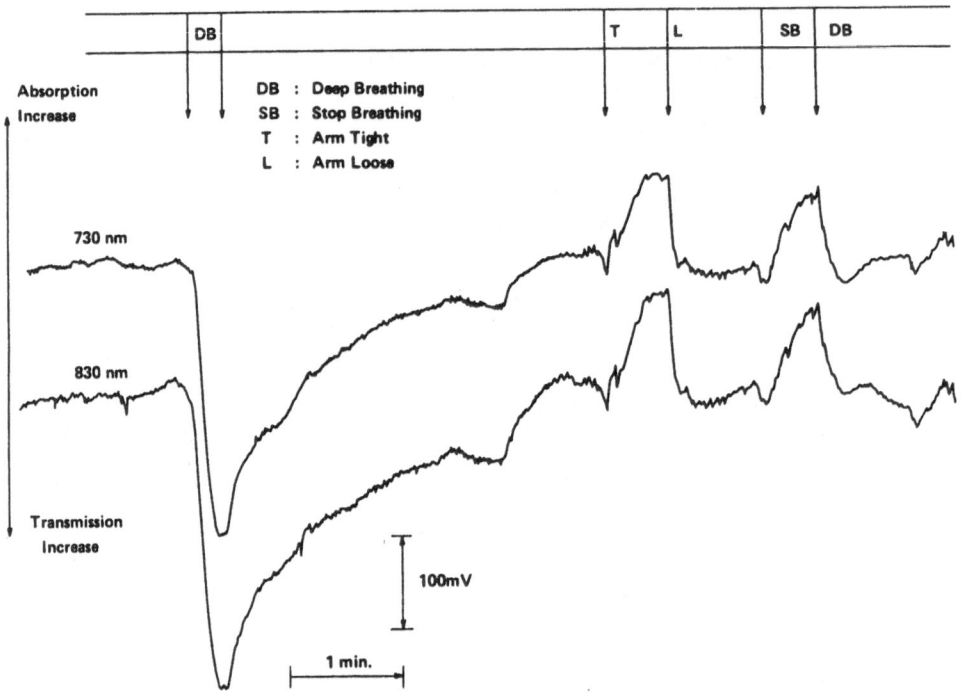

Figure 2. Transmission measurements of Human Palm.

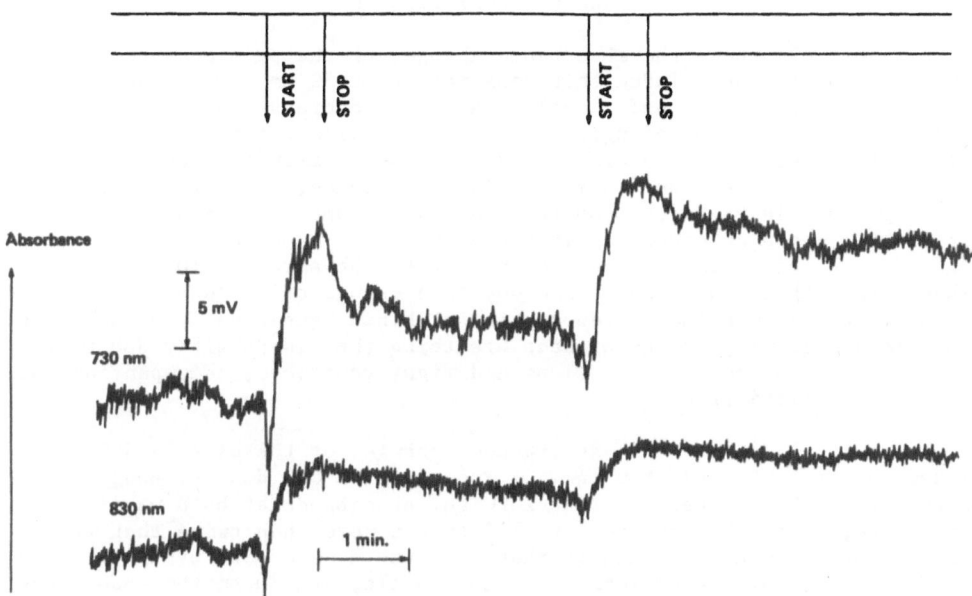

Figure 3.   Absorbance measurements of Forearm
(under physical exercises)

As shown in Figure 1 a white light from the 100 W halogen lamp illuminated the tissue through the optical fibre bundle (8 mm diameter). The measuring positions in this study were palm, wrist, forearm and upper arm. In each position except for the forearm, the absorbance changes were measured during deep breathing, choking and arm-artery compression. The effects on the forearm, of bending and stretching the fingers were also measured. Light transmitted through the tissue was collected by the second optical fibre bundle (2.5 mm diameter) and introduced into the detector. We selected the wavelengths at 730 and 830 nm. The absorption change at 730 nm is in the opposite direction to that at 830 nm for the same degree of oxy-deoxy transition of haemoglobin. On the other hand, a change of blood volume produces the same absorption change at both 730 and 830 nm.

RESULTS AND DISCUSSION

Figure 2 shows the absorbance changes of the palm measured at 730 and 830 nm. The tissue thickness was about 30 mm. When the subject (22 years old) breathed deeply, a large decrease in absorbance was observed at both wavelengths. Both optical signals recovered their original levels within 3 min. Upper arm compression (indicated by T in the figure) caused an increase in absorbance, which recovered very rapidly on release of the compression (L). Stopping breathing (SB) for about 40 sec also caused an increase in the absorbance. The direction and degree of the absorbance changes were the same at 730 and 830 nm. Thus, the observed changes are due to a change of blood volume rather than a change in the degree of haemoglobin oxygenation. The decrease in blood volume produced by deep breathing (DB) is possibly due to the increase in blood in the lung and right ventricle, accompanying the decrease in blood pressure.

Figure 3 shows the absorbance changes of the exercised forearm measured at 730 and 830 nm. The exercise was done by bending and stretching the fingers. On exercise, absorbance at both wavelengths increased, but the change at 730 nm was more than twice that at 830 nm. The reason for this is that exercise caused both hyperaemia and deoxygenation of haemoglobin in the muscle; at 730 nm the absorbance changes produced were additive while at 830 nm they were subtractive.

Though the data presented here are rather qualitative, the usefulness of near-infrared transmission spectroscopy is well documented. Quantitative analysis is now being attempted by several authors including us (Hazeki, Seiyama and Tamura, 1987).

REFERENCES

Jobsis, F.F. (1977). Non-invasive, infrared monitoring of cerebral and myocardial oxygen sufficiency and circulatory parameters. Science, 198, 1264-1267.

Gianni, I., Ferrari, M., Carpi, A. and Fasella, P. (1982). Rat brain monitoring by near-infrared spectroscopy: an assessment of possible clinical significance. Physiol. Chem. Phys. 14, 295-305.

Hazeki, O., Seiyama, A. and Tamura, M. Near-infrared spectrophotometric monitoring of haemoglobin and cytochrome a,a3 in situ. This volume.

Kariman, K. and Burkhart, D.S. (1985). Non-invasive in vivo spectrophotometric monitoring of brain cytochrome aa3 revisited. Brain Res. 360 203-213.

THE EFFECT OF INTRACELLULAR OXYGEN CONCENTRATION ON VENTRICULAR FIBRILLATION IN PERFUSED RAT HEART

M. MAKIGUCHI, H. KAWAGUCHI, H. YASUDA and M. TAMURA[*]

Department of Cardiovascular Medicine, School of Medicine, and [*]Biophysics Division, Research Institute of Applied Electricity, Hokkaido University, Sapporo 060, Japan

INTRODUCTION

From the clinical viewpoint, the patients with hypoxaemia have a high incidence of ventricular arrhythmias. However, there have been diverse results in animal models (Turnbull et al., 1965; Szekeres and Papp, 1967; Rogers et al., 1973; Murnaghan, 1975) possibly due to the effects of extracardiac factors such as neurohumoral reflex. To exclude such factors, we used an isolated perfused heart preparation to examine the direct effects of hypoxia on the arrhythmia. In this study, measuring the ventricular fibrillation threshold (VFT) in the isolated perfused rat heart, we examined the oxygen dependence of the susceptibility to ventricular fibrillation quantitatively under various conditions.

METHODS

Hearts of male Wistar rats were perfused by the Langendorff method. Aerobic and anoxic conditions were obtained in the perfusates by equilibration with 95% O_2 + 5% CO_2, and 95% N_2 + 5% CO_2, respectively. Hypoxia was obtained by equilibrating the perfusates with a mixture of these gases. The myocardial oxygen concentration was monitored by the absorbance measurement of myoglobin (Mb) at 580 - 620 nm (Araki, Tamura and Yamazaki, 1983). Thin platinum wire electrodes were inserted at the apex and base of the left ventricle for delivery of electric stimuli. Electrocardiograms were recorded from one electrode on the aortic cannula and the other on the left ventricle. The ventricular fibrillation threshold was estimated by applying stimuli of 5 ms duration within the vulnerable period after the R wave (Fig. 1). Current strength was raised from 1 mA in 0.25 mA steps until ventricular fibrillation occurred. Ventricular fibrillation usually reverted to normal sinus rhythm spontaneously within 10 - 20 seconds but when it persisted, electrical defibrillation was used. The VFT was defined as the minimal current required to induce at least six consecutive ventricular arrhythmias. After determining the VFT during aerobic perfusion, hypoxic perfusion was started and then VFT was determined. When VFT became lower, the tissue became more susceptible to fibrillation.

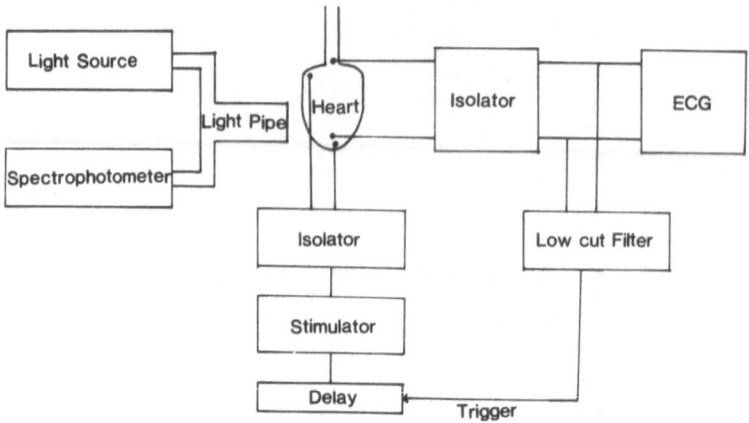

Figure 1. Schematic representation of the apparatus.

RESULTS

Effect of ventricular fibrillation on cardiac performance
    Figure 2 shows a typical example of transient ventricular fibrillation, where left ventricular pressure (LVP), MbO_2 saturation and coronary flow were also recorded.

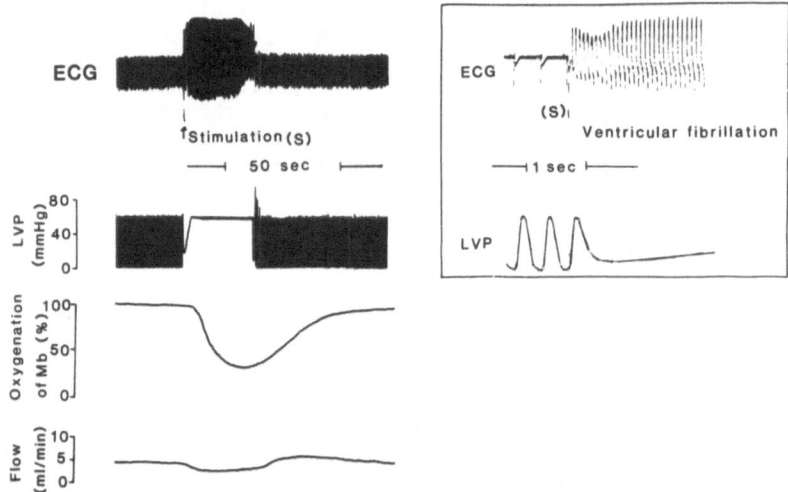

Figure 2. Effects of ventricular fibrillation on cardiac performance.

When ventricular fibrillation was induced by electrical stimulation, left ventricular pressure decreased and MbO_2 saturation fell gradually. After fibrillation stopped MbO_2 saturation recovered within a few minutes. Coronary flow decreased slightly during fibrillation and then increased again. The fall of MbO_2 saturation became more pronounced the longer the duration of fibrillation.

Effects of intracellular oxygen concentration on VFT
    To exclude the effects on VFT of heart rate, VFT during hypoxia was determined within the period where heart rate changed little. Figure 3 shows the relationship between intracellular oxygen concentration and VFT. The VFT remained unchanged above an intracellular oxygen concentration of 10 μM but fell below this concentration.

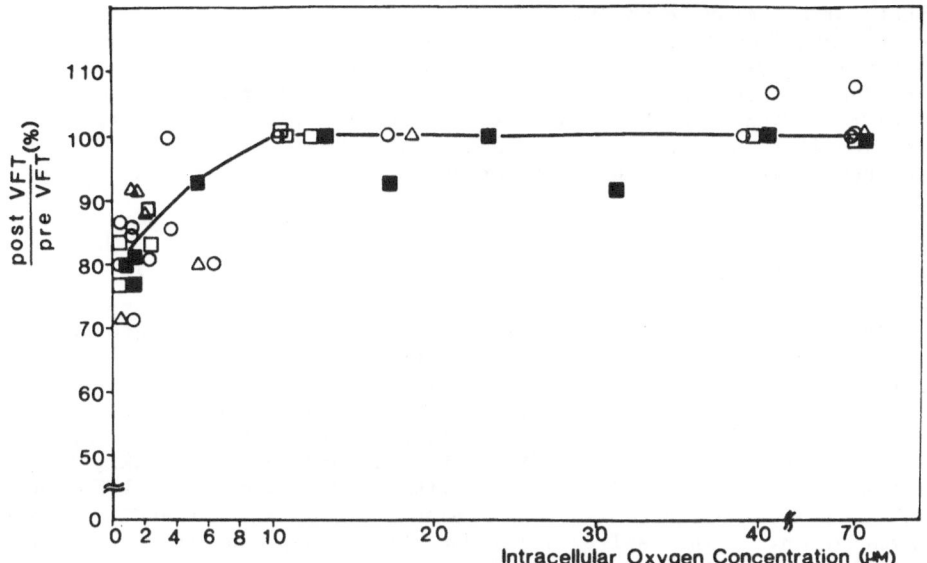

Figure 3. The relationship between intracellular oxygen concentration
and VFT. Pre VFT = the VFT during aeorbic perfusion; Post
VFT = that during hypoxic perfusion. Different symbols
indicate results from different hearts.

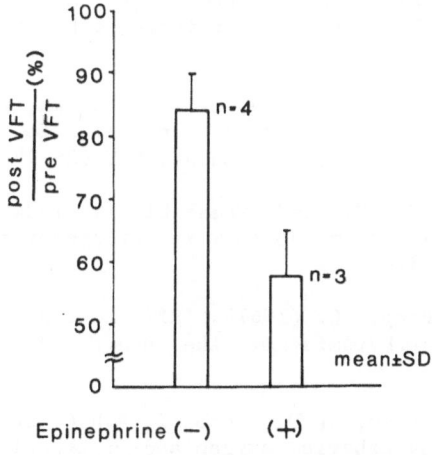

Epinephrine (−)   (+)

Intracellular O_2 Concn. (5~7uM)

Figure 4. VFT change induced by epinephrine infusion during aerobic
perfusion. Left: heart perfused with hypoxic medium without
epinephrine; $[O_2]$ adjusted to 5-7 μM. Right:heart perfused
with aerobic medium + epinephrine; in this state $[O_2]$
decreased to 5-7 μM due to increased cardiac work. Pre VFT =
VFT during aerobic perfusion or before epinephrine infusion
+ aerobic perfusion; Post VFT = VFT during hypoxic perfusion
or during epinephrine infusion + aerobic perfusion.

Effects of epinephrine on VFT

Epinephrine infusion during aerobic perfusion increased both left ventricular pressure and heart rate. The increase in oxygen consumption due to increase in cardiac work caused a decrease in the intracellular oxygen concentration (Araki et al., 1983). However, at the same level of intracellular oxygen concentration (5-7 μM), the VFT fall was much greater with the epinephrine infusion than with hypoxia alone (Fig. 4).

DISCUSSION

As seen in Figure 3, VFT starts to fall at an oxygen concentration below 10 μM. This oxygen concentration is also the critical oxygen concentration for the reduction of pyridine nucleotide, and decrease in oxygen consumption rate (Araki et al., 1983). Thus changes in the critical energy state of cardiac tissue are also reflected in the electrophysiological process (Sagisaka, Tamura and Yamazaki, 1984). Epinephrine, known to induce arrhythmia, decreased the VFT under aerobic perfusion (Fig. 4), and also caused a decrease in oxygen concentration. The fall of VFT induced by epinephrine is much greater than that caused by hypoxia. Thus, another factor such as c-AMP may be a candidate for the increase in vulnerability.

REFERENCES

Araki, R., Tamura, M. and Yamazaki, I. (1983). The effect of intracellular oxygen concentration on lactate release, pyridine nucleotide reduction, and respiration rate in the rat cardiac tissue. Circ. Res. 53, 448-455.

Murnaghan, M.F. (1975). The effect of anoxia on the ventricular fibrillation threshold in the rabbit isolated heart. Br.J.Pharmac. 54, 413-420.

Rogers, R.M., Spear, J.F., Moore, E.N., Horowitz, L.H. and Sonne, J.E. (1973). Vulnerability of canine ventricle to fibrillation during hypoxia and respiratory acidosis. Chest, 63, 986-994.

Sagisaka, K., Tamura, M. and Yamazaki, I. (1984). The effect of K^+ concentration on energy metabolism in perfused rat heart. J. Biochem. (Tokyo), 95, 1091-1103.

Szekeres, L. and Papp, G. (1967). Effect of arterial hypoxia on the susceptibility to arrhythmia of the heart. Acta Physiol. Hung. 32, 143-162.

Turnbull, A.D., MacLean, L.D., Dobell, A.R.C. and Demers, R. (1965). The influence of hyperbaric oxygen and of hypoxia on the ventricular fibrillation threshold. J. Thorac. Cardiovasc. Surg. 50, 842-847.

COMPARISON OF INTRACELLULAR PO$_2$ AND CONDITIONS FOR BLOOD-TISSUE O$_2$ TRANSPORT IN HEART AND WORKING RED SKELETAL MUSCLE

C.R. HONIG and T.E.J. GAYESKI

The University of Rochester, School of Medicine and Dentistry, 601 Elwood Avenue, Rochester, NY 14642, U.S.A.

INTRODUCTION

Recent measurements (Honig et al., 1984; Gayeski, Connett and Honig, 1985; Gayeski and Honig, 1986) and mathematical models (Hellums, 1977; Clark et al., 1985; Federspiel and Popel, in press; Groebe and Thews, 1986) indicate that the principal resistance to O$_2$ mass transfer is at the capillary under conditions of high O$_2$ flux. Federspiel and Popel (in press) have described this behaviour by relating the transcapillary O$_2$ flux to a driving force. The factor of proportionality is a mass transfer coefficient homologous to a conductance:

$$J = C \times [PcapO_2 - PcellO_2] \qquad (1)$$

J is the O$_2$ flux for a short length of capillary. In the steady state J is equal to the rate of O$_2$ consumption ($\dot{V}O_2$) in the surrounding tissue. C is the mass transfer coefficient, and the difference between capillary PO$_2$ (PcapO$_2$) and intracellular PO$_2$ is the driving force. PcapO$_2$ cannot yet be measured. However, coronary venous PO$_2$ affords a minimum estimate of PcapO$_2$ along the red cell flow path. PcellO$_2$ can be determined by cryomicrospectroscopy of myoglobin (Mb) (Gayeski, 1982; Honig et al., 1984; Gayeski et al., 1985; Gayeski and Honig, 1986).

The experiments to be described were undertaken to determine whether increases in cardiac $\dot{V}O_2$ are accompanied by an increase in transcapillary conductance for O$_2$ (C). To this end we compared PO$_2$ in individual myocytes in heart and red skeletal muscle. Results indicate that increase in C is essential for cardiac adaptation to stress.

METHODS

Preparative procedures and fast freezing

Cats and hound-type mongrel dogs were anaesthetized with pentobarbital, 30 mg/kg body weight. Preparative procedures for gracilis muscles have been described in previous reports (Losse, Schuchhardt and

Niederle, 1975; Clark and Clark, 1983; Gayeski et al., 1985). The left ventricular free wall was exposed through a left thoracotomy. FIO_2 was 0.35. Peak inflation pressure and respiratory frequency were adjusted to set $PaCO_2$ and pHa within normal limits. In dogs, cardiac output, left ventricular end-diastolic pressure and left ventricular work and power output were determined.

When all variables were in a steady state the hearts were frozen in situ with a copper heat sink at $-196^{\circ}C$. No attempt was made to synchronize freezing with systole or diastole since the O_2 saturation of Mb is constant throughout the cardiac cycle (Fabel, 1968; Makino et al., 1983). Approximately 10 ms were required to reach $0^{\circ}C$ 100 μm from the surface (Clark and Clark, 1983; Gayeski and Honig, 1986). Mb saturation and a derived optical parameter which depends on ice crystal size were constant between 20 and 600 μm from the pericardial surface. The calculated error in Mb saturation due to O_2 loading or unloading during freezing is <0.1% (Gayeski and Honig, 1986).

Determination of saturation and PO_2

Mb saturation was computed with a four-wavelength method for a non-isosbestic point at 578 nm, as described previously (Gayeski, 1982; Gayeski et al., 1985). The method takes account of the effect of scattered light. Reference spectra of Mb, cytochrome a,a_3, and cytochrome c indicate that cytochrome redox does not have a detectable effect on light absorption at the above wavelengths (Fabel, 1968). Since the molar ratio of Mb to cytochrome a,a_3 is ~8:1 in rat heart, the contribution of cytochrome should be small. In accord with this expectation cytochrome peaks could not be identified in reflection spectra between 530 and 620 nm from subcellular volumes in frozen normoxic or anoxic rat hearts. Absence of cytochrome absorption at our measuring wavelengths has been reported by others (Fabel, 1968; Makino et al., 1983).

The PO_2 in equilibrium with Mb ($PmbO_2$) was calculated from the oxymyoglobin dissociation curve using a Mb P_{50} of 5.3 torr at $37^{\circ}C$. The overall error of the Mb saturation determination was <5% (Gayeski and Honig, 1986). The error in $PmbO_2$ which results from error in saturation depends on the absolute value of saturation because of the influence of the hyperbolic dissociation curve. The standard deviation for saturation corresponds to an error of 0.3 torr when saturation is 20%. The minimum Mb saturation which can be distinguished from zero is 5% corresponding to 0.25 torr. $PmbO_2$ can be interpreted as the PO_2 to which cytochrome a,a_3 is exposed, since the ΔPO_2 between the cytosol and the surface of a mitochondrion is less than 0.05 torr (Clark and Clark, 1985; Connett, Gayeski and Honig, 1986).

Tissue preparation and sampling

Heart. All observations were made on the superficial bulbospiral muscle-band which lies immediately beneath the pericardium. The left ventricular free wall was cleaved into 2 blocks ~ 1.0 x 0.5 x 0.3 cm under liquid N_2. One was from the apex, the other from the base. Blocks were transferred to a cold stage of a Leitz MPV I microscope regulated at $-110^{\circ}C$. A cell 50-200 μm from the pericardium was selected at one edge of a block, and the measuring diaphragm was positioned at the centre of the cell profile. Each cell sampled subsequently was separated from other measuring sites by at least 10 cell diameters. The position of the measuring diaphragm was checked visually before and after each scan.

Gracilis. Each muscle was fractured into 5 blocks. Ten cells were selected in each block; each cell was at least 500 μm from any other. Probability distributions therefore take account of heterogeneity of O_2 delivery and $\dot{V}O_2$ on scales of centimetres and microns. In both heart and gracilis probability distributions for saturation and $PmbO_2$ were based on 50 cells per muscle.

Spatial resolution
Ventricular muscle fibres averaged 16 μm in diameter, so high spatial resolution is necessary to distinguish between Hb and Mb. Spatial resolution in a frozen specimen is determined by the size of the ice crystals and the size of the measuring diaphragm. Internal reflection sharply limits the amount of light emerging from an ice crystal (Gayeski, 1982). In a Mb-free muscle (chicken breast) a 2 x 3 μm measuring diaphragm centred 3 μm from a capillary or venule does not 'see' a Hb signal. The centre of a 2 x 3 μm measuring diaphragm is ~ 8 μm from a capillary when centred in a typical cardiac myocyte. Thus spatial resolution is sufficient.

RESULTS AND DISCUSSION

Myoglobin saturation and intracellular PO_2 in myocardium
Representative PO_2 distributions are shown in Figure 1. Dog Number 8 had normal blood pressure, heart rate and cardiac output. Median $PmbO_2$ was 6.2 torr; the inner quartile range was 5.3-6.8 torr. Dog Number 2 had a spontaneous supraventricular tachycardia which doubled its power output and reduced by about 1/3 the time available for diastolic coronary blood flow. Median saturation was 10% lower at the faster rate. $PmbO_2$ was only 1.6 torr lower, partly because Mb functioned on the steep slope of its oxydissociation curve. The minimum $PmbO_2$ found was 3.7 torr, no different from that in the unstressed heart. Stability of the lower tails of the $PmbO_2$ distributions during tachycardia was also observed in cat. The median Mb saturation was near 50% in all hearts studied to date. In this range change in saturation produces relatively little change in PO_2. Thus cardiac Mb is an efficient PO_2 buffer. Small cell size, the buffer function of Mb, and high functional capillary segment density (FCSD) together account for the remarkable uniformity of $PmbO_2$ in the normal epicardium.

Figure 2 compares PO_2 distributions obtained by Mb spectroscopy in cat subepicardium with distributions for cat subepicardium obtained with O_2 electrodes. The spectroscopic data (filled circles) represent 100 cells, 50 in each of 2 cats. The distribution is not different from that obtained in dogs. Whalen, Nair and Buerk (1973) used microelectrodes with tip diameter 1-2 μm; see Figure 2, open circles. The electrodes could also record intracellular action potentials. Only sites at which normal action potentials were obtained were included in summary statistics. Values < 5 torr could not be resolved. Mean PO_2 was 9.6 torr; the median was > 5 and < 10 torr. These values are only slightly greater than those obtained by Mb spectroscopy. The long upper tail reflects the influence of Hb within the catchment volume of the intracellular electrode and/or the uncontrolled effects of impaling a moving myocyte. Whalen et al. (1973) observed a gradient in PO_2 from epicardium to endocardium. Their mean value for subendocardium was 1.4 torr.

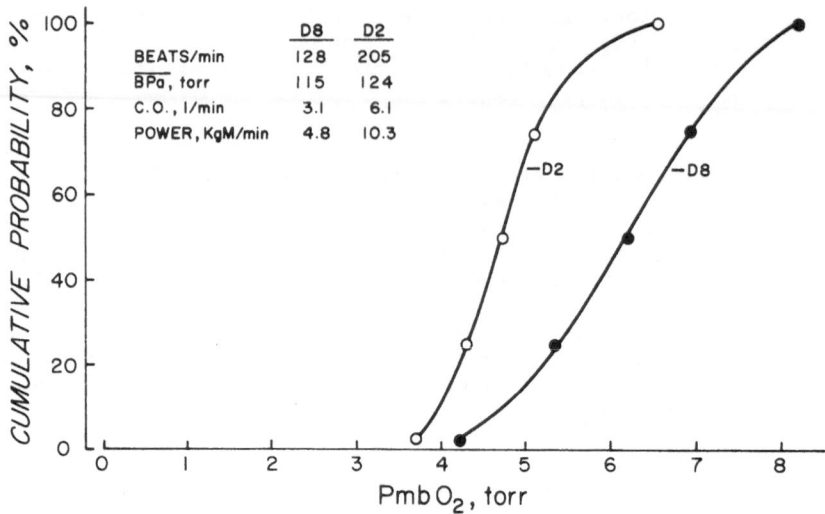

Figure 1. Probability distributions for PO_2 in equilibrium with myoglobin ($PmbO_2$) in two dog hearts (D2 and D8).

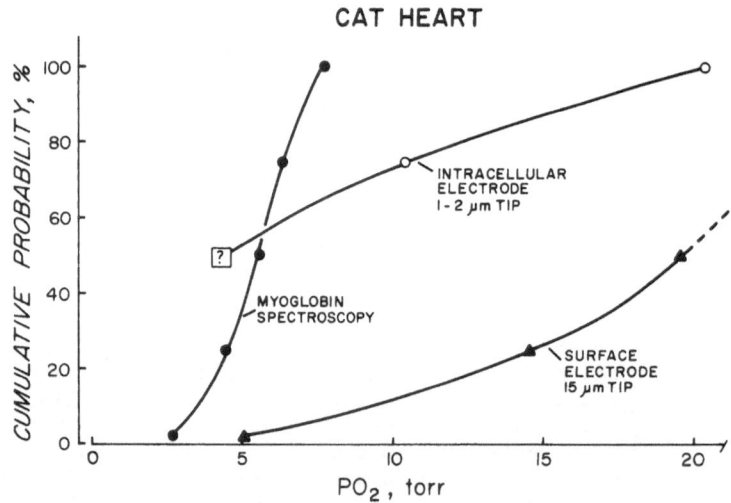

Figure 2. Comparison of PO_2 determined by Mb spectroscopy with PO_2 determined with O_2 electrodes.

The corresponding Mb saturation is 21%. Kirk and Honig (1964) observed a comparable transmural PO_2 gradient in dogs. O_2 tensions of 1-2 torr are highly adaptive because all the transport functions of Mb are enhanced at low Mb saturation (Honig et al., 1984; Gayeski et al., 1985; Gayeski and Honig, 1986): oxygen tension is well buffered, the Mb-facilitated O_2 flux is large, and diffusional shunting of O_2 in the closely packed capillary array is virtually eliminated. Most importantly, PO_2 in the range 0.5 - 1.0 torr maximizes the driving force for O_2 release from capillaries (Equation 1, and Gayeski et al., 1985; Gayeski and Honig, 1986).

Are the O_2 tensions determined spectroscopically high enough to run the mitochondria? The critical PO_2 for cytochrome turnover ($PcritO_2$) can be defined as the PO_2 at which a just detectable decrease in $\dot{V}O_2$ is observed. $PcritO_2$ is less than 0.3 torr in dog gracilis at the same $\dot{V}O_2$ observed in the unstressed dog heart (Connett et al., 1985). Measurements of ubiquinone content (Beyer, Noble and Hirschfeld, 1962) and succinic dehydrogenase activity indicate that O_2 flux per mol of cardiac cytochrome is about one quarter the flux per mol of skeletal cytochrome at equal $\dot{V}O_2$. Consequently $PcritO_2$ should be less than 0.1 torr in the unstressed heart, and should not exceed 0.5 torr at cardiac $\dot{V}O_{2max}$. Judging from the data of Whalen et al. (1973), even in subendocardium PO_2 should be sufficient for maximal $\dot{V}O_2$.

Oxygen distributions obtained with surface electrodes are qualitatively and quantitatively different from those based on intracellular measurements (Losse et al., 1975; Skolasinska et al., 1978). In the example shown in Figure 2 median $PmbO_2$ is 21 torr, three times the maximum $PmbO_2$ determined spectroscopically. Values in the upper half of the distribution (not shown) are comparable to PO_2 in blood. Data obtained with surface and other extracellular electrodes register a quantity which has been termed 'O_2 availability' i.e. a summation of PO_2 from all O_2 sources within the catchment volume. This volume is of the order of 1500 μm^3 for a cathode 15 μm in diameter (Silver, 1973). Though surface electrodes cannot resolve cell PO_2, they can provide valuable information about large scale O_2 gradients, such as those around ischaemic regions.

Relation between $PmbO_2$ and PO_2 in coronary veins

The upper tails of $PmbO_2$ distributions furnish information about the relation between $PcapO_2$ and $PmbO_2$. The maximum $PmbO_2$ found was less than 10 torr in all normal hearts studied to date. Henquell and Honig (1976) reported the mean PO_2 in veins draining the canine left ventricle as 26.8 torr $\pm$ 1.6 S.E.M., n=11. The ΔPO_2 between end-capillary blood and tissue is therefore of the order of 15 torr in the unstressed heart. This large difference could be accounted for by diffusional shunting of O_2.

Countercurrent blood flow promotes diffusional shunting. Countercurrent capillary flow has been observed by in vivo microscopy (Chang et al., 1982), and a network model predicts that almost 40% of cardiac capillary segments function in the countercurrent mode (von Restorff, Holtz and Bassenge, 1977). Diffusional shunting of heat and metabolically inert, lipid-soluble tracers has been demonstrated convincingly by tracer techniques (Bassingthwaighte, Yipintsoi and Knopp, 1984), and a large O_2 shunt is predicted by Grunewald and Sowa's (1978) partial countercurrent model of coronary microcirculation. However, Grote and collaborators found no evidence of a diffusional O_2 shunt in perfused rat heart (Huhmann, Niesel and Grote, 1967). The difference between the behaviour of non-metabolized gases and O_2 can be accounted for by Mb, which serves as a low-tension O_2 'trap' between capillaries. Since Mb buffers $PmbO_2$ well below PvO_2 it makes no difference whether flow is co-current or countercurrent, nor how large the difference in PO_2 between adjacent capillaries. Thus

Mb permits close packing of cardiac capillaries without the ineffic-
iency of a large O_2 diffusion shunt. It is significant that
Grunewald and Sowa's model did not include O_2 binding by Mb.

We conclude that red cells do not equilibrate with cardiac tissue,
and that the transcapillary conductance for O_2 is signficantly lower
than one would predict from the high capillary density and capillary
surface area in myocardium. To put this surprising result in
perspective we compare, below, tissue O_2 transport in heart and red
skeletal muscle during maximal exercise.

Comparison of O_2 transport in canine heart and gracilis muscle

$\dot{V}O_2$ and O_2 delivery. Cardiac O_2 extraction (Henquell and
Honig, 1976), cardiac output, BP, left ventricular dP/dt and external
cardiac work were the same as reported by others for unstressed hearts
(Feigl, 1983). This allows us to relate $PmbO_2$ to values of
myocardial $\dot{V}O_2$ and coronary flow taken from the literature. We chose
the data set of von Restorff et al. (1977) because it includes
measurements during maximal treadmill exercise; see Tables 1 and 2.

Flow and $\dot{V}O_2$ are about the same in the unstressed heart as in
dog gracilis at 4-6 Hz. However, the unstressed heart and the gracilis
at 4-6 Hz function at about 15 and 65% of their respective aerobic
capacities. Minimum resistance to flow in dog gracilis is observed at
8-10 Hz. Higher stimulus frequencies evoke fusion of tension and
obstruction to flow due to steady vascular compression. Resistance to
flow is 3 times greater in gracilis than in heart when both muscles are
at aerobic capacity. The heart's remarkable ability to vasodilate
reflects the fact that vascular compression during systole occupies a
progressively smaller fraction of each cycle as heart rate rises. The
coronary vascular resistance in maximal exercise reported by von
Restorff et al. (1977) is ~33% that in the unstressed heart, despite
a 350% increase in rate! Moreover, a further increase in coronary flow
could be induced by brief arterial occlusion during maximal exercise,
indicating that the coronary flow reserve was not exhausted. The
stability of $PmbO_2$ at near-maximal heart rate helps explain the
ability of the normal heart to sustain the tachycardia of heavy
exercise. The high O_2 extraction of both the stressed and unstressed
heart does not indicate limited capacity for O_2 delivery, nor insufficient O_2 at mitochondria. Instead, high extraction reflects the
heart's capacity for O_2 mass transfer from capillary to cell.

O_2 extraction. Increased O_2 extraction accounted for ~ 40%
of the rise in $\dot{V}O_2$ in canine hearts during maximal treadmill exercise (von Restorff et al., 1977). Coronary venous PO_2 and hence
minimum $PcapO_2$, dropped to 12 torr to achieve this. Even if $PmbO_2$
were zero, the ΔPO_2 across the coronary capillaries would not be
greater than at rest, yet $\dot{V}O_2$ increased four-fold! In contrast, the
small increase in gracilis $\dot{V}O_2$ when twitch frequency rose from 4 Hz
to 8-10 Hz was achieved entirely by increased flow. In some dogs O_2
extraction per decilitre actually fell significantly between 75 and
100% $\dot{V}O_{2max}$ (Honig et al., 1984). The minimum driving force for
transcapillary O_2 flux can be approximated by the difference between
coronary venous PO_2 and $PmbO_2$. This estimated driving force in
gracilis was at least double that in maximally working myocardium, but
gracilis $\dot{V}O_{2max}$ was only one-third cardiac $\dot{V}O_{2max}$. Substituting
these relationships in Equation (1) demonstrates that the mass transfer
coefficient for maximally working heart is about six times that of

Table 1.

	$\dot{V}O_2$	Flow ml/100 g·min	O_2 Delivery ml/100 g·min	Resistance PRV	$\dot{V}O_2$ %max	Flow % change	Functional capillary density % change
Dog Heart							
Unstressed	9	64	12	1.6	16	–	–
Max. exercise	53	256	59	0.5	90	400	+200
Dog Gracilis							
4–6 Hz Twitch	9.8	68.3	14.2	2.5	60	–	–
S.D. (n=7)	0.7	9.3	1.3	0.5			
8–10 Hz	16.4	122	25.5	1.4	90–100	180	–30
S.D. (n=7)	1.3	13	2.9	0.2			

Table 2.

	$\dot{V}O_2$ %max	CaO_2 ml/dl	SvO_2 %	PvO_2 torr	E ml/dl	E %change	$PmbO_2$ 5th Percentile	$PmbO_2$ 50th	PvO_2–Median $PmbO_2$ torr
Dog Heart									
Unstressed	16	20	23	17	14.1	–	1.5	10	8
Max. exercise	90	23	9	10	20.8	+ 149	(>0.5)	–	<9
Dog Gracilis									
4-6 Hz Twitch	60	21.0	27.5	21.7	14.3	–	4.4	11.4	10.3
S.D. (n=7)		2.3	9.4	6.6			1.0	1.7	
8-10 Hz	90-100	21.0	32.3	24.4	13.3	– 9	1.3	4.5	19.9
S.D. (n=7)		1.5	9.1	3.5			1.8	2.6	

maximally working gracilis. These comparisons indicate that the trans-capillary conductance for O_2 as well as driving force must be con-sidered in evaluating O_2 mass transfer. Martin et al. (1979) reached the same conclusion from experiments in which the Hb P_{50} was lowered in isolated blood-perfused rat hearts. Flow, coronary sinus O_2 content, O_2 extraction and $\dot{V}O_2$ remained constant, but coronary sinus PO_2 fell 18 torr. They suggested an increase in FCSD was responsible for the increase in O_2 conductance. von Restorff et al. (1977) and Mildenberger et al. (1977) drew the same inference from the relation- ship between coronary blood flow and O_2 extraction. Evidence for adaptive change in FCSD has been obtained in hearts of various species (Duran, Marsicano and Anderson, 1977; Honig, Frierson and Gayeski, in press).

Determinants of transcapillary conductance for O_2

The overall transcapillary conductance for O_2 depends on the time necessary for O_2 release, and the O_2 flux per red cell (Honig et al., 1984; Groebe and Thews, 1986; Federspiel and Popel, in press). This latter depends, for a given $\dot{V}O_2$, on the mitochondrial volume served by a unit length of capillary. Canine cardiac and gracilis myocytes are ~ 16 μm and 60 μm in diameter, respectively. The number of red cell-containing capillary profiles around the perimeter of a fibre is about the same in heart and gracilis. Therefore a unit length of gracilis capillary supplies ~ 10 times as much tissue volume. On the other hand, respiratory activity per tissue volume is ~ 4 times greater in heart (Beyer et al., 1962). Thus O_2 demand per unit length of capillary is ~ 2.5 times greater in gracilis.

Oxygen flux per red cell depends on transit time through the capillary network as well as on tissue O_2 demand. Transit time is defined as the ratio of length of flow path to red cell velocity. The linear velocity of a red cell is directly proportional to volume flow and inversely proportional to the aggregate cross-sectional area of the local capillary network. Note in Table 1 that volume flow is about the same in unstressed heart as in gracilis at 4-6 Hz. Under those conditions, the ratio of red cell velocities in the two muscles should be inversely proportional to the ratio of functional capillary segment densities. Ashikawa et al. (1984) measured red cell velocity in capillaries of dog left ventricular epicardium by use of high speed cinematography. Mean velocity over the cardiac cycle was 2.5 mm/s. Calculated red cell velocity in dog gracilis at 4 Hz is 3.5 - 4.0 mm/s (Honig and Odoroff, 1981). Then:

$$\frac{\text{FCSD heart}}{\text{FCSD gracilis}} \sim \frac{\text{Red cell velocity gracilis}}{\text{Red cell velocity heart}}$$

$$\frac{1500}{975} = 1.54 \qquad \frac{3.5\text{-}4.0}{2.5} = 1.4\text{-}1.6$$

We conclude that estimates of FCSD in Table 1 can be used as scaling factors to estimate red cell velocities in maximal exercise. When twitch frequency increases to 8-10 Hz gracilis flow increases 80% but FCSD falls 30%. Therefore red cell velocity in gracilis capill-aries becomes: (1.8/0.7) x 3.8 mm/s, or ~ 9.8 mm/s at $\dot{V}O_{2max}$. Mean capillary length in gracilis is 1150 μm. Taking account of capillary branching, red cell path length should be ~ 10% longer, or 1265 μm. Mean transit time is therefore 1265 mm/9.8 mm$\cdot$s^{-1}, ~ 130 ms. The dispersion of lengths and velocities about the mean should have little influence on the above estimate (Honig and Odoroff, 1981). For parameters appropriate for dog gracilis the calculated minimum time required to decrease Hb saturation from 80 to 25% is at least double the above estimate of mean transit time in gracilis capillaries (Honig et al., 1984). The O_2 which cannot be released from gracilis capillaries during the red cell's passage appears in venous blood. Note in Table 2 that venous saturation at 8-10 Hz was at least as high as at 4-6 Hz, and that the O_2 extraction of skeletal muscle did not increase between 60 and 100% of $\dot{V}O_{2max}$. Thus, O_2 extraction and $\dot{V}O_{2max}$ in gracilis appear to have been limited by capillary conductance for O_2. The capacities of oxidative enzymes are matched to capacity for O_2 mass transfer. The $\dot{V}O_2$ predicted from gracilis succinic dehydrogenase activity is about the same as the $\dot{V}O_2$ observed in vivo at 8-10 Hz (Connett, Gayeski and Honig, 1983). Greater enzyme capacity could not be utilized, since more O_2 could not be released from capillary blood.

Conditions are strikingly different in the canine heart. Coronary venous saturation fell from 23% in dogs at rest to 9% during maximum treadmill exercise despite more than a 4-fold increase in flow (von Restorff et al., 1977)! Coronary venous saturations as low as 5% have been observed during steady work in unanaesthetized dogs. Because of the fall in $PcapO_2$ the driving force for O_2 release from cardiac capillaries must be less in maximal exercise than at rest, even if $PmbO_2$ were zero. Therefore the extraordinary extraction of the maximally working dog heart reflects an increase in O_2 conductance. Coronary flow increased 4-fold and FCSD increased 2-fold, so the mean velocity of the red cells should have doubled (4/2) x 2.5 $mm \cdot s^{-1}$, or ~ 5 $mm \cdot s^{-1}$). This velocity is half the estimated red cell velocity in dog gracilis at $\dot{V}O_{2max}$. Because of the circuitous route by which red cells traverse the complex coronary capillary network (Fabel, 1968; Bassingthwaighte, Yipintsoi and Harvey, 1974; Chang et al., 1982; Wieringa et al., 1982) the mean path length for red cells is 2-3 times longer in heart than gracilis. Therefore mean red cell transit time in cardiac capillaries is: 2.5-3.0 $mm/5$ $mm \cdot s^{-1}$, or ~ 500-600 ms at $\dot{V}O_{2max}$. Such transit times are consistent with the calculated minimum time for O_2 saturation to fall from 80% − 10% (Honig et al., 1984).

Transit time and O_2 flux per red cell are not the only determinants of capillary conductance for O_2. Calculations indicate that red cell shape and the length of diffusion path between red cell membrane and the sarcolemma are also important (Feigl, 1983; Honig et al., 1984). Unfortunately, no data are presently available as to whether these parameters change in response to stress. Differences in red cell transit time appear to account for a major portion of the difference in the mass transfer coefficients of heart and gracilis. This follows from the fact that the ratio of transit times and the ratio of mass transfer coefficients are approximately equal in the two muscles.

SUMMARY AND CONCLUSIONS

1. Neither anoxic nor hypoxic cells were found in epicardium of anaesthetized dogs, cats, rabbits and rats despite heterogeneity of flow (Wieringa et al., 1982) and haematocrit (Honig et al., in press) in the coronary capillary network.

2. Median PO_2 in unstressed dog heart and cat heart are 4.8 and 5.2 torr, respectively. These values are close to the P_{50} of the oxymyoglobin dissociation curve, and well above $PcritO_2$.

3. A dense, interconnected capillary network and high capillary haematocrit appear essential to achieve high O_2 extraction at flows characteristic of maximally working myocardium.

4. Mb promotes O_2 transport in myocardium by: a) maximizing the driving force for transcapillary diffusion, b) minimizing spatial variability in $PmbO_2$, c) facilitating O_2 diffusion in myocytes and, d) permitting close capillary packing without a diffusion shunt for O_2.

5. The O_2 conductance of the red cell-capillary system is a major determinant of O_2 mass transfer in red muscle.

ACKNOWLEDGEMENTS

Discussions with A. Clark, P. Clark, W. Federspiel and R. Connett helped clarify our thinking. Our research is supported by Grant HLB 03290 from the United States Public Health Service.

REFERENCES

Ashikawa, K., Kanatsuka, H., Suzuki, T. and Takishima, T. (1984). A new microscope system for the continuous observation of the coronary microcirculation in the beating canine left ventricle. Microvasc. Res. 28, 387

Bassingthwaighte, J.B., Yipintsoi, T. and Harvey, R.B. (1974). Microvasculature of the dog left ventricular myocardium. Microvasc. Res. 7, 229

Bassingthwaighte, J.B., Yipintsoi, T. and Knopp, T.J. (1984). Diffusional arterio-venous shunting in the heart. Microvasc. Res. 28, 233

Beyer, R.E., Noble, W.N. and Hirschfeld, T.J. (1962). Alterations of rat-tissue coenzyme Q (ubiquinone) levels by various treatments. Biochim. Biophys. Acta, 57, 376

Chang, B.-L., Yamakawa, T., Nuccio, J., Pace, R. and Bing, R.J. (1982). Microcirculation of left atrial muscle, cerebral cortex and mesentery of the cat. Circ. Res. 50, 240

Clark, A., Jr and Clark, P.A.A. (1983). Capture of spatially homogeneous chemical reactions in tissue by freezing. Biophys. J. 42, 25

Clark, A., Jr and Clark, P.A.A. (1985). Local oxygen gradients near isolated mitochondria. Biophys. J. 48, 931

Clark, A., Jr, Federspiel, W., Clark, P.A.A. and Cokelet, G.R. (1985). Oxygen delivery from red cells. Biophys. J. 47, 71

Connett, R.J., Gayeski, T.E.J. and Honig, C.R. (1983). Lactate production in a pure red muscle in absence of anoxia: mechanisms and significance. In: Oxygen Transport to Tissue-IV. Eds Bicher, H.I. and Bruley, D.F., Plenum Press, New York and London, (Adv. Exp. Med. Biol. 159, 327).

Connett, R.J., Gayeski, T.E.J. and Honig, C.R. (1985). An upper bound on the minimum PO_2 for O_2 consumption in red muscle. In: Oxygen Transport to Tissue-VII. Eds Kreuzer, F., Cain, S. M., Turek, Z. and Goldstick, T.K., Plenum Press, New York and London, (Adv. Exp. Med. Biol. 191, 291).

Duran, W.N., Marsicano, T.H. and Anderson, R.W. (1977). Capillary reserve in isometrically contracting dog hearts. Am. J. Physiol. 233, H276

Fabel, H. (1968). Normal and critical O_2-supply of the heart. In: Oxygen Transport in Blood and Tissue. Eds Lubbers, D.W., Luft, U.C., Thews, G. and Witzleb, E., Thieme Verlag, Stuttgart, pp. 159-171.

Federspiel, W.J. and Popel, A.S. A theoretical analysis of the effect of the particulate nature of blood on oxygen release in capillaries. Microvasc. Res., (in press).

Feigl, E.O. (1983). Coronary physiology. Physiol. Rev. 63, 1

Gayeski, T.E.J. (1982). A cryogenic microspectrophotometric method for measuring myoglobin saturation in subcellular volumes: Application to resting dog gracilis muscle. Ph.D. Dissertation, University of Rochester, Rochester, NY, U.S.A.

Gayeski, T.E.J., Connett, R.J. and Honig, C.R. (1985). Oxygen transport in rest-work transition illustrates new functions for myoglobin. Am. J. Physiol. 248, H914

Gayeski, T.E.J. and Honig, C.R. (1986). O_2 gradients from sarcolemma to cell interior in a red muscle at maximal VO_2. Am. J. Physiol. 251, H789

Groebe, K. and Thews, G. (1986). Theoretical analysis of oxygen supply to contracted skeletal muscle. In: Oxygen Transport to Tissue-VIII. Ed. Longmuir, I.S., Plenum Press, New York and London, (Adv. Exp. Med. Biol. 200, 495).

Grunewald, W.A. and Sowa, W. (1978). Distribution of the myocardial tissue PO_2 in the rat and the inhomogeneity of the coronary bed. Pflugers Arch. 374, 57

Hellums, J.D. (1977). The resistance of oxygen transport in the capillaries relative to that in the surrounding tissue. Microvasc. Res. 13, 131

Henquell, L. and Honig, C.R. (1976). O_2 extraction of right and left ventricles. Proc. Soc. Exp. Biol. Med. 152, 52

Honig, C.R., Frierson, J.L. and Gayeski, T.E.J. Red cell spacing and recruitment of red cell-containing capillary segments in rat heart: role in O_2 transport. Am. J. Physiol., (in press).

Honig, C.R., Gayeski, T.E.J., Federspiel, W., Clark, A., Jr and Clark, P. (1984). Muscle O_2 gradients from hemoglobin to cytochrome: new concepts, new complexities. In: Oxygen Transport to Tissue-V. Eds Lubbers, D.W., Acker, H., Leniger-Follert, E. and Goldstick, T.K., Plenum Press, New York and London, (Adv. Exp. Med. Biol. 169, 23).

Honig, C.R. and Odoroff, C.L. (1981). Calculated dispersion of capillary transit times: significance for oxygen exchange. Am. J. Physiol. 240, H199

Huhmann, W., Niesel, W. und Grote, J. (1967). Untersuchungen uber die Bedingungen fur die Sauerstoffversorgung des Myokards an perfundierten Rattenherzen. Pflugers Arch. 294 250

Kirk, E.S. and Honig, C.R. (1964). Nonuniform distribution of blood flow and gradients of oxygen tension within the heart. Am. J. Physiol., 207, 661

Losse, B., Schuchhardt, S. and Niederle, W. (1975). The oxygen pressure histogram in the left ventricular myocardium of the dog. Pflugers Arch. 356, 121

Makino, N., Kanaide, H., Yoshimura, R. and Nakamura, M. (1983). Myoglobin oxygenation remains constant during the cardiac cycle. Am. J. Physiol. 245, H237

Martin, J.L., Duvelleroy, M., Teisseine, B. and Durable, M. (1979). Effect of an increase in HbO_2 affinity on the calculated capillary recruitment of an isolated rat heart. Pflugers Arch. 382, 57

Mildenberger, R.R., L'Abbate, A., Zborowska-Sleus, D.T. and Klassen, G.A. (1977). The relationship between oxygen extraction and mean transit time in the canine heart. Can. J. Physiol. Pharmac. 55, 478

Silver, I.A. (1973). The oxygen microelectrode. In: Oxygen Transport to Tissue. Eds Bicher, H.I. and Bruley, D.F., Plenum Press, New York and London, (Adv. Exp. Med. Biol. 37A, 478).

Skolasinska, K., Harbig, K., Lubbers, D.W. and Wodick, R. (1978). PO_2 and microflow histograms of the beating heart in response to changes in arterial PO_2. Basic Res. Cardiol. 73, 307

von Restorff, W., Holtz, J. and Bassenge, E. (1977). Exercise induced augmentation of myocardial oxygen extraction in spite of normal coronary dilatory capacity in dogs. Pflugers Arch. 372, 181

Whalen, W.J., Nair, P. and Buerk, D. (1973). Oxygen tension in the beating cat heart in situ. In: Oxygen Supply. Eds Kessler, M., Bruley, D.F., Clark, L.C., Jr, Lubbers, D.W., Silver, I.A. and Strauss, J., Univ. Park Press, Baltimore, pp. 199

Wieringa, P.A., Spaan, J.A.E., Stassen, H.G. and Laird, J.D. (1982). Heterogeneous flow distribution in a three dimensional network simulation of the myocardial microcirculation - a hypothesis. Microcirculation, 2, 195

DYNAMIC STRUCTURE OF PHOSPHOLIPID BILAYERS ON THE PATH FOR OXYGEN DIFFUSION IN THE OX LUNG

T. KOYAMA and T. ARAISO

Division of Physiology, Research Institute of Applied Electricity, Hokkaido University, 060 Sapporo, Japan

SUMMARY

The dynamic properties of the phospholipid bilayer of cell membranes were studied with a nanosecond fluorometer to obtain information on the microstructure of the path for oxygen diffusion in the ox lung. The viscosity in membranes of pneumocytes, endothelial cells from pulmonary artery, and erythrocytes was 47, 54 and 162 mPa·sec respectively. The wobbling angle representing the oscillation of phospholipid molecules was 47, 44 and 38 degrees, respectively.

INTRODUCTION

Inhaled oxygen molecules diffuse through the alveolar and endothelial membranes, the blood plasma and the erythrocyte membrane. Several phospholipid bilayers, therefore, can be distinguished on the path for oxygen diffusion in the lung. Measurements of the dynamic structure of phospholipid bilayers are required for understanding oxygen diffusion barriers in the lung. In the present study the viscosity (η) and range of oscillation (θ) of phospholipid molecules in the lipid bilayers were measured for isolated pneumocytes, endothelial cells and erythrocyte ghost membranes by the nanosecond fluorescence depolarization technique .

METHODS

Pulmonary artery and pieces of lung tissue of oxen were excised within 30 min after slaughter and transported to the laboratory in ice-cooled Dulbecco's PBS(−) solution containing antibiotics. The pulmonary artery was longitudinally incised and fixed on a plate with the interior surface open upwards. The interior surface was gently scraped with a razor blade (Ryan, Mortara and Whitaker, 1980). The mucus-like fluid on the blade was placed in PBS(−) containing 0.1% collagenase and agitated gently with a magnetic stirrer at 20°C for 30 min. The mixture was then filtered with a nylon mesh (30 μm) and the filtrate was centrifuged at 4000 r.p.m. for 5 min. The collagenase solution was removed by aspiration and the pellet of endothelial cells was resuspended in a small quantity of PBS(-). The cells were washed three times. Fragments of the pavement-like monolayer of small polygonal (endothelial) cells (Fig. 1A) were dispersed into individual spherical cells of a similar size (Fig. 1B).

Isolated pneumocytes have been studied in rabbits and rats (Clements, Jones and Felts, 1972; Kikkawa and Yoneda, 1974; Douglas and Smith, 1982). The lungs of these small animals were readily washed and degassed by tracheal cannulation. This method, however, was impossible in oxen and the following procedure for degassing was applied to minced lung tissues. The lung tissue was minced with scissors in ice-cooled PBS(-). Six ml of PBS(-), suspending about 2 ml of the minced lung tissue, were placed in a 10 ml syringe. The tip of the needle was sealed with a rubber stopper. The piston was pulled, so as to make air bubbles come out from the alveolar lumen. Then air bubbles were expelled from the syringe. This procedure was repeated several times to degas the lung tissue and to wash free cells off the alveolar surface into the PBS(-). The degassed lung mince was washed with PBS(-) and filtered through a nylon mesh (30 μm), so that free cells could be eliminated. Two ml of the lung mince were placed in a 10 ml syringe. Six ml of trypsin solution (10 mg/ml in PBS(-)) were sucked into this syringe and the procedure of degassing was repeated three times so as to introduce trypsin solution into the alveolar lumen. Two ml of air were sucked into the syringe and the mixture was incubated at 37°C in a rotating tonometer. After a 30 min incubation the procedure for degassing was repeated again to disperse isolated cells from the alveolar lumen. The dispersed cells were filtered through nylon mesh and centrifuged with a swing rotor at the rate of 2000 r.p.m. for 30 min. The trypsin solution was removed and the cells were washed three times with PBS(-) at room temperature and used for measurements without any further separation of cell types. Erythrocytes could be easily discarded, because they formed a red pellet at the bottom of the centrifuge tube.

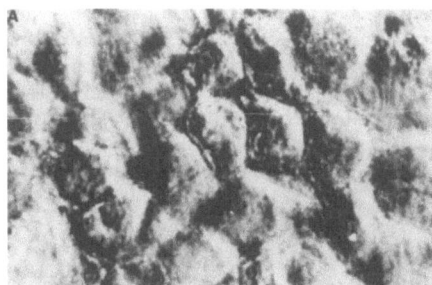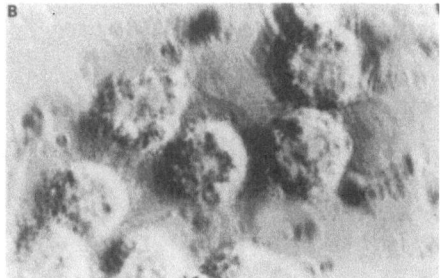

Figure 1. A. Pavement-like structure of the endothelium.
B. Dispersed endothelial cells.

For measurements on erythrocytes, ghost membranes were used because the haemoglobin contained in erythrocytes strongly quenches the fluorescence of the fluorophore, 1,6-diphenyl-1,3,5-hexatriene (DPH).

The nanosecond time-resolved fluorometer has been described previously (Kinosita, Kawato and Ikegami, 1977; Kawato, Kinosita and Ikegami, 1977; Koyama et al., 1984; Araiso et al., 1986) and its application to measurements in erythrocyte membrane ghosts in several

animal species has been reported (Koyama, Araiso and Mochizuki, 1986).
Briefly, the fluorophore DPH, taken into the phospholipid bilayer of
cells, was excited repeatedly with a polarized pulsed light of a
nanosecond duration at a wavelength of 340 nm. The excited
DPH-molecules emit anisotropic fluorescent light after the end of each
light pulse. The anisotropy of the fluorescent light decays as a
function of the viscosity in the micro-region of the phospholipid
array. The value of the anisotropy ratio at infinite time (20 to 30
nsec after pulsed illumination with parallel-polarized light) reflects
the wobbling angle of DPH, i.e. the range of oscillation of the lipid
chains of phospholipid molecules. To get the anisotropy decay curve,
the parallel and perpendicular components of the fluorescent light,
($I\|(t)$ and $I\perp(t)$), were measured at 428 nm with a single photon
counting method. The anisotropy decay curve, $r(t)$, was calculated with
a personal computer from these curves and displayed on an X-Y plotter
together with intensity decay curves of the fluorescent light. The
anisotropy decay curve is expressed by the following equation:
$r(t) = (r_0 - r_\infty) \cdot \exp(-t/\phi) + r_\infty$, where r_0, r_∞ and ϕ represent the
initial value of $r(t)$, the steady value of $r(t)$ at infinite time and
the rotational correlation time, respectively. The viscosity, η, and
wobbling angle, ϕ, were calculated using the values for ϕ and r_∞ with a
personal computer as described in detail previously (Araiso et al.,
1986). The wobbling angle represents the range of oscillation of
phospholipid molecules constituting cell membranes (T. Araiso and T.
Koyama, to be published).

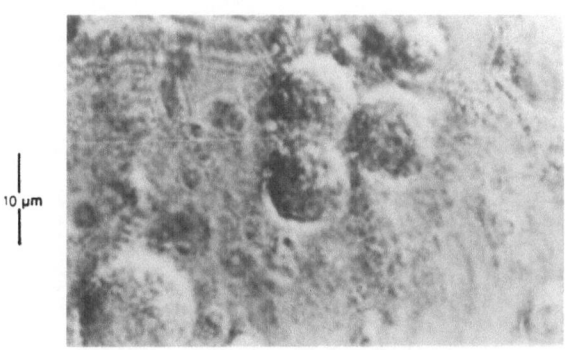

Figure 2. Dispersed pneumocytes.

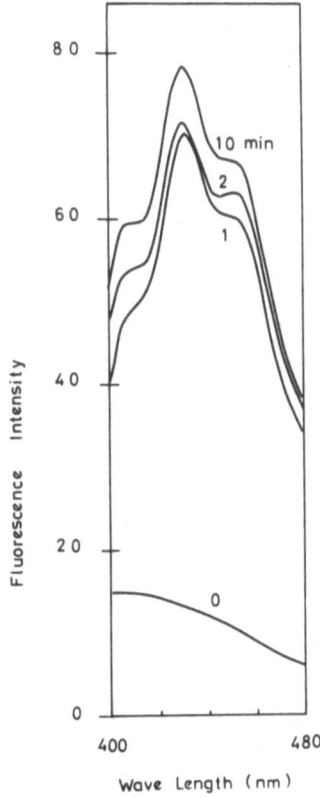

Figure 3. Increase in DPH
fluorescent light.

The anisotropy ratio (rs) of the DPH fluorescence excited with continuous illumination, was measured with a commercially available spectrofluorophotometer (Shimadzu RF-540). Approximately 10^4 cells were suspended in 3 ml saline and placed in a non-fluorescent quartz cuvette and used in measurements with both pulsed and continuous illumination. The spectrum of DPH fluorescence was measured with the spectrofluorometer in the spectrum mode and an excitation wavelength of 340 nm. All measurements were made at 37°C.

RESULTS AND DISCUSSION

A photomicrograph of dispersed pneumocytes in PBS(-) is shown in Figure 2. Pneumocytes of different types with different diameters can be seen; the biggest cell was 12.2 μm in diameter, while the smallest was 5.5 μm. The other, medium-sized, cells ranged in diameter from 7.8 to 10.4 μm. Seventy-two per cent of randomly photographed cells were medium-sized, and their mean diameter $\pm$ S.D. was 9.3 $\pm$ 0.8 μm. All cells seemed spherical. Ninety per cent of the cells remained unstained by Trypan blue application. The intensity of DPH fluorescence increased quickly after the addition of the cell suspension to the DPH dispersion and reached a steady state in 30 min. An example obtained in endothelial cells is shown in Figure 3.

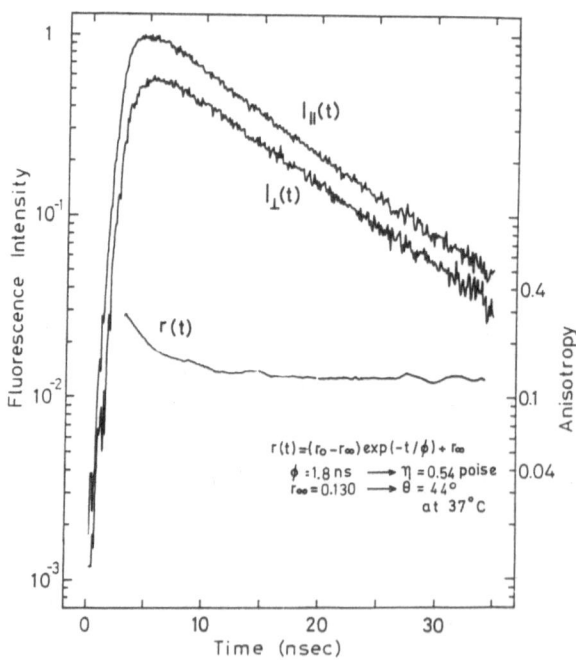

Figure 4. Recordings of anisotropic decay curves of DPH fluorescence in endothelial cells. The fluorescence intensity on the ordinate is normalized with the maximal value in I‖(t).

An example of recordings of parallel and perpendicular components of DPH fluorescence in endothelial cells is shown in Figure 4.

Both I‖(t) and I⊥(t) decreased to less than 1/100th of their initial intensity in 50 nsec. The two curves were less separated in endothelial cells than in erythrocyte ghost membranes as shown in Figure 5.

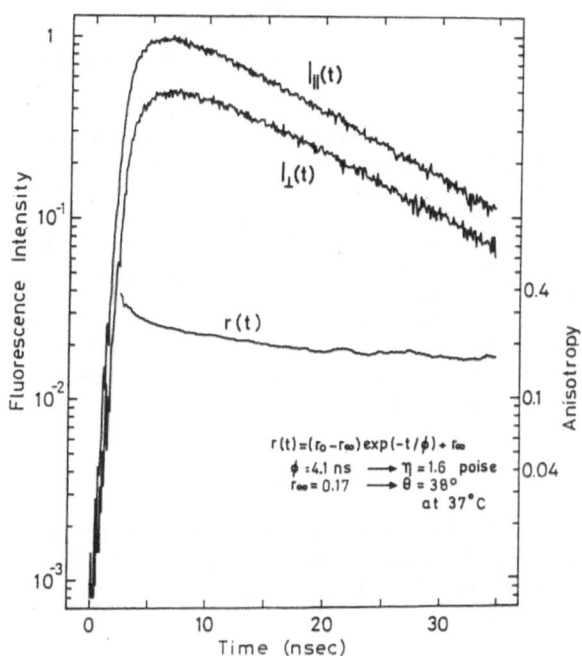

Figure 5. Anisotropic decay curves of DPH fluorescence in erythrocyte ghost membranes. The scale for the fluorescence intensity is the same as in Figure 4.

The anisotropy decay curve decreased in endothelial cells more steeply than in erythrocyte ghost membranes. The variables calculated from the anisotropy decay curves obtained with pulsed illumination, and rs are listed in Table 1.

Table 1. Parameters of fluorescence anisotropy for DPH in ox lung at 37°C

Cell type	rs	r	Θ degr	Φ nsec	*D 10^{-7}/sec	η mPa.sec
Pneumocytes	0.15	0.11	47	1.5	7.6	47
Endothelial cells	0.17	0.13	44	1.8	7.9	54
Erythrocytes	0.21	0.17	38	4.1	2.7	162

*D, rotational diffusion coefficient.

The wobbling angle was smaller and the viscosity greater, in erythrocyte membranes than in pneumocytes and endothelial cells. The following biological interpretation seems possible. Erythrocytes are strongly deformed when they are forced to flow through capillaries. To permit a reversible deformation, a viscous and tight membrane is required for erythrocytes. The endothelial cells are exposed to continuous shear stress from blood flow, but no deformation is required

The pneumocytes, especially alveolar Type I cells are repeatedly subjected to slight stretching and relaxation with respiratory movements, but they probably suffer a relatively small stretch force. Thus, the results listed in Table 1 seem to be reasonable.

The viscosity is inversely proportional to the diffusion coefficient, while the wobbling angle represents the range of oscillation of phospholipid molecules. These two factors are of great importance with respect to permeation of oxygen and other substances across the cell membranes.

Both the endothelial cells of the pulmonary artery, and the pneumocytes became spherical when enzymatically dispersed. This spherical shape is different from their stretched shape in vivo. It is therefore possible that the measured viscosity may be higher than that of the actual cell membranes through which oxygen molecules diffuse down the partial pressure gradients in vivo. The difference, however, seems to be small, since the viscosity of phospholipid vesicles decreased only slightly when they were swollen by exposure to a hypotonic solution and their surface area was increased (M. Kinjyo, personal communication). The present results provide information on the relative importance of each step in the diffusion of oxygen in the lung.

ACKNOWLEDGEMENTS

The authors wish to express their thanks to Professors M. Wakita, School of Dentistry, Hokkaido University and T. Akino, Sapporo Medical College for their valuable advice.

REFERENCES

Araiso, T., Shindo, Y., Arai, T., Nitta, J., Kikuchi, Y., Kakiuchi, Y and Koyama, T. (1986). Viscosity and order in erythrocyte membranes studied with nanosecond fluorometery. Biorheology, 23, 467–483.

Clements, J.A., Jones, A.L. and Felts, J.M. (1972). Dispersal of rabbit lung into individual viable cells: a new model for the study of lung metabolism. Science, 178, 1209–1210.

Douglas, W.H.J. and Smith, S.S. (1982). Lung cell culture systems. In: Lung Development; Biological and Clinical Perspectives 1. Ed. Farrell, P.M., Academic Press Inc., New York, pp. 151–163.

Kawato, S., Kinosita K., Jr and Ikegami, A. (1977). Dynamic structure of lipid bilayers studied by nanosecond fluorescence techniques. Biochemistry, 16, 2319–2324.

Kikkawa, Y. and Yoneda, K. (1974). The type II epithelial cell of the lung, I. Method of isolation. Lab. Invest. 30, 76–84.

Kinosita, K., Jr, Kawato, S. and Ikegami, A. (1977). A theory of fluorescence polarization decay in membranes. Biophys. J. 20, 289–305.

Koyama, T., Araiso, T. and Mochizuki, M. (1986). Oxygen diffusion coefficient of cell membranes. In: Oxygen Transport to Tissue-VIII. Ed. Longmuir, I.S., Plenum Press, New York and London, (Adv. Exp. Med. Biol. 200, 99–106).

Koyama, T., Araiso, T., Shindo, Y. and Arai, T. (1984). A nanosecond fluorometer for studies of rheological aspects of biomembranes. Biorheology, Suppl. I, 323-329.

Ryan, U.S., Mortara, M. and Whitaker, C. (1980). Methods for microcarrier culture of bovine pulmonary artery endothelial cells avoiding the use of enzymes. Tissue & Cell, 12, 619-635.

EFFECTS OF TEMPERATURE ON OXYGEN TRANSFER CONDUCTANCE OF HUMAN RED BLOOD CELLS

J. PIIPER, K. YAMAGUCHI and P. SCHEID

Department of Physiology, Max Planck Institute for Experimental Medicine, Gottingen, F.R.G.

SUMMARY

The influence of temperature (varied from 37 to 7°C; average pH = 7.4) on the kinetics of O_2 uptake and release by human red blood cells under stopped-flow conditions was investigated by double-beam spectrophotometry. The kinetics were characterized by the specific transfer conductance for O_2, G. The temperature coefficient of G, $Q_{10}(G)$, for O_2 uptake averaged 1.17, and activation energy, $E_a(G)$ = 2.9 kcal/mol O_2. The average values for O_2 release were: $Q_{10}(G)$ = 1.30, and $E_a(G)$ = 4.8 kcal/mol O_2. The G values for release of O_2 from oxyhaemoglobin solution, G_{sol}, yielded $Q_{10}(G_{sol})$ = 2.06, $E_a(G_{sol})$ = 13.4 kcal/mol O_2. Comparison of these Q_{10} and E_a values with those for diffusion of O_2 and haemoglobin in aqueous media leads to the conclusion that the kinetics of O_2 uptake and release by red blood cells in the stopped-flow condition is mainly limited by diffusion of O_2 and haemoglobin in the red cell interior and by diffusion of O_2 in the medium, and to a lesser degree by chemical reaction kinetics.

INTRODUCTION

The effects of temperature have been used to attempt to quantify the limiting roles of diffusion and chemical reaction in O_2 transfer kinetics of red blood cells (RBC) using rapid-mixing devices. Thus Sirs (1963) and Lawson, Holland and Forster (1965) obtained Q_{10} values of about 2.2–2.5 for O_2 release from RBC, suggesting that the deoxygenation of RBC was mainly determined by reaction limitation, since the temperature effect was consistent with that expected for the dissociation reaction of oxyhaemoglobin (HbO_2). In contrast, the results of recent studies by Olson and his associates (Coin and Olson, 1979; Vandegriff and Olson, 1984a, b) and by our group (Yamaguchi et al., 1985) appeared to indicate that O_2 release from RBC at a sufficiently high concentration of dithionite was mainly limited by intracellular O_2 diffusion and that O_2 uptake by RBC was additionally limited by O_2 diffusion in the medium surrounding the RBC, i.e. in both cases by diffusion in an aqueous medium. Therefore, the present study was undertaken to reinvestigate the effects of temperature variations on both oxygenation and deoxygenation of human RBC in the stopped-flow condition.

METHODS

The measurements were made on freshly drawn venous blood samples (from two male subjects) which were immediately cooled to $4^{\circ}C$ for storage and used within 24 hours.

The O_2 transfer kinetics were investigated using the stopped-flow technique with double-beam spectrophotometry described previously (Yamaguchi et al., 1985). The RBC suspension and a buffer solution were rapidly mixed and the time-course of O_2 saturation changes in the mixture after flow stop was recorded. The equilibration system and the stopped-flow apparatus were thermostatically controlled, within $0.1^{\circ}C$, at 7, 17, 27, or $37^{\circ}C$. The pH of the RBC suspension medium and of the buffer solution was kept at 7.10, 7.40 or 7.70 by choice of appropriate PCO_2. The following three series of experiments were performed:

(1) Oxygen uptake into RBC
A red cell suspension was prepared by diluting 1 ml whole blood with 99 ml of isotonic saline-phosphate-bicarbonate buffer. This suspension was fully deoxygenated with a gas mixture containing only CO_2 and N_2. Cell-free isotonic buffer solution was equilibrated with gas mixtures containing the appropriate O_2 and CO_2 concentrations balanced with nitrogen. The changes of O_2 saturation after mixing of these liquids were recorded.

(2) Oxyyen release from RBC
An isotonic buffer solution containing 80 mmol/l sodium dithionite was prepared as described before (Yamaguchi et al., 1985). The reaction between oxygenated RBC, equilibrated with a gas mixture containing 20% O_2, and deoxygenated dithionite solution was investigated.

(3) Oxygen release from haemoglobin solutions
To estimate the apparent dissociation velocity constant (k) of oxyhaemoglobin, 1 ml of fresh whole blood was lysed by addition of 3 mg saponin, and 99 ml of a buffer formulated to resemble the intraerythrocyte electrolyte concentration were added to the lysate. The haemoglobin solution thus prepared was equilibrated with a gas mixture containing 20% O_2 and 4-6% CO_2 at the desired temperature. A fully deoxygenated solution of 80 mmol/l dithionite was prepared using the same intracellular-like buffer equilibrated with 4-6% CO_2. The pH in both solutions was 7.20.

CALCULATIONS

Specific oxygen transfer conductance (G)
To analyse the RBC O_2 transfer kinetics, the specific O_2 transfer conductance, G, was calculated in accordance with Yamaguchi et al. (1985) as the ratio of the effective transfer rate of O_2 per unit RBC volume (obtained from the measured rate of change of O_2 saturation, dS_{O_2}/dt, and the RBC O_2 capacity, C) and the difference between medium PO_2 (P_m) and the haemoglobin equilibrium PO_2 (P_{eq}):

$$G = \frac{C \cdot dSO_2/dt}{P_m - P_{eq}} \qquad (1)$$

A G value can be calculated for any SO_2, and represents an overall O_2 transfer conductance in the process of O_2 transport between the surrounding medium and the haemoglobin in the cells. Although formally calculated as a diffusive conductance, G may be determined by other rate-limiting steps in the O_2 exchange process, e.g. by reaction of O_2 with haemoglobin.

The haemoglobin equilibrium PO_2, P_{eq}, is determined by SO_2 and by the O_2 dissociation curve for the given pH and temperature. Assuming temperature invariance of the shape of the O_2 dissociation curve but temperature dependence of O_2 affinity, O_2 dissociation curves were constructed as isomorphic with the standard curve according to Severinghaus (1979). For the effect of temperature on O_2 affinity, the half-saturation PO_2, P_{50}, was spectrophotometrically determined at 37, 27, 17 and $7^\circ C$ from the Hill plots established for the SO_2 range 0.3–0.8.

Temperature coefficient (Q_{10}) and activation energy (E_a)
These were calculated according to their definitions:

$$Q_{10}(x) = \frac{x_{T+10}}{x_T} \qquad (2)$$

$$E_a(x) = -2.303 \ R \cdot \frac{\Delta \log x}{\Delta (1/T)} \qquad (3)$$

where T is the absolute temperature, in Kelvin; x, the temperature-dependent variable under study with the values x_T and x_{T+10} at temperature T and T+10 Kelvin; Δ, difference, between values at higher T at lower T; R, the gas constant [= 1.987 cal/(mol·Kelvin)]; 2.303 = natural logarithm of 10.

The Q_{10} and E_a values are interconvertible according to the relationship resulting from combination of Equations (2) and (3):

$$E_a(x) = 2.303 \ RT \cdot \frac{T + 10}{10} \cdot \log Q_{10}(x) \qquad (4)$$

Dissociation kinetics of oxyhaemoglobin solutions
For haemoglobin solution in the presence of 40 mmol/l dithionite (after mixing), the decrease of SO_2 was nearly exponential and could thus be characterized by a (nearly) SO_2-independent dissociation velocity constant, k:

$$k = -\frac{dSO_2/dt}{SO_2} \qquad (5)$$

$E_a(k)$ was obtained from a plot of log k against $1/T$ calculated according to Equation (3). Since the specific O_2 reaction conductance of haemoglobin solution, G_{sol}, is proportional to k and inversely proportional to P_{eq}, the temperature dependence of G_{sol} results from the temperature dependence of k and of P_{50}:

$$\frac{\Delta \log G_{sol}}{\Delta(1/T)} = \frac{\Delta \log k}{\Delta(1/T)} - \frac{\Delta \log P_{50}}{\Delta(1/T)} \tag{6}$$

RESULTS

The mean Q_{10} and E_a values are shown in Table 1. The Q_{10} values were averaged for the temperature steps $37°/27°$, $27°/17°$ and $17°/7°$. The E_a values were calculated from Q_{10} for the temperature step $37°/27°$ using Equation (4).

Table 1. Mean experimental Q_{10} and E_a values

	Q_{10}	E_a (kcal/mol O_2)
G for O_2 uptake by RBC	1.17	2.9
G for O_2 release by RBC	1.30	4.8
G for O_2 release by haemoglobin solution	2.06	13.4
Half-saturation PO_2 of haemoglobin, P_{50}	1.81	11.0

DISCUSSION

The Q_{10} and E_a values measured for the O_2 transfer conductance, G, should yield information on the rate-limiting effects of the individual processes involved in O_2 transfer of RBC: chemical reaction with haemoglobin (association or dissociation of O_2), diffusion within the RBC, diffusion across the RBC membrane, diffusion in the medium surrounding the RBC. In the analysis, the experimental Q_{10} and E_a values (Table 1) will be compared with the corresponding values for the determinants of diffusive O_2 transfer shown in Table 2.

Reaction limitation

Since the value measured for O_2 release from RBC, $E_a(G) = 4.7$ kcal/mol, was much lower than that obtained for O_2 release from oxyhaemoglobin solution, $E_a(G_{sol}) = 13.4$ kcal/mol, reaction limitation probably was not of major importance. Moreover, the value of G for O_2 transfer of RBC revealed no significant changes with pH in the range between 7.1 and 7.7 (unpublished measurements). Since in this pH range the reaction velocity constants are expected to be sensitive to changes of pH (Bauer et al., 1973), this finding may be considered as further evidence against a predominant limitation by chemical reaction.

Table 2. Q_{10} and E_a values for determinants of diffusive O_2 transport, from various data in the literature (see Discussion). E_a calculated from Q_{10} for the temperature step $37^\circ/27^\circ$

	Q_{10}	E_a (kcal/mol O_2)
Solubility in RBC	0.85	-3.0
Krogh's diffusion constant for RBC and plasma	1.10	1.8
Haemoglobin diffusion coefficient in RBC	1.30	4.8
Krogh's diffusion constant for RBC membrane	1.47	7.1

Diffusion limitation in RBC membrane

The diffusive conductance is expected to be proportional to the Krogh O_2 diffusion constant (which is equal to the product of O_2 solubility and O_2 diffusivity). From the data sets published by Power and Stegall (1970) and by Fischkoff and Vanderkooi (1975), the following mean estimates for the Krogh diffusion constant of the RBC membrane are obtained: $E_a = 7.1$ kcal/mol; $Q_{10} = 1.47$. Since these Q_{10} and E_a values are considerably higher than those derived from RBC O_2 transfer kinetics in our experiments, particularly for O_2 uptake, an important limitation of O_2 transfer by the RBC membrane appears improbable.

Diffusion limitation in the RBC

The diffusion coefficients for O_2 and for haemoglobin within the RBC were assumed to follow the temperature dependence of viscosity of water and aqueous media (Longsworth, 1954), yielding an activation energy of 4.8 kcal/mol. The temperature dependence of the physical solubility of O_2 in the extracellular medium and in RBC, as derived from the values reported by Altman and Dittmer (1971), yields an activation energy of -3.0 kcal/mol. Thus, E_a of the Krogh diffusion constant of free O_2 is estimated at 1.8 kcal/mol. For haemoglobin diffusion, assumed to be responsible for facilitated diffusion of O_2, the E_a and Q_{10} values expected are those for the diffusion coefficient of haemoglobin. Hence, for diffusion of O_2 in RBC, $E_a(G)$ values are expected in the range of 1.8 to 4.8 kcal/mol, corresponding to Q_{10} of 1.10 to 1.30. This range covers the Q_{10} and E_a values for O_2 transfer of RBC calculated from our experimental data.

Diffusion limitation in medium surrounding RBC

In measurements of O_2 uptake by RBC, diffusion limitation in the medium appears to play a major role (Coin and Olson, 1979; Vandegriff and Olson, 1984a,b; Yamaguchi et al., 1985). The Q_{10} and E_a values for O_2 uptake by RBC (Table 1) are in good agreement with the predicted values. The higher Q_{10} and E_a values for O_2 release as compared with O_2 uptake may point to the relatively more important role of haemoglobin diffusion in the former.

CONCLUSION

The temperature dependence of the kinetics of O_2 uptake and O_2 release by human red blood cells in stopped-flow conditions suggests that the major limitation is imposed by diffusion in aqueous media (red cell interior and surrounding extracellular medium), whereas diffusion across the lipid cell membrane and the chemical reaction appear to be of minor importance. In physiological conditions, when there is probably little limitation by the external medium, the main limitation appears to be provided by free O_2 diffusion inside the RBC with some contribution by diffusion of haemoglobin, and a minor contribution by reaction kinetics.

REFERENCES

Altman, P.L. and Dittmer, D.S. (Eds.) (1971). Biological Handbooks: Respiration and Circulation, Fed. Am. Soc. Exp. Biol., Bethesda, MD, pp. 17-18..

Bauer, C., Klocke, R.A., Kamp, D. and Forster, R.E. (1973). Effect of 2,3-diphosphoglycerate and H^+ on the reaction of O_2 and hemoglobin. Am. J. Physiol. 224, 838-847.

Coin, J.T. and Olson, J.S. (1979). The rate of oxygen uptake by human red blood cells. J. Biol. Chem. 254, 1178-1190.

Fischkoff, S. and Vanderkooi, J.M. (1975). Oxygen diffusion in biological and artificial membranes determined by the fluorochrome pyrene. J. Gen. Physiol. 65, 663-676.

Lawson, W.H., Jr, Holland, R.A.B. and Forster, R.E. (1965). Effect of temperature on deoxygenation rate of human red cells. J. Appl. Physiol. 20, 912-918.

Longsworth, L.G. (1954). Temperature dependence of diffusion in aqueous solutions. J. Phys. Chem. 58, 770-773.

Power, G.G. and Stegall, H. (1970). Solubility of gases in human red blood cell ghosts. J. Appl. Physiol. 29, 145-149.

Severinghaus, J.W. (1979. Simple, accurate equations for human blood O_2 dissociation computations. J. Appl. Physiol. Respirat. Environ. Exercise Physiol. 46, 599-602.

Sirs, J.A. (1963). Uptake of O_2 and CO by hemoglobin in sheep erythrocytes at various temperatures. J. Appl. Physiol. 18, 166-170.

Vandergriff, K.D. and Olson, J.S. (1984a). The kinetics of O_2 release by human red blood cells in the presence of external sodium dithionite. J. Biol. Chem. 259, 12609-12618.

Vandergriff, K.D. and Olson, J.S. (1984b). Morphological and physiological factors affecting oxygen uptake and release by red blood cells. J. Biol. Chem. 259, 12619-12627.

Yamaguchi, K., Nguyen-Phu, D., Scheid, P. and Piiper, J. (1985). Kinetics of O_2 uptake and release by human red blood cells studied by a stopped-flow technique. J. Appl. Physiol. 58, 1215-1224.

OXYGEN TENSION AND SPERM MIGRATION IN THE FEMALE BED BUG

A. RUKNUDIN* and I.A. SILVER

Department of Pathology, Medical School, Bristol, U.K.
*On leave from the Department of Zoology, Jamal Mohamed
College, Tiruchirapalli, India

INTRODUCTION

Common bed bugs (Cimex spp) have an uncommon way of insemination, haemocoelic insemination. Sperm deposited in a pouch, the spermalege (Fig. 1), in the abodomen of the female, migrate to sperm storing organs through the haemocoel which is filled with haemolymph. Highly motile sperm in the haemocoel exhibit directed migration towards their storage sites, the conceptaculae (Abraham, 1934). The sperm, being outside the reproductive organs and surrounded by visceral organs in the haemocoel, must orientate themselves towards the conceptaculae by some mechanism. Abraham (1934) proposed sperm chemotaxis although he did not get evidence that any part of the female reproductive tissue attracted sperm. By comparing the sperm migration in related groups of bugs, it was suggested that an oxygen gradient could be responsible for directed migration of sperm in bed bugs (Carayon, 1966). The aggregated sperm attach themselves to the wall of the lateral oviducts during their migration and drag themselves toward the conceptacula (Davis, 1966). During in vitro experiments, the aggregated sperm masses when introduced into a gradient of oxygen in buffer, move towards the higher concentration of oxygen (Rao and Davis, 1969). Thus the measurement of oxygen tension in the haemolymph has become necessary to establish whether there is an oxygen concentration gradient along the path of migrating sperm in the female bug.

Various methods are employed in the biomedical field to measure oxygen tension in both body fluids and tissues (Cobbold, 1974). Among them polarographic determination of oxygen concentration is convenient and is used in the present work in view of the small size of the insect. Since oxygen was first eletrolysed from an aqueous solution in Nernst's laboratory, the technique has been used by several workers. Flush-ended electrodes are generally used with platinum (Cater, Silver and Wilson, 1959; Lubbers et al., 1969) or gold (Whalen, Nair and Ganfield, 1973) cathodes. We have used both these types of metal cathodes for making microelectrodes to measure oxygen tension in the bed bug.

MATERIALS AND METHODS

One-week-old bugs (<u>Cimex hemipterus</u>), obtained from laboratory colonies, were used for all experiments. The oxygen microelectrodes were made as descibed by Silver (1965). A brass jacket formed the sleeve over the glass capillary, the two being held together by araldite. This assembly was fitted into a Josephson's electrode mount and then into a micromanipulator.

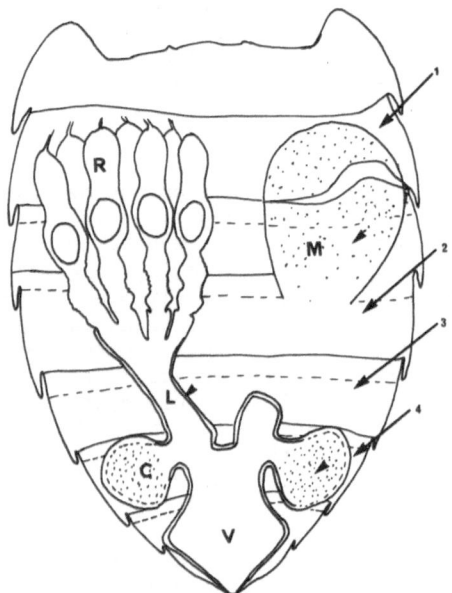

Figure 1. Diagram showing reproducive organs of bug and location of microelectrodes in haemolymph (numbered arrows) and in reproductive organs (arrow heads). C, conceptacula; L, lateral oviduct; M, mesospermalege; R, ovary; V, vagina.

The microelectrodes were tested for linearity in physiological saline containing different amounts of dissolved oxygen. The chosen ones were calibrated (Silver, 1973) in coconut milk buffer prepared from ripe coconuts. This buffer was selected since its chemical composition is similar to haemolymph and in it the sperm remained motile for some hours (Ruknudin, 1981).

During measurements of oxygen tension in a bug, the living insect was held firmly by crossed pins at the cephalothoracic junction and the microelectrode was carefully inserted through a puncture made with a fine needle in the intersegmental membrane.

Insertions into the haemolymph were done 1) between the 3rd and 4th abdominal segments where the anterior margin of the mesospermalege is situated, 2) between the 5th and 6th abdominal segments where the posterior region of the mesospermalege and middle region of ovary are found, 3) between the 6th and 7th abdominal segments where terminal oocytes are located, and 4) between the 7th and 8th abdominal segments where the conceptaculae are found. When measuring oxygen tension in the organs, the tergal cuticular plates were removed to expose the organs and the microelectrodes were inserted and measurement made quickly. All electrode insertions were done under a stereoscopic microscope and measurements were made in at least nine bugs for the calculation of mean and standard deviations.

RESULTS

The oxygen tension in various regions along the pathway of migrating sperm in the haemocoel of the bed bug was determined using gold microelectrodes (Table 1). The oxygen tension at the anterior margin of mesospermalege of a mated bug is low, 6.46 +2.3 kPa whereas in the region of the conceptacula it is higher, 10.99 +1.2 kPa. The mid and lower regions of the ovary showed oxygen concentrations intermediate between these two values.

Table 1. Oxygen tension in the haemolymph in various regions of mated bed bugs

Region	Oxygen tension kPa (Mean $\pm$ S.D.)	
	Using Au cathode	Using Pt cathode
Mesospermalege	6.46 $\pm$ 2.3	6.98 $\pm$ 1.13
Middle of the ovary	8.01 $\pm$ 1.9	Not measured
Base of the ovary	9.36 $\pm$ 1.57	9.51 $\pm$ 1.3
Conceptacula	10.99 $\pm$ 1.2	10.59 $\pm$ 0.83

The measurements using gold microelectrodes were compared with values obtained with platinum microelectrodes. As Table 1 shows the values did not differ significantly. From these data, it can be inferred that there is an oxygen gradient in the haemolymph between the mesospermalege and conceptacula. Similar measurements were also carried out in the body cavity of virgin bugs (Table 2).

Table 2. Oxygen tension in the haemolymph in various regions of virgin bed bugs

Region	Oxygen tension kPa (Mean ± S.D.)
Mesospermalege	5.26 ± 1.87
Middle of the ovary	6.42 ± 1.1
Base of the ovary	7.2 ± 1.7
Conceptacula	9.15 ± 1.2

The values in the corresponding regions were less than those in the mated bugs indicating that the oxygen gradient between the spermalege and conceptaculae became more pronounced after mating. When oxygen tension inside the three reproductive organs (arrow heads, Fig. 1) was measured in mated bugs, the corresponding values were higher (Table 3) than those obtained from the surrounding haemolymph (Table 1).

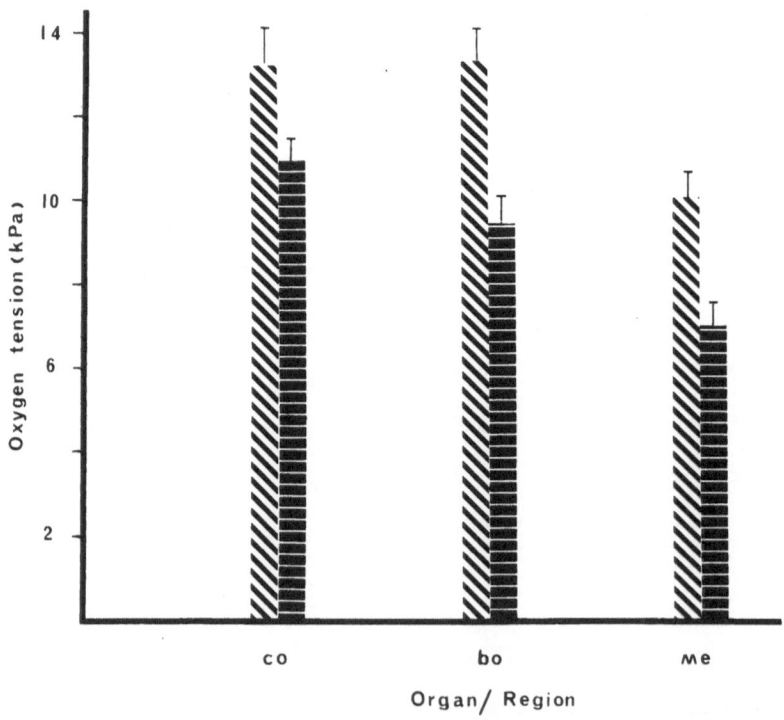

Figure 2. A comparison of oxygen tension inside the reproductive organs (hatched bars) and in surrounding haemolymph (horizontal bars). co, conceptacula; bo, base of ovary; me, mesospermalege. Mean ± S.D.

Table 3.   Oxygen tension within the different reproductive organs of
            mated bed bugs

Organ	Oxygen tension kPa (Mean $\pm$ S.D.)
Mesospermalege	10.12 $\pm$ 1.07
Wall of lateral oviduct	13.38 $\pm$ 1.3
Conceptacula	13.22 $\pm$ 1.7

The higher oxygen tension within the organs indicates that the source
of oxygen supply for the surrounding haemolymph could possibly be these
organs (Fig. 2). The distribution of tracheae supports this suggestion
since the reproductive organs having the higher oxygen tension were
richly tracheated (Ruknudin, 1981).

DISCUSSION

Oxygen tension measurements reveal haemolymph regions with
different oxygen tensions in the haemocoel of the female bed bug. The
regions investigated lie along the path of migratory sperm and results
indicate the existence of an oxygen gradient in the haemolymph with the
lowest concentration at the beginning, and the highest at the end, of
the pathway in the haemolymph. The sperm of the bed bug differ from
those of other insects in their need to migrate outside the
reproductive organs.   To avoid the sperm being lost in the wilderness
of the haemocoel, the oxygen gradient across the two regions seems to
direct sperm towards their destination. This chemotactic behaviour has
been suggested by a number of workers (Abraham, 1934; Carayon, 1965,
1977) who observed the sperm around the tracheae in the haemocoel of
the mated bug.   In their in vitro studies Rao and Davis (1969) found
that the sperm of the bed bug moved towards higher oxygen tension but
when the bugs were submerged in water or kept under nitrogen for a few
hours immediately after mating, migration did not occur during that
period.   The bugs survived the anoxia and when they were respiring
normally again, the sperm migration continued.

The difference in oxygen tension between the mesospermalege and
the conceptacula, which are seperated by 1 mm, is 4.53 kPa. Such a
steep oxygen gradient is not peculiar to bed bugs. Leichweiss et al.
(1969), using oxygen microelectrodes demonstrated that nearby regions
in the cortex of rat kidney have different oxygen tensions. In dog
kidney, a difference in the concentration of oxygen of 42 mmHg (5.5
kPa) existed over a distance of 10 μm at certain places indicating a
steep oxygen gradient in the gas diffusion of the renal tissue
(Baumgartl and Lubbers, 1973). The oxygen concentrations of blood of
mated bugs are higher than those of virgins and in accordance with
this, the inseminated Cimex lectularius females show an increase in the
rate of oxygen uptake when compared with virgins (Rao, 1973). This
increased uptake probably accentuates the oxygen gradient in the mated
bug so as to help the migating sperm.

Sperm chemotaxis as a mechanism for egg location is gaining
increasing recognition and has been demonstrated in a number of animal
groups   (Miller, 1985).   Substances released from the outer jelly of
the ova are known to cause phospholipid methylation, thus inducing a

chemotactic response in the starfish sperm (Tezon, Miller and Bardin, 1986). However, in the bed bug the chemotactic response is elicited by a gradient of oxygen rather than by any chemical substance released from the egg.

As a corollary to the measurement of oxygen tension in the haemolymph surrounding paragenital and reproductive organs, the oxygen tension in these organs themselves was determined. The results corroborated the findings on haemolymph gas tension. The oxygen supplied by the tracheae to the various organs diffuses into adjacent tissues (Weis-Fogh, 1964) and it is evident from the present study that oxygen from the organs measured diffuses into haemolymph. Also, tracheoles supplying the ovarioles are found to be restricted to the external ovariole sheath thus facilitating the diffusion into haemolymph (Bonhag and Arnold, 1961). The data presented here show that a gradient of oxygen tension exists in the haemocoel and this could possibly be responsible for directed migration of sperm in the female bug. However, the mechanism whereby the sperm responds to this chemotactic stimulus remains to be determined.

REFERENCES

Abraham, R. (1934). Das Verhalten der Spermien in der weiblichien Bettwanze (Cimex lectularius L) und der Verblib der uberschussigen Spermasse. Z. Parasit. 6, 560-591.

Baumgartl, H. and Lubbers, D.W. (1973). Platinum needle electrode for polarographic measurement of oxygen and hydrogen. In Oxygen supply:theoretical and practical aspects of oxygen supply and microcirculation of tissue. Eds Kessler, M., Bruley, D.F., Clark, L.C.Jr., Lubbers, D.W., Silver, I.A. and Strauss, J., Urban and Schwarzenberg, Munchen-Berlin-Wien, pp. 130-136.

Bonhag, P.F. and Arnold, W.J. (1961). Histology, histochemistry and tracheation of the ovariole sheath in the American cockroach, Periplanata americana(L). J. Morph. 108, 107-130.

Carayon, J. (1966). Traumatic insemination and paragenital system. In: Monograph of Cimicidae. Ed. Usinger, R.L., Thomas Say Foundation, Vol. 7, pp. 81-166.

Carayon, J. (1975). Insemination extragenitale traumatique et systeme paragenitale chez les Hemipteres Cimicoidea. Vol. 2, These, Doc. en Sc., Univ. de Paris.

Cater, D.B., Silver, I.A. and Wilson, G.M. (1959). Apparatus and technique for the quantitative measurement of oxygen tension in living tissues. Proc. Roy. Soc. Lond. B. 151, 256-276.

Cobbold, R.S.C. (1974). Transducers for biomedical measurements: principles and applications. John Wiley and Sons, New York.

Davis, N.T. (1966). Reproductive physiology. In: Monograph of Cimicidae. Ed. Usinger R.L., Thomas Say Foundation, Vol.7, pp.167-178.

Leichweiss, H.P., Lubbers, D.W., Weiss, C.H., Baumgartl, H. and Reschke, W. (1969). The oxygen supply of the rat kidney: measurements of internal pO_2. Pflugers Arch. 309, 328-349.

Lubbers, D.W., Baumgartl, H., Fabel, H., Huch, H., Kessler, M., Kunze, K., Riemann, H., Seiler, D. and Schuchhardt, S. (1969). Principle of construction and application of various platinum electrodes. Prog. Resp. Res. 3, 136-146.

Miller, R.L. (1985). Demonstration of sperm chemotaxis in Echinodermata: Asteroidea, Holothuroidea, Ophiuroidea. J. Exp. Zool. 234, 383-414.

Rao, H.V. (1973). Oxygen consumption in virgin and mated bed bugs. Curr. Sci. 42, 208-209.

Rao H.V. and Davis N.T. (1969). Sperm activation and migration in bed bugs. J. Insect Physiol. 15, 1815-1832.

Ruknudin, A. (1981). Studies on the reproductive physiology of the common bed bug, Cimex hemipterus (F) (Cimicidae: Heteroptera) with special reference to sperm physiology. Ph.D. Thesis, University of Madras, India.

Silver, I.A. (1965). Some observations on the cerebral cortex with an ultramicro-membrane covered, oxygen electrode. Med. Electron. Biol. Engng. 3, 337-387.

Silver, I.A. (1973). Problems in the investigation of tissue oxygen microenvironment. Adv. in Chem. 118, 343-351.

Tezon, J., Miller, R.L., and Bardin, C.W. (1986). Phospholipid methylation in starfish spermatozoa is linked to sperm chemoattraction. Proc. Natl. Acad. Sci., U.S.A. 83, 3589-3593.

Weis-Fogh, T. (1964). Diffusion in insect wing muscle, the most active tissue known. J. Exp. Biol. 41, 229-256.

Whalen, N.J., Nair, P. and Ganfield, R.A. (1973). Measurements of oxygen tension in tissues with a micro-oxygen electrode. Microvasc. Res. 5, 254-262.

SURFACTANT IN PULMONARY OXYGEN TOXICITY

S. ARMBRUSTER, J. KLEIN, E.M. STOUTEN, W. ERDMANN and
B. LACHMANN.

Department of Anaesthesia, Erasmus University Rotterdam,
The Netherlands

INTRODUCTION

Prolonged continuous exposure to high concentrations of oxygen can be lethal, due to respiratory failure. However, the cause of this respiratory failure is not known. The present study was designed to investigate the role of pulmonary surfactant in the development of respiratory failure induced by high concentrations of oxygen.

METHODS

Experiments were performed in Sprague Dawley rats weighing 250-300 g and maintained on a standard laboratory diet. Twenty-two rats were exposed to pure oxygen at 1 ATA (101 kPa) in a special polystyrene chamber. Oxygen concentration was measured continuously (oxygen monitor, Instrumentation Laboratories, Lexington, MA) and kept above 95%.

The CO_2 concentration was held constant at a level similar to that of room air (0.033%) by means of a high oxygen flow and placement of soda lime chip containers in the cage. The cage temperature was held between 23 and 26°C. Survival was monitored on an hourly basis. All oxygen-exposed rats died within 72 hours. Directly after death a thoracotomy was performed and the volume of pleural effusion measured. The lungs were dissected, the weight was determined and the trachea cannulated. Pressure-volume (p.v.) diagrams were then recorded with a maximum insufflation pressure of 40 cmH$_2$0. Following this procedure, broncho-alveolar lavage was performed by infusion and gentle aspiration of 9 ml physiological saline and the surface tension activity of the lavage fluid measured with a Wilhelmy balance (Biegler, Austria).

In 4 air-exposed control rats broncho-alveolar lavage was performed in vivo according to the method of Lachmann, Robertson and Vogel (1980) to produce a surfactant-deficient lung. After completion of this procedure the rats were ventilated for 15 minutes and then killed with an intraperitoneal (I.P.) injection of a barbiturate overdose.

Another group of six air-exposed control rats was killed, untreated, in the same way. Lung weight and surface tension in both groups was measured and p.v. curves were recorded following the same procedure as described for oxygen-exposed rats.

RESULTS

Pressure-volume studies

The lungs of oxygen-exposed rats appeared completely atelectic. The p.v. curves demonstrated an opening pressure in these lungs higher than 10 cmH$_2$O (Fig. 1). At an insufflation pressure of 40 cmH$_2$O, the inflated volume was less than 50% of that in normal lungs (air exposed and untreated lungs; Fig. 2).

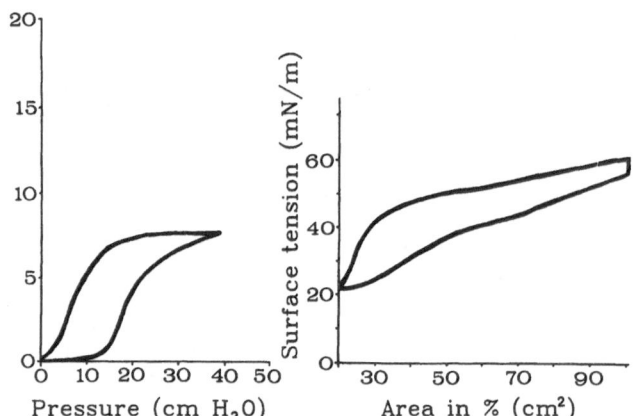

Figure 1. Left: Pressure/volume diagram (ordinate = volume in ml). Right: Surface tension/surface area diagram. Both diagrams show mean values from 22 rats exposed to pure oxygen at 1 ATA (101 kPa).

The characteristic deflation limb of the p.v. curves recorded in oxygen-exposed lungs suggests clearly that these lungs are over stabilized in comparison with the lungs of the air-exposed group. This was confirmed by the fact that these lungs did not collapse after inflation; they stayed completely aerated.

The lungs in which surfactant deficiency was produced by broncho-alveolar lavage had the same retractive forces as the oxygen-exposed lungs, with similar opening pressures, but the deflation curve was nearly linear, an obvious sign of stiffness of the lungs (Fig. 3).

Surface tension measurements
There were no significant differences between surface tension activity of the lavage fluid from oxygen-exposed lungs and air-exposed lungs (Figs 1 and 2); however, in the surfactant-depleted lungs the surface tension activity was very low (Fig. 3, Table 1).

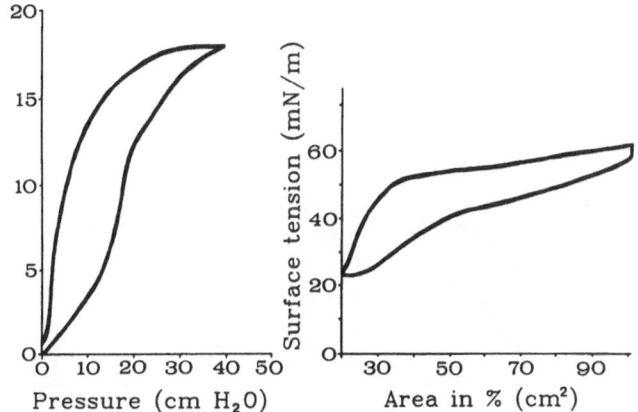

Figure 2. Left: Pressure/volume diagram (ordinate = volume in ml). Right: Surface tension/ surface area diagram. Both diagrams show mean values from 6 air-exposed control rats.

Table 1.   Surface tension measurements

	Gamma max. mN/m		Gamma min. mN/m	
	$\bar{x}$	$\pm$ S.D.	$\bar{x}$	$\pm$ S.D.
Air-exposed lungs	51.3	3.2	20.7	1.2
Oxygen-exposed lungs	51.5	1.6	22.9	3.5
Surfactant-depleted lungs	71.6	1.4	43.4	5.2

The wet weight of oxygen-exposed lungs was increased in comparison with the wet weight of air-exposed lungs: 3.05 + 0.15 vs. 1.9 + 0.18 g ($\bar{x}$ + S.D.). The surfactant-deficient lungs had a wet weight of 5.8 + 1.3 g.

Pleural effusion volume measurements
The amount of pleural effusion in the oxygen-exposed group was 12.8 + 2.6 ml ($\bar{x}$ + S.D.) consisting of a large fibrin component and a serum protein component of 4.1 + 0.2 g %.

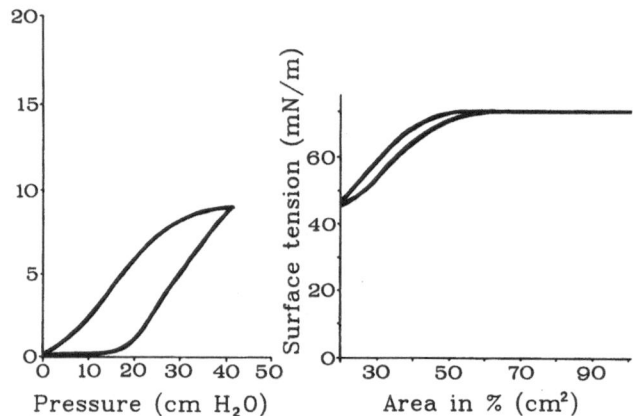

Figure 3. Left: Pressure/volume diagram (ordinate = volume in ml). Right: Surface tension/ surface area diagram. Both diagrams show mean values from 4 rats following lung lavage to produce surfactant-depleted lungs.

DISCUSSION

Prolonged exposure to high concentrations of oxygen can lead to death from respiratory failure. Interestingly enough, this study shows that lung failure induced by oxygen exposure is not the primary cause of death since:

1) The oxygen-exposed lungs showed normal surface tension activities;

2) the total wet weight of the oxygen-exposed lungs was only slightly increased compared with the wet weight of the air-exposed lungs;

3) although normal surface tension activity was measured in oxygen-exposed lungs, the p.v. curves showed high retractive forces. This might have been due to changes in anatomical structure, but this is not supported by previous pathological examinations of oxygen-exposed lungs. Another explanation could be that the high retractive forces are induced by a dysfunction of the pulmonary surfactant system probably due to an inhibitor which is not yet detectable by surface tension activity measurements in vitro. The stabilization of the deflation limb in these lungs can be explained by the presence in small airways of fluid with high surface tension which leads to air trapping.

4) in spite of the fact that the p.v. curves of the oxygen-exposed group showed changes in lung mechanics when compared with the control group, we do not assume that these changes can be deemed responsible for the respiratory failure leading to death. In previous experiments performed in our department it was demonstrated that rats with similar p.v. curves could survive.

Thus, like Smith et al. (1981) we think the large amount of pleural effusion found in the oxygen-exposed rats which resulted in severe compression of the lungs is the only possible explanation left to be invoked as the primary cause of death from respiratory failure in rats exposed to 100% O_2.

REFERENCES

Lachmann, B., Robertson, B. and Vogel, J. (1980). In vivo lung lavage as an experimental model of the respiratory distress syndrome. Acta Anaesth. Scand. 24, 231.

Smith, R.M., Rogers, R.M., Horton, F.O. and McCallum, R.E. (1981). Lung mechanics and a simultaneous comparison of alveolar pleural and peritoneal phagocytic cells lavaged from fasted infection-free oxygen toxic rats. Chest, 80, (July Suppl) 15-185.

ACUTE RESPIRATORY FAILURE INDUCED BY TRACHEAL INSTILLATION OF XANTHINE

OXIDASE, ITS PREVENTION AND THERAPY BY EXOGENOUS SURFACTANT

INSTILLATION

B. LACHMANN, O.D. SAUGSTAD[*], J. KLEIN and W. ERDMANN

Department of Anaesthesia, Erasmus University Rotterdam, The Netherlands and [*]Department of Pediatrics, National Hospital Norway, Oslo, Norway

INTRODUCTION

Free oxygen radicals play an important role in a variety of diseases. The hypoxanthine-xanthine oxidase system, which generates the superoxide radical, exerts a damaging effect on several organs, including the lung (Review: Saugstad 1985). It has been shown that intravenous hypoxanthine in rats breathing 100% oxygen can cause lung damage, in contrast to hypoxanthine or oxygen alone (Saugstad et al., 1984a). Further, it has been shown (Johnson et al., 1981) that this system acutely induces increased capillary permeability in the rat lung. We have demonstrated that xanthine oxidase (XO) applied to the trachea of guinea-pigs induces dramatic changes in lung-thorax compliance, in the course of a few minutes, by destroying the functional integrity of the bronchial and alveolar surfactant system, probably by formation of free oxygen radicals (FOR). This effect could be partly prevented by superoxide dismutase (SOD) which is a superoxide radical scavenger (Saugstad et al., 1984b). The purpose of this study was to investigate whether it is possible to influence the functional changes induced by FOR by tracheal instillation of natural surfactant (NS).

MATERIALS AND METHODS

Preparation and in vivo characterization of surfactant

The surfactant used in these experiments was a natural surfactant isolated from adult rabbit lungs in basically the same manner as previously described (Metcalfe, Enhorning and Possmayer, 1980) with some modifications. In brief, immediately after the animals were killed with an overdose of pentobarbitone sodium, the lungs were minced and the tissue washed with saline (100 ml/20-25 g tissue) for 30 min, filtered and the filtrates were centrifuged at 500 $\underline{g}$ at room temperature to remove cell debris. The supernatant was then centrifuged for one h at 8000 $\underline{g}$ at 2-4°C. The surfactant phospholipids were extracted from the resulting white pellet with chloroform: methanol, ethanol and acetone. The latter procedure was repeated at least twice.

The resulting surface active material was resuspended in saline so that a concentration of 60 mg total phospholipids/ml resulted. The activity of this preparation was tested in vivo (Lachmann 1986) in immature rabbit fetuses from day 27 of gestation. In these premature rabbits the instilled surfactant increased lung-thorax compliance more than 11-fold.

Experimental procedure

The experiments were performed on young guinea-pigs, weight range 330-390 g. Following anaesthesia with pentobarbitone sodium (60 mg/kg body weight) the animals were tracheotomized. After relaxation with pancuronium (2 mg/kg body weight) the animals were ventilated in parallel (6 animals simultaneously) with a Servo ventilator 900 B (Siemens Elema, Solna, Sweden) according to Lachmann (1985). The respiratory rate was 20 per min, peak pressure 15 cmH_2O, inspiratory time 50% and inspiratory oxygen concentration 100%.

One unit XO (Sigma Co.) dissolved in 1 ml saline was instilled into the trachea of the animals after a ventilation period of 10 min. Simultaneously with the application of the fluid, the peak pressure of the ventilator was raised to 35 cmH_2O and positive end expiratory pressure to 4 cmH_2O; this was necessary to keep the animals alive. This working pressure was maintained in all animals throughout the experimental period, except for the few minutes when pressure-volume diagrams were recorded. Pressure-volume diagrams were measured in each animal by placing them alternately into a specially constructed pressure-constant body plethysmograph connected to a Servo ventilator 900 B (Siemens Elema). Volume changes were recorded with a Fleisch tube connected to the body plethysmograph via a differential transducer (Siemens Elema EMT 34) and amplifier (EMT 31), and an integrator unit (Siemens Elema Mingograph 81).

Five groups of animals were studied :
1) animals receiving 1 U XO plus, 35-40 min later, 1.5 ml surfactant
2) animals receiving 1 U XO plus, 35-40 min later, 1.5 ml saline
3) animals receiving nothing, only ventilated
4) animals receiving 1 U XO followed 1-2 min later by 1.0 ml surfactant plus, 35-40 min later, additional 1.0 ml surfactant
5) animals receiving 1 U XO followed 1-2 min later by 1.0 ml saline plus, 35-40 min later, 1.0 ml surfactant.

RESULTS AND DISCUSSION

Table 1 shows that the decreased lung-thorax compliance after XO instillation could be almost completely restored by tracheal instillation of surfactant (Group 1).

Table 1. Lung-thorax compliance $(ml/cmH_2O \cdot kg^{-1})$ from the deflation curve of the P-V diagrams

x ± SD	Initial values	30' after XO	15' after surfact. or saline
Group 1, (n = 7) XO + 1.5ml surf. 35-40'later.	1.02 ± 0.27	0.32 ± 0.11	0.86 ± 0.19
Group 2, (n = 8) XO + 1.5ml sal. 35-40'later.	1.10 ± 0.18	0.37 ± 0.14	0.28 ± 0.09

This effect must be related to the surfactant lipids and not to fluid administration since the lung-thorax compliance deteriorated further in those animals receiving saline (Group 2) instead of surfactant (Table 1). In those healthy animals which were only artificially ventilated (Group 3, n = 4) there was no effect on lung-thorax compliance (initial values, 1.11 $\pm$0.23; after 30' ventilation, 1.23 $\pm$0.22; after 45' ventilation, 1.21 $\pm$0.14).

Tracheal instillation of surfactant 1-2 min after XO instillation partly prevents the effect of XO instillation (Table 2). One possible explanation for the latter result could be that the lipid peroxidation by free oxygen radicals takes place with exogenous surfactant lipids, thus preventing the peroxidation of the lipids of the surfactant system, which lines the alveoli and airways. While surfactant replacement partly prevented and restored the functional changes after tracheal XO instillation, it can be concluded from the results that free oxygen radicals, generated by the xanthine oxidase system, also destroy the functional integrity of the bronchial and alveolar surfactant system.

Table 2. Lung-thorax compliance $(ml/cmH_2O.kg^{-1})$ from the deflation curve of the P-V diagrams

	Initial values	30' after 1st instill. (surf. or sal.)	15' after 2nd instill (surf.)
Group 4 (n = 7) XO, 2' later 1 ml surf.; 35-40' later 1 ml surf.	100% $\pm$16	77.5% $\pm$21	92% $\pm$12
Group 5 (n = 5) XO, 2' later 1 ml sal.; 35-40' later 1 ml surf.	100% $\pm$13	35.3% $\pm$17	84% $\pm$14

In earlier investigations we demonstrated that the combination of fluid with artifical ventilation itself leads to damage, mainly of the bronchial surfactant system (Lachmann, 1985). Similarly, in this study, one factor of functional disturbance can be attributed to the fluid administration. However, the combination of fluid and artificial ventilation during this observation period mainly led only to an increased opening pressure, without significant restrictive lung volume changes, at a pressure of 30 cmH_2O. In contrast, after XO instillation we observed both an increased opening pressure and a significant decrease in volume, at a pressure of 30-35 cmH_2O.

CONCLUSION

The general application of radical scavengers, in patients who are at risk of developing acute respiratory distress syndrome (ARDS) might be one important therapeutic approach in the future. However, in all cases where one of the primary causes of respiratory failure are free oxygen radicals, additional 'local' surfactant replacement might be an important step for a successful treatment of ARDS.

REFERENCES

Johnson, K.J., Fanton, J.C., Kaplan, J. and Ward, P. (1981). In vivo damage of rat lungs by oxygen metabolites. J. Clin. Invest. 67, 983-993.

Lachmann, B. (1985). Possible function of bronchial surfactant. Eur. J. Resp. Dis. 67, 49-61.

Lachmann, B. (1986). New aspects in pathophysiology and therapy of respiratory distress syndrome : some criteria for characterisation of exogenous surfactant in surfactant deficient animal models. In: Selected Topics in Perinatal Medicine. Eds Cosmi, E.V. and Di Renzo, G., C.I.C. International Publishers, Rome, pp. 177-198.

Metcalfe, I.L., Enhorning, G. and Possmayer, F. (1980). Pulmonary surfactant-associated proteins: their role in the expression of surface activity. J. Appl. Physiol. 49, 34-41.

Saugstad, O.D. (1985). Oxygen radicals and pulmonary damage. Pediatr. Pulmonol. 1, 167-175.

Saugstad, O.D., Hallman, M., Abraham, J., Cochrane, G.G., Epstein, B. and Gluck, L. (1984a). Hypoxanthine and oxygen induced lung injury: a possible basic mechanism of tissue damage? Pediatr. Res. 18, 501-504.

Saugstad, O.D., Hallman, M., Becher G., Oddoy, A. and Lachmann, B. (1984b). Protective effect of superoxide dismutase (SOD) on severe lung damage caused by xanthine oxidase (XO). Pediatr. Res. 18, 802.

ENDOTOXIN PROTECTION AGAINST PULMONARY OXYGEN TOXICITY AND PLASMA

PROSTAGLANDIN LEVELS IN THE RAT

J. KLEIN, A. TROUWBORST and W. ERDMANN

Department of Anaesthesia, Erasmus University Rotterdam
The Netherlands

SUMMARY

Exposure of rats to high concentrations of oxygen (> 95%) at 1 ATA pressure (101 kPa) is lethal within three days. Rats treated with a small dose of endotoxin are protected against these lethal effects of hyperoxia. Recently, we found that the lysine salt of acetylsalicylic acid antagonises this protective action of endotoxin. This suggests that prostaglandin metabolism plays an important role in the protective action of endotoxin against pulmonary oxygen toxicity. Therefore, we measured the plasma levels of $6KPGF_{1\alpha}$, a stable degradation product of prostacyclin (PGI_2), PGE_2 and thromboxane B_2, the stable degradation product of thromboxane A_2, in rats exposed to air or > 95% oxygen for 48 hours. We compared these with the plasma levels of rats treated with endotoxin (Salmonella typhimurium lipopolysaccharide 1 mg/kg) and exposed to air or > 95% oxygen for 48 hours. We found that exposure of rats to > 95% oxygen for 48 hours leads to a significant rise in the $6KPGF_{1\alpha}$ levels. Rats exposed to > 95% oxygen for 48 hours and treated with endotoxin had significantly higher PGE_2 and significantly lower $6KPGF_{1\alpha}$ plasma levels than saline-treated rats exposed to > 95% oxygen for 48 hours.

INTRODUCTION

Despite the potential dangers of hyperoxia in the lung, administration of oxygen at above ambient tensions is necessary for treatment of severe hypoxaemia caused by respiratory failure or acute lung injury. Small doses of bacterial endotoxin markedly increase the survival rate of adult rats exposed to 98% oxygen for periods that are normally lethal (60–72 h) (Frank and Roberts, 1979; Frank, Summerville and Massaro, 1980).

Details concerning the mechanism and even the cellular site(s) of action of endotoxin are not yet known. We recently reported the reversal of endotoxin-induced protection against pulmonary oxygen toxicity by acetylsalicylic acid (Klein, Trouwborst and Salt, 1986). This suggested that prostaglandin metabolism may play an important role in the protective action of endotoxin during hyperoxia.

In a pilot study we investigated the changes in plasma levels of i) thromboxane B_2 (TxB$_2$), the stable degradation product of thromboxane A_2, ii) 6 keto-prostaglandin $F_{1\alpha}$ (6KPGF$_{1\alpha}$), the stable degradation product of prostacyclin (PGI$_2$) and iii) prostaglandin E_2, (PGE$_2$) in rats exposed to hyperoxia and compared them with the levels in rats exposed to hyperoxia and treated with endotoxin.

METHODS

Animals

Male Sprague-Dawley rats (TNO-Rijswijk, NL) weighing 250-399 g, and maintained on a standard laboratory diet (Hope Farms rat food, Woerden, NL) were assigned at random to four different groups: one hyperoxia-exposed group (n = 19), one endotoxin-treated hyperoxia-exposed group (n = 7), one air-exposed group (n = 7) and one endotoxin-treated air-exposed group (n = 7).

Endotoxin treatment

Just before exposure to air or oxygen, endotoxin treated rats were given 1 mg/kg endotoxin (Salmonella typhimurium lipopolysaccharide, phenol water extraction: Sigma Chemical Co. London, U.K.) dissolved in normal saline. Control rats received an equal volume of I.P. normal saline.

Oxygen and air exposure

Rats were exposed to > 95% oxygen or compressed air at 1 ATA in special airtight cases with an overflow hole, each cage containing 6 or 7 rats. The oxygen concentration was continuously measured (oxygen monitor, Instrumentation Laboratories, Lexington, MA) and was constantly higher than 95% in the oxygen-perfused cages. The CO_2 concentration was held constant at a level similar to that of room air (0.033%) by means of a high oxygen flow (7 to 8 complete gas changes per hour) and by placing containers of soda lime chips in the cage. The cage temperature was held constant between 23 and 26°C. Water and food were provided ad libitum. The cages were opened once a day for 10 to 15 min to facilitate replenishment of food and water and waste removal.

Blood sampling

All rats were exsanguinated during ether narcosis by puncture of the abdominal aorta. Blood samples were collected in polypropylene tubes, containing 20 µl of heparin (500 U/ml Thromboliquine, Organon, NL) and 50 µl indomethacin (0.1 mg/ml in 0.1 M phosphate buffer, pH 8.0). Samples were centrifuged immediately at 1400 g for 10 min and plasma stored at -20°C until assay.

Radioimmunoassay method

One ml of plasma was applied to a Seppak C_{18} Cardridge (Waters Ass. Inc., NL) which had previously been washed with 10 ml absolute ethanol and 10 ml distilled water. The column was rinsed with 2 ml distilled water and the prostaglandin-like compounds were eluted with 2 ml absolute ethanol. Radioimmunoassay kits were obtained from New England Nuclear (Boston).

Statistical analysis

Significant differences in median prostaglandin concentrations of the different groups were estimated using the Mann Whitney U test. Level of significance $p \leq 0.05$.

356

RESULTS

Changes in PGE_2, $6KPGF_{1\alpha}$ and T_xB_2 plasma levels after exposure to hyperoxia for 0, 24 and 48 hours are shown in Figure 1.

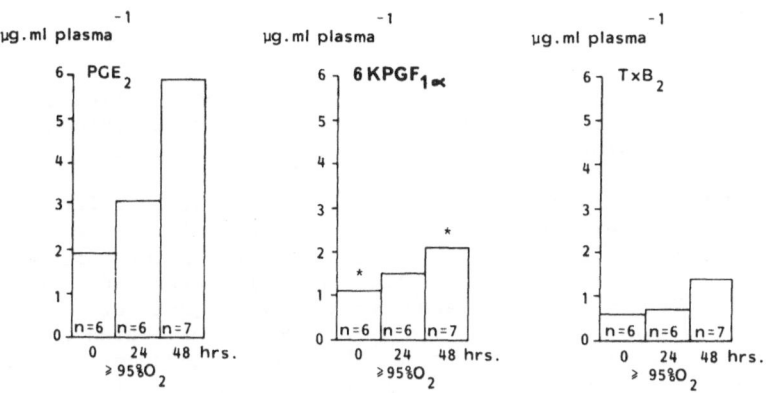

* p < 0.05; MANN WHITNEY U-test

(median concentrations)

Figure 1. Effects of exposure to hyperoxia for 0, 24 and 48 h on plasma levels of PGE_2, $6KPGF_{1\alpha}$ and TxB_2 in rats.

Exposure to hyperoxia results in increases in the levels of PGE_2, $6KPGF_{1\alpha}$ and T_xB_2, only the rise in $6KPGF_{1\alpha}$ being significant. Plasma prostaglandin and thromboxane B_2 levels of rats treated with endotoxin or saline and exposed to > 95% oxygen or compressed air for 48 hours are shown in Figure 2.

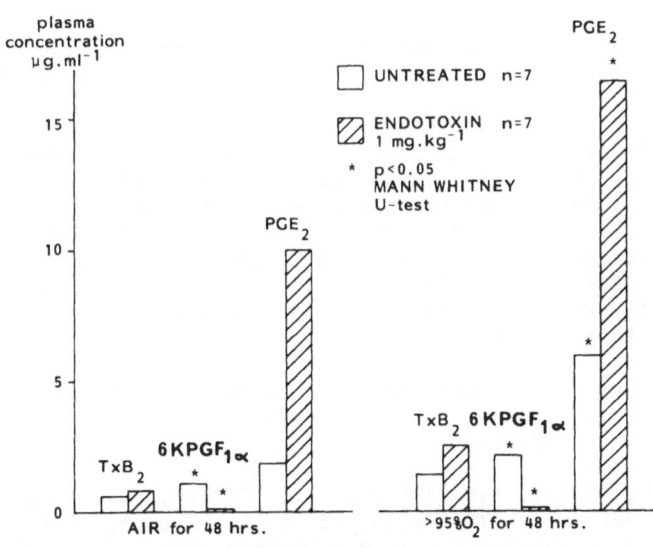

(median concentrations)

Figure 2. Effect of endotoxin on plasma levels of TxB_2, $6KPGF_{1\alpha}$ and PGE_2 in rats exposed to air or hyperoxia for 48 h. Untreated animals received saline.

357

Rats exposed to > 95% oxygen for 48 hours and treated with endotoxin had significantly higher PGE_2 and significantly lower $6KPGF_{1\alpha}$ plasma levels than the saline-treated rats exposed to > 95% oxygen for the same period.

DISCUSSION

In this pilot study we found that endotoxin-treated, oxygen-exposed rats had significantly lower $6KPGF_{1\alpha}$ and significantly higher PGE_2 levels than saline-treated rats exposed to hyperoxia.

Although it cannot be assumed that these differences are responsible for the protective action of endotoxin, we think the rise in PGE_2 level could possibly provide the basis for this protective effect. It has been shown that PGE_2 inhibits the release of leukotriene B_4 from activated neutrophils (Ham et al., 1983). Leukotriene B_4 is a metabolite of arachidonic acid which has a potent chemotactic activity for polymorphonuclear leukocytes (PMN) and promotes their accumulation and adherence to the vascular endothelium. PMNs play an important role in the genesis of oxygen toxicity (Fox et al., 1981). Recently, Taniguchi et al. (1986) found that inhibition of leukotriene B_4 production by a lipoxygenase inhibitor decreased mortality in rats exposed to hyperoxia for 72 hours. Thus, endotoxin could possibly exert its protective effect by the induction of PGE_2 synthesis which inhibits the release of leukotriene B_4 from PMNs. However, further experiments are needed to confirm this.

REFERENCES

Fox, R.B., Hoidal, J.R., Brown, D.M. and Repine, J.E. (1981). Pulmonary inflammation due to oxygen toxicity: involvement of chemotactic factors and polymorphonuclear leukocytes. Am. Rev. Respir. Dis., 123, 521-523.

Frank, L. and Roberts, R.J. (1979). Oxygen toxicity: protection of the lung by the bacterial lipopolysaccharide (endotoxin). Toxicol. Appl. Pharmac. 50, 371-380.

Frank, L., Summerville, J. and Massaro, D. (1980). Protection from oxygen toxicity with endotoxin: role of the endogenous antioxidant enzymes of the lung. J. Clin. Invest. 65, 1104-1110.

Ham, E.A., Soderman, D.D., Zanetti, M.E., Dougherty, H.W., McCauley, E. and Kuehl, F.A., Jr (1983). Inhibition by prostaglandins of leukotriene B_4 release from activated neutrophils. Proc. Natl. Acad. Sci. USA, 80, 4349-4353.

Klein, J., Trouwborst, A. and Salt, P.J. (1986). Endotoxin protection against oxygen toxicity and its reversal by acetylsalicylic acid. Crit. Care Med. 14, 32-33.

Taniguchi, H., Taki, F., Takagi, K., Satake, T., Sugiyama, S. and Ozawa, T. (1986). The role of leukotriene B_4 in the genesis of oxygen toxicity in the lung. Am. Rev. Respir. Dis., 133, 805-808.

SURVIVAL TIME OF RATS EXPOSED TO DIFFERENT HIGH OXYGEN PRESSURES AFTER ADMINISTRATION OF PERFLUOROCHEMICALS

M. STARK and J. LUTZ

Physiologisches Institut der Universitaet Wuerzburg
Roentgenring 9, D-8700 Wuerzburg, F.R.G.

INTRODUCTION

Hyperbaric oxygen administration is considered useful in the treatment of anaerobic infections (Cline and Turnbull, 1985) as well as in tumour radiation therapy (Fischer, Rockwell and Martin, 1986) where very high tissue partial pressures of oxygen are needed. The most commonly used oxygen pressure of 300 kPa (3 ATA) may not be sufficient to reach this goal. There must be sufficient oxygen capacity in the blood to maintain an elevated partial pressure even at the venous end of capillaries when oxygen is consumed. More effective treatment could be accomplished if haemoglobin were to remain in the oxygenated form. The administration of perfluorochemicals (PFC) which increase the oxygen capacity under elevated oxygen partial pressure can help in attaining this goal. However, as judged from previous results obtained by Geyer (1983), hyperbaric oxygen treatment combined with PFC administration can be expected to augment oxygen toxicity considerably. To test this supposition hyperbaric oxygen was applied and survival times were measured in rats as an index of the overall tolerance of the organism at various oxygen pressures. Survival time was found to give a more precise endpoint than the determination of the onset of convulsions. It was the goal of this study to establish the relationship between oxygen tension and lethal duration of exposure in both control and PFC-treated animals.

MATERIALS AND METHODS

Male white Wistar rats with body weights of 200-240 g were used throughout these studies. They were housed in groups of four in macrolon cages, kept on a standard diet (Altromin, Lage, F.R.G.) and had access to water ad libitum. They were exposed to hyperbaric oxygen, in groups of four, within a cylindrical pressure chamber (30 cm length; 20 cm internal diameter) closed with 20 mm thick steel-armoured acrylic glass plate. Pressurization was produced with compressed oxygen from a commercial cylinder. Pressures of 2.6, 3.3, 4.3, 5.3 and 6.8 x 100kPa (ATA) were applied. To prevent accumulation of carbon dioxide a dish with soda lime (Draegersorb 650, Draeger, Luebeck F.R.G.) was placed on the bottom of the chamber; for the longer experiments (at the two lower pressures) siccagel spheres (Kali-Chemie, Hannover, F.R.G.) were added to avoid an increase in humidity. PFC was administered in the form of

Fluosol-DA (Green Cross Corp., Osaka, Japan) in a dose of 8.0 g.kg^{-1} body weight, by intravenous injection into the dorsal vein of the penis under ether anaesthesia. This dose is up to twice that used in previous clinical trials (Mitsuno, Ohyanagi and Naito, 1982; Tremper et al., 1982).

RESULTS AND DISCUSSION

The relationship between survival time (T) and oxygen pressure (P) in control groups (Fig. 1) can be best described by the power function:

$$T = 74 \cdot P^{-2.55}$$

After logarithmic transformation of all values a linear regression was found:

$$\log T = -2.61 \log P + 1.91$$

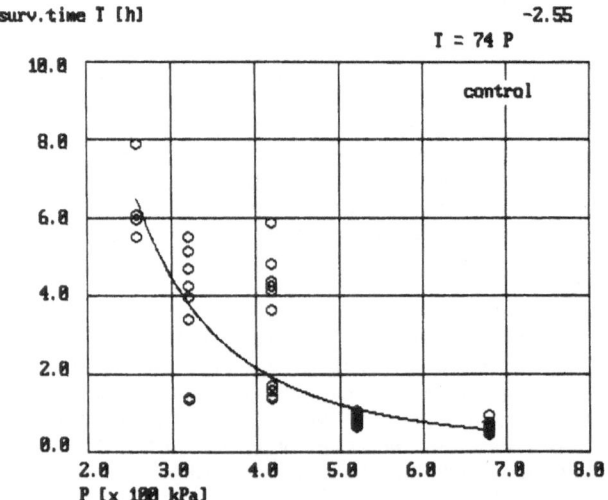

Figure 1. Survival times of Wistar rats in the presence of elevated oxygen pressures. The symbols (o) represent the time to respiratory arrest.

The standard deviation of the regression coefficient (sb) and the y-intercept (sa) could be determined as sb = ± 0.26 and sa = ± 0.154; the correlation coefficient was r = − 0.823, ($p < 0.001$). This function is shown in Figure 2.

A corresponding curve was found for PFC-treated animals with the equation:

$$\log T = - 2.24 \log P + 1.56$$

with sb = ± 0.24 and sa = ± 0.155. The correlation coefficient was r = - 0.816 (p < 0.001). Thus both curves showed a good fit to the data. The regression line for the PFC-animals lies below that of the controls though by only a small amount (Fig. 3).

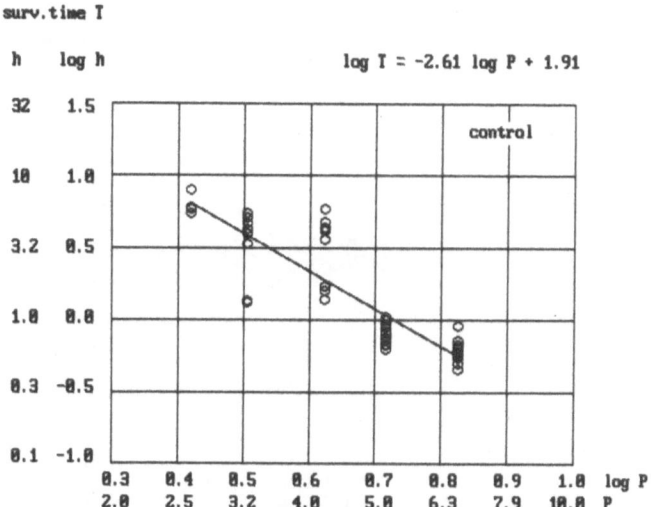

Figure 2. Logarithmic transformation of both oxygen pressure and survival time to rectify the power curve depicted in Figure 1. Absolute values are also plotted for clarification. The abscissa is P[x 100 kPa].

As can be seen from Figure 4 the values of PFC-treated rats generally lie within the 95% confidence band of the controls though predominantly in the lower one. Figure 5 shows that differences between the survival in control and PFC-treated animals are variable at different oxygen tensions. However, depression of the survival time under PFC could not be statistically confirmed by different tests (U-test, Fisher's LSD-test, analysis of covariance), even though a total of 91 animals was used.

Whereas PFCs seem to provide a suitable method of increasing the amount of expired nitrogen in the situation of decompression from hyperbaric air application (Lutz and Herrmann, 1984) up to now little has been known about their influence under hyperbaric oxygen conditions. Previous studies on survival times at high oxygen tension demonstrated that these could be altered by as much as 90% by various substances (Gershman, Gilbert and Caccamise, 1958; Clark, 1984). In comparison the PFC-induced shortening of survival found in this study was distinctly less. This may turn out to be important for combination of hyperbaric oxygen treatment with the administration of PFC under the protection of oxygen radical scavengers.

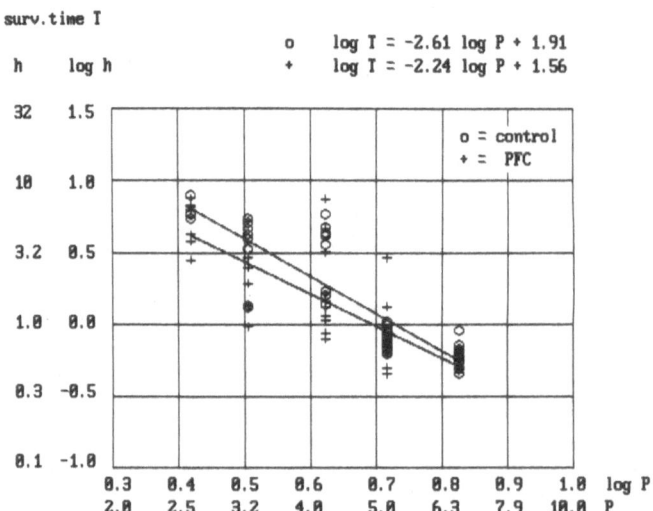

Figure 3.   Linear regressions of survival time on oxygen pressure after logarithmic transformation in control (o) and PFC-treated rats (+). The lower line represents the regression obtained after PFC administration. The abscissa is P[x 100 kPa].

CONCLUSION

To our knowledge this constitutes the first experimental study of survival time with PFC under elevated oxygen pressures. Our findings demonstrate a relatively weak life-shortening effect of PFC under these conditions. This is particularly remarkable considering the marked reduction of survival time under hyperbaric oxygen produced by some other substances. On the basis of these data it is concluded that the use of PFC under increased oxygen pressure does not compound oxygen toxicity as severely as previously expected.

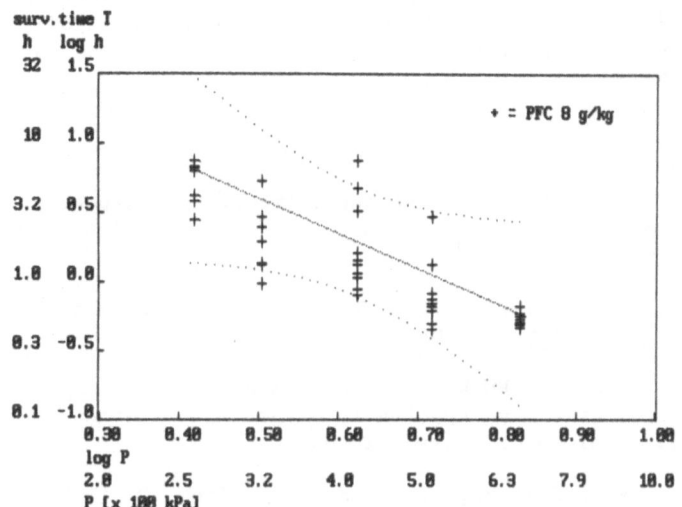

Figure 4. Linear regression and 95% confidence band of the logarithmically-transformed data from control rats. Data for PFC-treated rats are depicted by the symbols (+). All but two of the observations on PFC-treated animals lie within the confidence band of the controls, but usually in the band below the regression line.

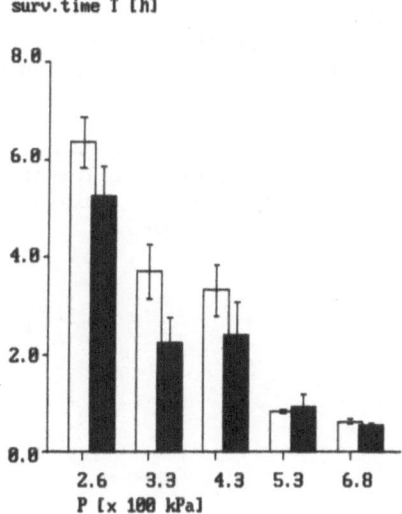

Figure 5. Survival times of rats at different oxygen pressures. The empty columns represent the mean values of the controls, the black ones those of the PFC-treated animals. The error bars give SEM. There were 12 to 20 animals at each pressure setting.

ACKNOWLEDGEMENTS

The authors wish to express their gratitude to Dr A. Sebald and Dipl. Phys. C. Stoehr for the production of plotting programs. They also thank Dr H. Beisbarth and Pfrimmer & Co., Erlangen, F.R.G. for supplying the Fluosol-DA from the Green Cross Corporation.

REFERENCES

Clark, J.M. (1984). Interacting effects of hypoxia and acute hypercapnia on oxygen tolerance in rats. J. Appl. Physiol. 56, 1191-1198.

Cline, K.A. and Turnbull, T.L (1985). Clostridial myonecrosis. Ann. Emerg. Med. 14, 459-466.

Fischer, J.J., Rockwell, S. and Martin, D.F. (1986). Perfluorochemicals and hyperbaric oxygen in radiation therapy. Int. J. Radiol. Oncol. Biol. Phys. 12 ,95-102.

Gerschman, R., Gilbert, L. and Caccamise, D. (1958). Effect of different substances on survival times of mice exposed to different high oxygen tensions. Amer. J. Physiol. 192, 563-571.

Geyer, R. (1983). PFC as a blood substitute - an overview. In: Advances in Blood Substitutes Research. Eds Bolin, R.B., Geyer, R.P. and Nemo, G.J., Alan R. Liss, New York, pp. 157-168.

Lutz, J. and Herrmann, G. (1984). Perfluorochemicals as a treatment of decompression sickness in rats. Pflugers Arch. 401, 174-177.

Mitsuno, T., Ohyanagi, H. and Naito, R. (1982). Clinical studies of perfluorochemical whole-blood substitutes (Fluosol-DA). Ann. Surg. 195, 60-69.

Tremper, K.K., Friedman, A.E., Levine, E.M., Lapin, P.R. and Camarillo, D. (1982). The preoperative treatment of severely anemic patients with a perfluorochemical oxygen-transport fluid, Fluosol-DA. N. Engl. J. Med. 307, 277-283.

EFFECT OF ANTIOXIDANTS ON THE BIOCHEMICAL RESPONSE IN THE OXYGEN-EXPOSED RAT

N. R. WEBSTER AND C. TOOTHILL[*].

Clinical Research Centre, Division of Anaesthesia
Watford Road, Harrow, Middlesex HA1 3UJ, and
[*]Department of Chemical Pathology, University of Leeds
Leeds, U.K.

INTRODUCTION

Our aerobic existence provides overwhelming energetic advantages over that of anaerobes, the oxidation of glucose yielding approximately 15 times more energy than the anaerobic conversion of glucose to lactate. However, the four electron reduction of oxygen to water necessarily progresses through toxic intermediate free radical species. All aerobically respiring cells are consequently under constant threat of oxidation from these highly reactive free radical intermediates (Fridovich, 1982).

Apart from cells of the lung only the red cell is exposed to oxygen tensions comparable to those of atmospheric air. In addition to this oxidative stress, the red cell is at a disadvantage because of its inability to replace damaged constituents by resynthesis. Oxidation of cellular components with consequent haemolysis is a constant threat to the red cell.

The successive four electron reduction of oxygen to water with release of energy during oxidative phosphorylation results in the formation of superoxide anion ($O_2^-\cdot$), hydrogen peroxide (H_2O_2), hydroxyl radical ($OH\cdot$) and finally water (H_2O). These intermediates are normally 'hidden' from cell components but may under some circumstances result in damage (Fisher and Forman, 1985). In addition there are other potential sources of the reduction products of oxygen within cells, such as xanthine oxidase, catechols and cytochrome-P 450.

Although many of these enzyme systems are saturated at normoxia and are therefore unable to reduce oxygen at a faster rate in the presence of hyperoxia, many others are not, and therefore yield increased concentrations of oxygen-derived free radicals when exposed to a high PO_2. Another explanation for the increased concentrations of oxygen-derived free radicals when exposed to a high PO_2 is the increased $O_2^-\cdot$ generation from mitochondria (Turrens, Freeman and Crapo, 1982; Turrens et al., 1982).

The response of the cell to oxidant stress is manifest in terms of detoxifying mechanisms (SOD, CAT and GPX), and by the generation of naturally occurring antioxidant compounds e.g. reduced glutathione. Such mechanisms act in two ways: i) preventative, or primary antioxidants which reduce the rate of initiation of free radical chain reaction (for example by reducing the free radical concentrations within the cell); and ii) chain-breaking, or secondary antioxidants which trap the chain-propagating peroxyl radicals.

There are thus several defence mechanisms to give protection from the deleterious effects of free radicals. It is considered that only when these are exhausted does significant free radical-induced cellular damage occur (Slater, 1972). The concept of antioxygenic potential of tissues was first suggested in 1960 by Bieri and Anderson who proposed a balance between factors tending to promote autoxidation of lipids opposed by factors exerting an antioxidant action. Their hypothesis was related specifically to vitamin E but can be extended to include all other protective systems.

Oxygen therapy is widely used in clinical practice often at toxic concentrations. The lung (as the organ involved in gaseous exchange) and the erythrocyte (as the organ of oxygen transport) are unique. These tissues are exposed to the highest oxygen tensions and are, therefore, more vulnerable to damage by oxygen. Failure of oxygen supply, uptake or delivery may require the administration of high partial pressures of oxygen. Prophylactic treatment would be a most useful adjunct to oxygen therapy and would allow the safe use of high concentrations of oxygen without the appearance of pathophysiological changes.

This study was therefore undertaken to determine changes in the antioxidant status of rats exposed to oxygen and to examine possible pharmacological approaches to the treatment of oxygen toxicity.

METHODS

Young adult male Wistar (SPF) rats were exposed to a mixture of 80% oxygen: 20% nitrogen for 5 days. Daily I.M. injections were given of either αtocopherol (40 mg/kg); β carotene (60,000 I.U./kg); reduced glutathione (100 mg/kg); methylprednisolone (25 mg/kg) or ascorbic acid (200 mg/kg). Two groups without antioxidant treatment were included where one received 80% oxygen: 20% nitrogen alone, whilst the other was an air control group. Food (Labsure – which contains vitamin A 8000 I.U./kg and vitamin E 60 I.U./kg) and water were allowed ad libitum.

At the end of the exposure period the animals were anaesthetized, blood withdrawn from the abdominal aorta and the lungs washed with 0.9% (w/v) saline. One lung was inflated to 5 cm water with formalin solution for histological examination, whilst the other was homogenized in 10ml 50 mmol/l phosphate buffer, pH 7.5. A sample of liver was also taken from the same animals.

The following assays were performed:

1. Superoxide dismutase (SOD) which catalyses the reaction :-
$$O_2^- \cdot + O_2^- \cdot + 2H^+ \longrightarrow H_2O_2 + O_2$$
was assayed by the method of Crapo, McCord and Fridovich (1978).

2. Catalase (CAT) which catalyses the reaction :-
$$2H_2O_2 \longrightarrow 2H_2O + O_2$$
was assessed by the method of Beutler (1975).

3. Glutathione peroxidase (GPX) which catalyses the reaction:-
$$2GSH + R-O-O-H \longrightarrow GSSG + H_2O + R-OH$$
was assessed by the method of Beutler (1975) following the oxidation of NADPH after the addition of t-butyl hydroperoxide as substrate:
$$GSSG + NADPH + H^+ \longrightarrow 2GSH + NADP^+$$

4. Lipid peroxidation was assessed by the method of Buege and Aust (1978) which measures the formation of malondialdehyde in the thiobarbituric acid reaction.

5. Protein content of the lung and liver homogenates was determined by the Lowry method (Lowry et al., 1951).

6. Hydrogen peroxide haemolysis demonstrates a propensity of the red cell to undergo further peroxidative damage and was first described by Nitowsky (1956).

7. Histological examination was undertaken on haematoxylin and eosin stained lung sections by a pathologist who was unaware of the treatment schedule. An oedema score was allocated to each sample as follows: 0 = no oedema; 1 = mild oedema (interstitial); and 2 = severe oedema (alveolar flooding). The mean score for each treatment group was then determined.

8. Statistical analysis was by the Kruskal-Wallis test.

RESULTS

Table 1 gives the oedema score, and Figures 1-4 show the changes in biochemical parameters, produced by each of the seven treatments tested. 1) air control; 2) 80% oxygen:20% nitrogen for 5 days; 3-7) 80% oxygen:20% nitrogen for 5 days with daily injection of: 3) αtocopherol (40 mg/kg); 4) βcarotene (60,000 I.U./kg); 5) reduced glutathione (100 mg/kg); 6) methylprednisolone (25 mg/kg); 7) ascorbic acid (200 mg/kg). Results are mean and S.E.M. Statistical significance (treatment group versus 80% oxygen control group) is shown as *, P< 0.05; **, P< 0.01; ***, P< 0.001.

Table 1. Oedema resulting from different treatments

Treatment	No.	Oedema score
1 Air control	11	0.0
2. 80% oxygen control	6	1.6
3. 80% oxygen + αtocopherol	6	2.0
4. 80% oxygen + βcarotene	6	1.7
5. 80% oxygen + reduced glutathione	6	1.8
6. 80% oxygen + methylprednisolone	6	1.3
7. 80% oxygen + ascorbic acid	6	0.7

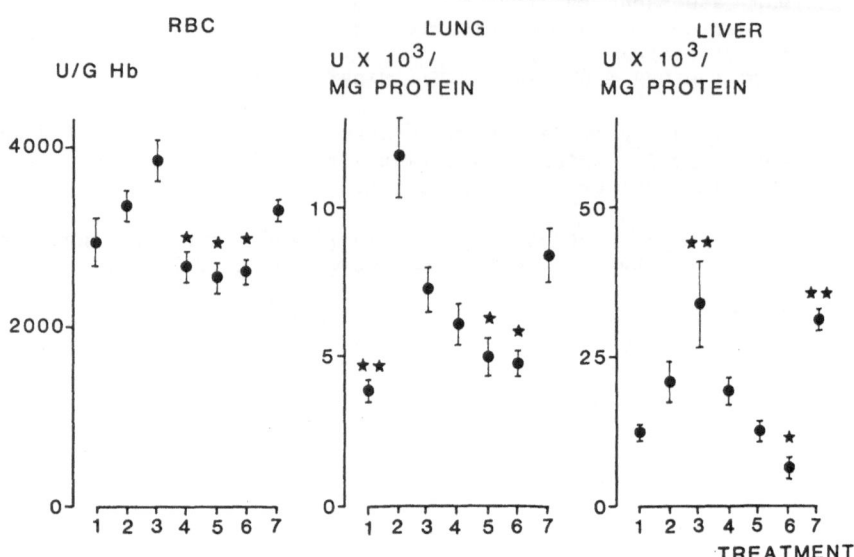

Figure 1.

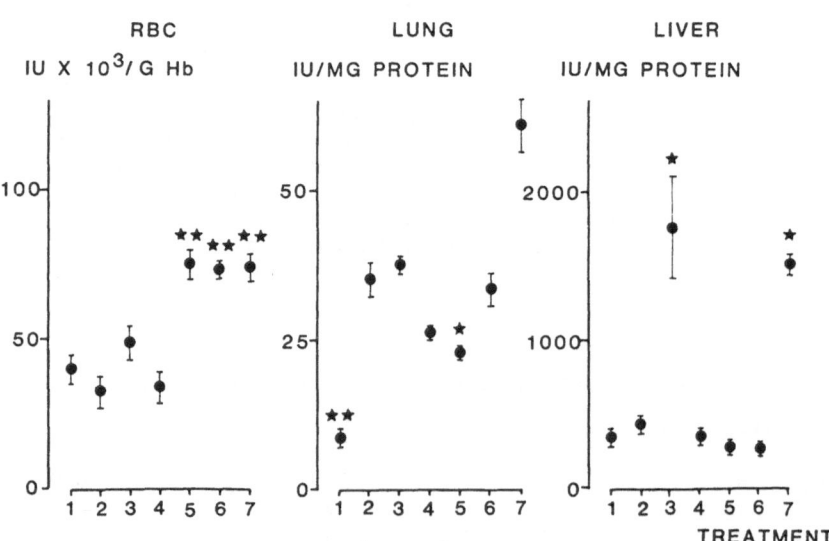

Figure 2.

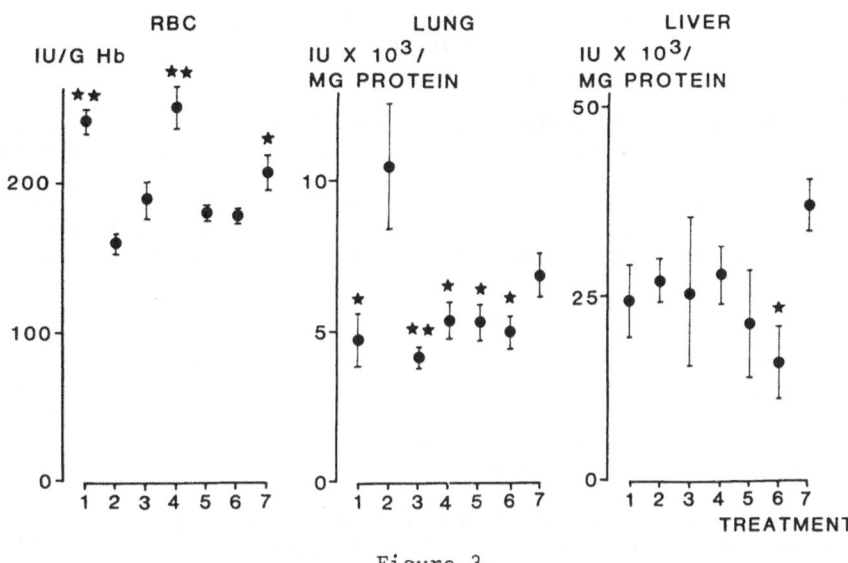

Figure 3.

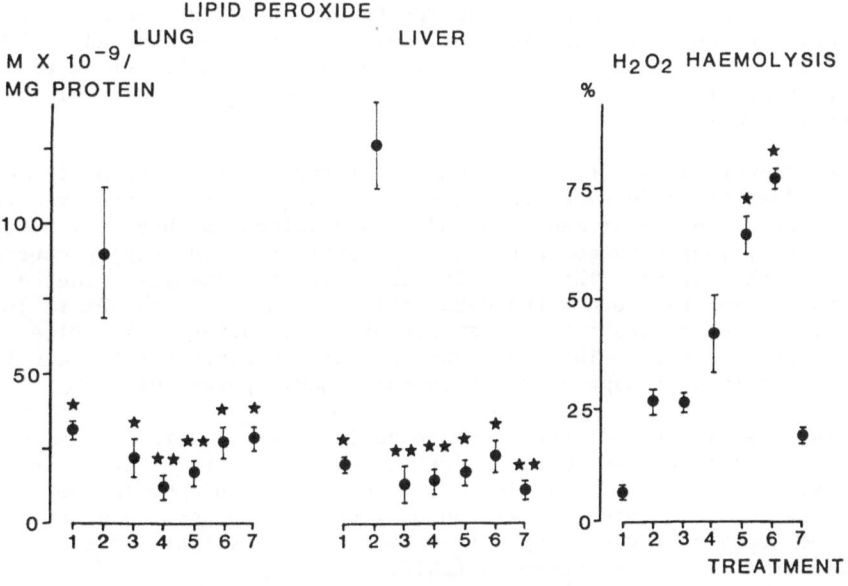

Figure 4.

DISCUSSION

No experimental animal died as a result of either oxygen exposure or antioxidant treatment. All compounds investigated reduced the response to oxygen as assessed by SOD and GPX activity in the lung. However, only ascorbic acid appeared to improve the histological changes of oxygen toxicity. Ascorbic acid would, therefore, appear to be worthy of further investigation for the prevention of the pathophysiological effects of oxygen.

Several groups of workers have investigated the relative ability of a number of antioxidant compounds to reduce the pulmonary damage caused by oxygen to different animal species, with differing results. Wide interspecies variation in the susceptibility to oxygen-induced damage is well recognized (Clark and Lambertsen, 1971) and it is not surprising therefore that results from different animal species are not comparable.

Matzen (1957) examined the effect of ascorbic acid and hydrocortisone on pulmonary oedema and mortality produced by ozone (up to 10 ppm) in mice. Ten mg of ascorbic acid (per animal) were found to be ineffective, however, 20 mg provided some protection, with mortality after two weeks being reduced. However, neither compound altered the total water content of these lungs.

In an experiment on lambs breathing 100% oxygen, Hansen, Hazinski and Bland (1982) found that 20 mg/kg of ∝tocopherol intra-muscularly produced no change in pulmonary microvascular permeability when compared with oxygen-only controls. All animals died in both groups by the fourth day. However, Wender et al. (1981) found that higher doses (60 mg/kg) were capable of preventing some of the pathological changes in the newborn rabbit as well as decreasing the biochemical response to oxygen. No such protection was found when 20 mg/kg was used. In an investigation of survival of mice exposed to 100% oxygen at 5 atmospheres absolute Jamieson and Van Den Brenk (1964) found that ascorbic acid and ∝tocopherol yielded similar survival times to control animals.

A reduction in the biochemical response to oxygen is most probably due to reduced concentrations of oxygen-derived free radical species, as a result of enhanced endogenous defence mechanisms. In the present study lung catalase activity was similar to the oxygen control values in all cases. This is in contrast to changes produced by antioxidants in the other two detoxifying enzymes SOD and GPX in lung homogenates, where return to almost pre-oxygen treatment activities was a consistent feature. This suggests that antioxidants act by reducing the concentrations of superoxide anion and organic peroxides.

In the liver a different response is seen to antioxidants, most probably reflecting the more important role of the liver in the metabolism of the compounds investigated. Marked changes in the three detoxifying enzyme systems were seen only with ∝tocopherol and ascorbic acid, the remainder producing either no change or slight reversal from the activities in oxygen control animals.

The changes seen in red cell enzyme activities are most likely to be due to changing cell populations. It is known that young red cells have higher activities of SOD and lower activities of CAT than older cells (Sass, Vorsanger and Spear, 1964; Bartosz et al., 1978).

The increase in hydrogen peroxide haemolysis following oxygen therapy is probably due to peroxidation of the lipid membrane component and the higher levels seen following treatment with β carotene, reduced glutathione and methylprednisolone could be due to increased antioxidant radical concentration within the membrane.

In conclusion, red cell enzyme activities would not appear to be useful markers of oxidant activity within the lung, and as such could not be used to indicate the occurrence of pulmonary oxidant injury in humans. Ascorbic acid was the only compound investigated which decreased the histological changes induced by oxygen although all compounds studied decreased the concentrations of malondialdehyde in the lung homogenates. This compound may therefore be of use in the prevention of oxidant injury in humans.

REFERENCES

Bartosz, G., Tannert, Ch., Fried, R. and Leyko, W. (1978). Superoxide dismutase activity decreases during erythrocyte aging. Experientia, 34, 1464.

Beutler, E. (1975). Red Cell Metabolism: a Manual of Biochemical Techniques. Grune and Stratton, New York.

Bieri, J.B. and Anderson, A.A. (1960). Peroxidation of lipids in tissue homogenates as related to vitamin E. Arch. Biochem. Biophys. 90, 105-110.

Buege, J.A. and Aust, S.D. (1978). Microsomal lipid peroxidation. In: Methods in Enzymology, 52, Academic Press, New York, pp. 302-310.

Clark, J.M. and Lambertsen, C.J. (1971). Pulmonary oxygen toxicity - a review. Pharmac. Rev. 23, 37-133.

Crapo, J.D., McCord, J.M. and Fridovich, I. (1978). Preparation and assay of superoxide dismutases. In: Methods in Enzymology, 53, Academic Press, New York, pp. 382-393.

Fisher, A.B. and Forman, H.J. (1985). Oxygen utilization and toxicity in the lungs. In: Handbook of Physiology, Section 3: Respiration, Volume 1. Eds Fishman, A.P. and Fisher, A.B., American Physiological Society, Bethesda, pp. 231-254.

Fridovich, I. (1982). Chemical and biological aspects of the superoxide radical and of superoxide dismutase. In: Membranes, Molecules, Toxins and Cells. Eds Bloch, K., Bolis, L. and Tosteson, D.C., PSG Inc., Boston, pp. 105-122.

Hansen, T.N., Hazinski, T.A. and Bland, R.D. (1982). Vitamin E does not prevent oxygen induced lung injury in newborn lambs. Pediatr. Res. 16, 583-587.

Jamieson, D. and Van Den Brenk, H.A.S. (1964). The effects of antioxidants on high pressure oxygen toxicity. Biochem. Pharmac. 13, 159-164.

Lowry, O.H., Rosenbrough, N.J., Farr, A.L. and Randall, R.J. (1951). Protein measurement with the folin phenol reagent. J. Biol. Chem. 193, 265-275.

Matzen, R.N. (1957). Effect of vitamin C and hydrocortisone on the pulmonary edema produced by ozone in mice. J. Appl. Physiol. 11, 105-109.

Nitowsky, H.M. (1956). Tocopherol deficiency in infants and children: II. Plasma tocopherol and erythrocyte haemolysis in hydrogen peroxide. Am. J. Dis. Child. 92, 164-174.

Sass, M.D., Vorsanger, E. and Spear, P.W. (1964). Enzyme activity as an indicator of red cell age. Clin. Chim. Acta, 10, 21-26.

Slater, T.F. (1972). Free Radical Mechanisms in Tissue Injury. Pion, London.

Turrens, J.F., Freeman, B.A., and Crapo, J.D.. (1982). Hyperoxia increases hydrogen peroxide release by lung mitochondria and microsomes. Arch. Biochem. Biophys. 217, 411-421.

Turrens, J.F., Freeman, B.A., Levitt, J.G. and Crapo, J.D.. (1982). The effect of hyperoxia on superoxide production by lung submitochondrial particles. Arch. Biochem. Biophys. 217, 401-410.

Wender, D.F., Thulin, G.E., Smith, G.J.W. and Warshaw, J.B. (1981). Vitamin E affects lung biochemical and morphological response to hyperoxia in the newborn rabbit. Pediatr. Res. 15, 262-268.

TUMOUR BLOOD FLOW FOLLOWING LOCAL ULTRASOUND HEATING COMPUTED FROM

THERMAL CLEARANCE CURVES

R.D. BRAUN, T.K. GOLDSTICK, M. KLUGE[*],
F. KALLINOWSKI[*], P. VAUPEL[*] and S.C. GEORGE

Chemical Engineering Department, Northwestern
University, Evanston, IL 60201, U.S.A. and [*]Department
of Applied Physiology, University of Mainz, D-6500
Mainz, F.R.G.

SUMMARY

Thermal clearance curves following termination of ultrasound-
induced hyperthermia in human mammary carcinomas implanted into the
flanks of nude rats were studied. They were found to be monoexponential in form, both with and without blood flow. From the
difference between the inverse time constants with and without flow,
the tumour blood flow rate could be calculated. Blood flow was found
to increase with very short exposure times at the therapeutic hyperthermia temperature and subsequently decrease as the exposure time
increased. A higher therapeutic hyperthermia temperature augmented this
effect.

INTRODUCTION

Tumour blood flow plays the major role in the determination of the
microenvironment of tumour cells. It is especially vital for oxygen
transport to tumour tissue and therefore is a paramount factor both for
tumour growth and for the efficacy of nonsurgical therapies. Among
these therapies, hyperthermia is of particular interest because, in
vivo, it preferentially destroys tumour cells (Dethlefsen and Dewey,
1982; Hahn, 1982; Storm, 1983; Vaupel and Kallinowski, 1986). The heat
sensitivity of tumours is explained by their chronically hypoxic,
acidic and generally nutritionally deprived environment related to
their poor circulation (Vaupel et al., 1984). Hyperthermia may further
aggravate this situation by reducing blood flow, thus enhancing its own
therapeutic effectiveness (Vaupel and Kallinowski, 1986). Since it is
necessary to measure temperature during hyperthermia, a feasible way of
determining tumour blood flow rates may be derivable from thermal
clearance curves. In order to test this hypothesis, thermal clearance
curves were obtained from hyperthermia experiments on human breast

cancers transplanted into nude rats. Preliminary results from the analysis of two experiments on squamous cell mammary carcinomas are presented here.

MATERIALS AND METHODS

Tissue from human breast cancers, serially transplanted into nude mice in the 34th to 50th generations, was implanted subcutaneously into the flanks of 113 circa two-month old nude rats. The breeding and maintenance of the animals as well as the implantation techniques have been previously described (Vaupel et al., 1985). After a tumour growth period of four to six weeks, tumour weights were around two to three g.

Each tumour implant was subjected to a series of heating cycles using an ultrasound (1.7 MHz) feedback control heater. The series of heating cycles consisted of heating the tumour to and maintaining it at a particular temperature for a given time. Temperature in the tumour centre was monitored with a thermocouple with a shaft diameter of 250 μm. After each hyperthermia cycle, the power input was shut off and the subsequent decline of temperature with time (thermal clearance curve) was recorded on a standard strip chart recorder. The time-temperature protocol was as follows: 1) one minute at a relatively low hyperthermia temperature, 38 to 39°C, to determine the 'normothermia' blood flow; 2) one minute at a higher, therapeutic, hyperthermia temperature, 40, 42 or 44°C; 3) three consecutive 20 minute cycles at the therapeutic hyperthermia temperature (totals of 21, 41 and 61 minutes of therapeutic hyperthermia); 4) one minute at the relatively low hyperthermia temperature; 5) one minute at the therapeutic hyperthermia temperature (total of 62 minutes of therapeutic hyperthermia). At the end of each experiment, the animal was killed, the tumours were heated up once again to the therapeutic hyperthermia temperature for one minute (total of 63 minutes of therapeutic hyperthermia), and the temperature decay in the tumour without blood flow' was recorded. The experiments were performed in a temperature controlled environment of approximately 24°C with the animal always on a heated operating table to maintain the rectal temperature at normal body temperature, around 37°C. Because tumour blood flow is critically dependent on arterial blood pressure (Vaupel, 1975), curves were rejected when the mean arterial blood pressure during the clearance period dropped below 100 mmHg or varied more than 30 mmHg. We report here on the results of the analysis from the first two experiments from this study which were performed on squamous cell carcinomas of the breast weighing 2.2 and 2.3 g.

The analysis involved finding the best functional representation for the data which had been manually digitized at 30 sec intervals, for 15 minutes, from the strip chart records. Based on the considerations outlined by Braun (1986), a relationship was then developed between this functional representation and the tumour blood flow rate.

The thermal clearance curves were fitted to the monoexponential function:

$$T = (T_o - T_{inf})\exp(-t/tc) + T_{inf} \qquad (1)$$

where T = tissue temperature in °C, T_o and T_{inf} = initial and final values of T in °C, t = time in minutes after the ultrasound heating was turned off, and tc = time constant of the thermal clearance curve in minutes. Least squares nonlear regression was used to fit for the parameters tc, T_o, and T_{inf} (Braun, 1986).

Tumour blood flow was then calculated by the simple relationship:

$$Q_b = 100(1/tc - 1/tc_o) \qquad (2)$$

where Q_b = blood flow rate in ml blood/(100 g tissue x min), tc and tc_o = time constant from Equation (1) with flow and without flow, respectively, in minutes. This equation had been previously determined by an empirical fit of the results of thermal clearance experiments carried out on a sponge phantom, the details of which have been presented elsewhere (Braun, 1986).

RESULTS

The thermal clearance curves, both with and without blood flow, were fitted excellently by Equation (1). Typical fits are shown in Figures 1 and 2. Of the 16 curves obtained, 15 were acceptable according to the blood pressure criteria given above. All 15 were fitted well by Equation (1). All had root mean square (R.M.S.) errors (comparable to the standard deviation of a mean) less than $\pm0.06°C$, 13 being less than $\pm0.04°C$. The mean R.M.S. error was $\pm0.036°C$. The goodness of fit was independent of both the temperature level to which the tumour was heated, and whether or not there was blood flow. Finally, all the information in a thermal clearance curve appears to be represented by a simple monoexponential function.

Figure 3 shows the tumour blood flow rate as a function of heating time at hyperthermia temperature. In both experiments, an increase in flow at very short hyperthermia times and a decrease with increased therapeutic hyperthermia exposure times was found. With the higher therapeutic hyperthermia temperature (44°C), the tumour blood flow was eventually reduced to 23% of the initial value, whereas with the lower therapeutic hyperthermia temperature (42°C), blood flow dropped to only 62% of the starting value.

DISCUSSION

The monoexponential fit of the thermal clearance curves to Equation (1) was striking. All 15 curves fitted with an R.M S. error less than $\pm0.06°C$ and 13 less than $\pm0.04°C$. The mean R.M.S. for all the curves was $\pm0.036°C$, approximately the precision to which the data could be read off the strip chart record. In another experimental series reported elsewhere (Braun, 1986), the thermal clearance data from 14 sponge phantom experiments providing a total of 106 curves also fitted Equation (1) excellently. Only 8 out of the 106 had R.M.S. errors greater than $\pm0.04°C$ and the mean was $\pm0.033°C$. Finally, in a third, separate experimental series involving tissue-isolated kidney tumours (DS-carcinosarcoma) in rats (Vaupel, Ostheimer and Mueller-Klieser, 1980), we have analysed the thermal clearance data from 10 experiments with a total of 38 curves. All of these also fitted Equation (1) excellently. The mean R.M.S. error was $\pm0.036°C$. The striking agreement with Equation (1) in these three independent studies suggests that the thermal clearance curves following hyperthermia induced by ultrasound contain no more information than a first order time constant. It can be argued that both of the limiting solutions of the bioheat transfer equation, conduction only and convection only, are monoexponential in form. Thus one might expect the solution to the combined conduction and convection condition also to be a monoexponential function.

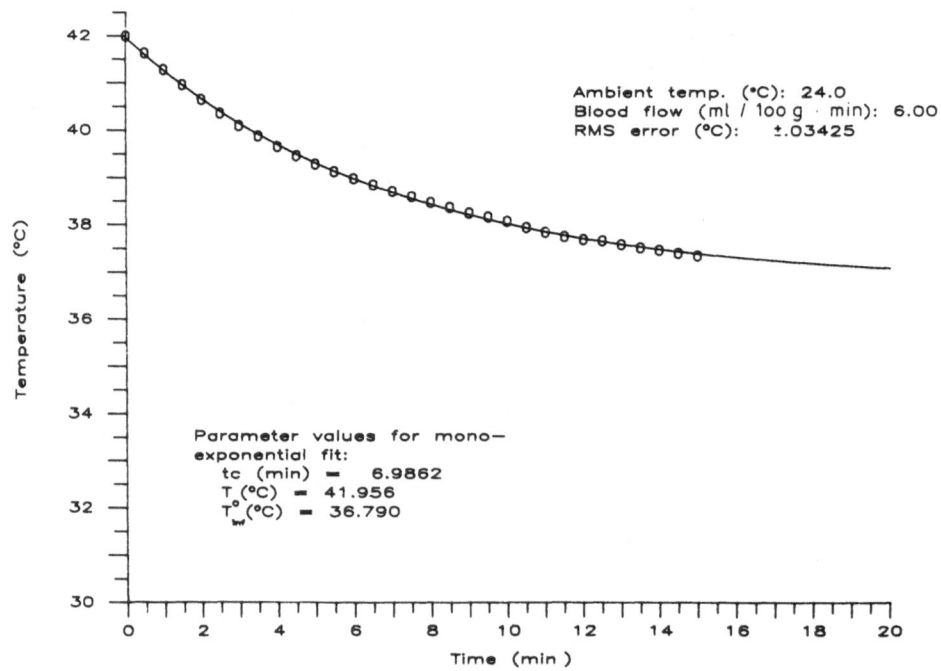

Figure 1.   Typical monoexponential fit (solid line) of the thermal washout data (points) from a human squamous cell mammary carcinoma transplanted into a nude rat. This particular curve (with blood flow) was obtained following the termination of ultrasound heating of the tumour at 42°C for 20 minutes (after a total hyperthermia time of 61 minutes).

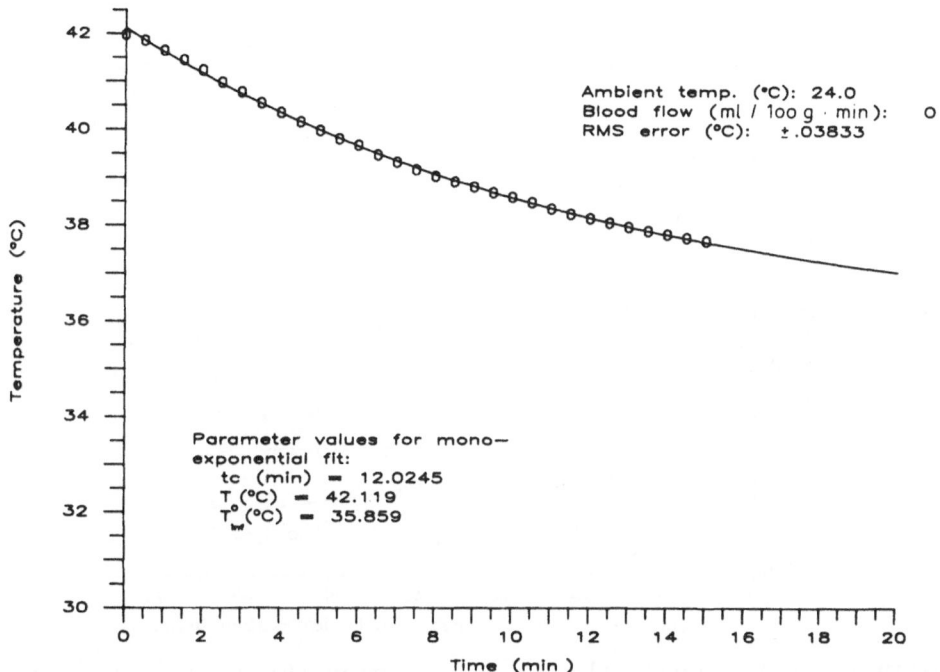

Figure 2. Typical monoexponential fit (solid line) of the thermal washout data (points) from a human squamous cell mammary carcinoma transplanted into a nude rat. This particular curve (without blood flow) was obtained on a dead rat following the termination of ultrasound heating of the tumour at 42°C for 1 minute (after a total hyperthermia time of 63 minutes).

Equation (2), which was based on the analysis of the phantom experiments, follows from the monoexponential character of the curves. In the limits, as tc_o approaches both infinity (no thermal conduction) and tc (no blood flow), it can be shown that Equation (2) is exact. Based on this, Equation (2) was assumed to give absolute values for Q_b. If the tumour could be very well insulated thermally, thus eliminating conduction, Q_b would become inversely proportional to $1/tc$ and this technique could be used without a no-flow curve, e.g., in humans during treatment in the clinic.

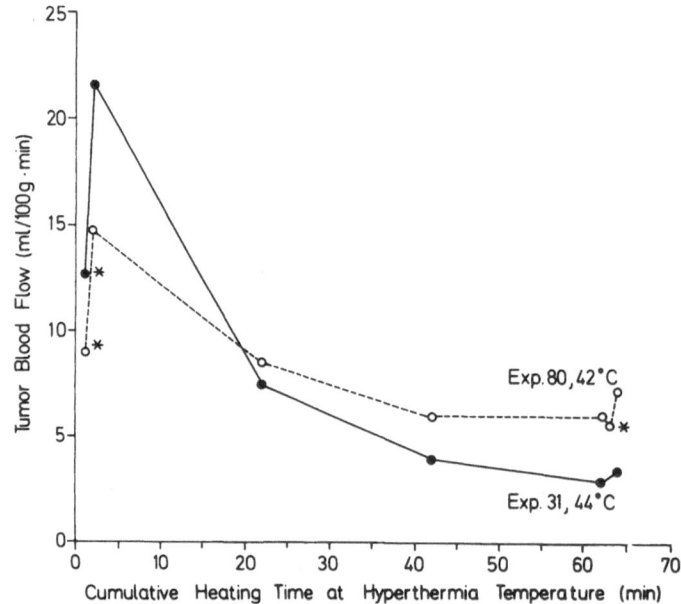

Figure 3. Tumour blood flow as a function of cumulative heating time for two different human squamous cell mammary carcinomas transplanted into nude rats. The two therapeutic hyperthermia temperatures were 42°C (dashed line) and 44°C (solid line). Tumour blood flow was measured after each heating cycle.
*Heating for 1 minute: to 38.5°C for the 42°C curve and to 39.0°C for the 44°C curve.

The method of analysis employing Equation (2) appears not to have been utilized before, but appears to be superior to other methods of analysis in both the simplicity of its application and its consistency with available experimental data. For example, the technique for Mueller-Schauenburg et al. (1975), which assumes a power curve for thermal clearance without blood flow (where we invariably found a monoexpeonential function), when applied to the present data gave different values of Q_b in different regions of the thermal clearance curve. Other aspects of the Mueller–Schauenburg technique have been discussed by Braun (1986).

Therapeutic hyperthermia initially increased, then decreased, Q_b. This is consistent with previous reports (Emami and Song, 1984; Jain and Ward–Hartley, 1984; Mueller–Klieser and Vaupel, 1984; Song, 1984; Vaupel and Kallinowski, 1986) and suggests that this form of nonsurgical treatment will be effective if appropriate combinations of hyperthermia levels and exposure times are used.

ACKNOWLEDGEMENTS

Part of the research reported herein was performed while R.D.B. was a visiting student on a Deutscher Akademischer Austauschdienst (DAAD) fellowship at the Department of Applied Physiology, University of Mainz, and T.K.G. was a Visiting Professor also on a DAAD grant at the same department. In addition, the authors gratefully acknowledge the research support provided by grants no. EY04085 from the National Eye Institute (USA) and no. 01 VF 034 from the Bundesministerium fuer Forschung and Technologie. Partial graduate student support for R.D.B. was provided by Northwestern University.

REFERENCES

Braun, R.D. (1986). Determination of tumor blood flow by thermal washout. M.S. thesis, Chemical Engineering Department, Northwestern University, Evanston, IL 60201, U.S.A.

Dethlefsen, L.A. and Dewey, W.C. (1982). Cancer therapy by hyperthermia, drugs and radiation. Bethesda National Cancer Institute Monograph, 61.

Emami, B. and Song, C.W. (1984). Physiological mechanisms in hyperthermia: a review. Int. J. Radiat. Oncol. Biol. Phys. 10, 289–295.

Hahn, G.M. (1982). Hyperthermia and Cancer. Plenum Press, New York and London.

Jain, R.K. and Ward–Hartley, K. (1984). Tumor blood flow characterization, modifications, and role in hyperthermia. IEEE Trans. Sonics Ultrasonics, SU–31, 504–526.

Mueller–Klieser, W. and Vaupel, P. (1984). Effect of hyperthermia on tumor blood flow. Biorheology, 21, 529–538.

Mueller–Schauenburg, W., Apfel, H.,Benzing, H. and Betz, E. (1975). Quantitative measurement of local blood flow with heat clearance. Basic Research in Cardiology, 70, 547–567.

Song, C.W. (1984). Effect of local hyperthermia on blood flow and microenvironment. Cancer Res. 44, Suppl, 4721s–4730s.

Storm, F.K. (1983). Hyperthermia in Cancer Therapy. Hall Publishers, Boston.

Vaupel, P. (1975). Interrelationship between mean arterial pressure, blood flow, and vascular resistance in solid tumor tissue of the DS-carcinosarcoma. Experientia, 31, 587-588.

Vaupel, P. and Kallinowski, F. (1986). Physiological effects of hyperthermia. Recent Results in Cancer Research (in press).

Vaupel,P., Kallinowski, F., Dave, S., Gabbert, H. and Bastert, G. (1985). Human mammary carcinomas in nude rats - A new approach for investigating oxygen transport and substrate utilization in tumor tissues. In: Oxygen Transport to Tissue-VII. Eds Kreuzer, F., Cain, S.M., Turek, Z. and Goldstick, T.K., Plenum Press, New York and London, (Adv. Exp. Med. Biol. 191, 737-751).

Vaupel, P., Mueller-Klieser, W., Otte, J. and Manz, R. (1984). Impact of various thermal doses on the oxygenation and blood flow in malignant tumors upon localized hyperthermia. In: Oxygen Transport to Tissue-V. Eds Lubbers, D.W., Acker, H., Leniger-Follert, E. and Goldstick, T.K., Plenum Press, New York and London, (Adv. Exp. Med. Biol. 169, 621-629).

Vaupel, P., Ostheimer, K. and Mueller-Klieser, W. (1980). Circulatory and metabolic responses of malignant tumors during localized hyperthermia. J. Cancer Res. Clin. Oncol. 98, 15-29.

OXYGEN CONSUMING REGIONS IN EMT60/RO MULTICELLULAR TUMOUR SPHEROIDS

DETERMINED BY NONLINEAR REGRESSION ANALYSIS OF EXPERIMENTAL PO_2

PROFILES

T.K. GOLDSTICK, W. MUELLER-KLIESER[*], B. BOURRAT[*] and
L.A. JURMAN

Chemical Engineering Department, Northwestern University,
Evanston, IL 60201, U.S.A. and [*]Department of Applied
Physiology, University of Mainz, D-6500 Mainz, F.R.G.

INTRODUCTION

Malignant cells can be studied in vitro, in a tumour-like microenvironment, by growing multicellular tumour spheroids in culture (Sutherland, McCredie and Inch, 1971). Franko and Sutherland (1979) utilized diffusion theory to explain the viable rim thicknesses of spheroids measured histologically. Without PO_2 profiles, however, an unequivocal interpretation of their results was not possible. Systematic studies of the PO_2 profiles in spheroids have since been made with oxygen microelectrodes by several groups (Carlsson et al., 1979; Kaufman et al., 1981; Mueller-Klieser and Sutherland, 1982a,b). Based on these measurements, new analyses utilizing diffusion theory are being developed to characterize oxygen transport and consumption in spheroids. Busch et al. (1982) proposed employing, piecewise, either first order or Michaelis-Menten oxygen consumption kinetics in the diffusion equation. They, however, gave no evidence that this was appropriate and their study contains insufficient data to enable evaluation of the methods. Grossmann, Carlsson and Acker (1983) solved the steady state diffusion equation with an oxygen consumption rate that was linear with the radius. This appears not to be generally applicable because a later analysis (Grossmann et al., 1984) shows a nonlinear dependence on radius. A semianalytical approach to the evaluation of PO_2 profiles in spheroids, including a diffusion-depleted 'stagnant' layer of medium, has been proposed (Mueller-Klieser, 1984). This features a constant, zero order oxygen consumption rate in the viable region and a uniform Krogh's diffusion constant throughout the spheroid. In a preliminary study, Groebe and Mueller-Klieser (1986) included this approach in a nonlinear regression analysis taking into account experimental errors during the PO_2 measurements in the spheroids. Jurman (1986) extended the nonlinear regression to include two compartments in the viable region (see below).

The objective of the present study was to determine oxygen consuming zones in spheroids by nonlinear regression analysis of experimental PO_2 profiles within spheroids. To avoid possible errors related to stirring in the diffusion-depleted zone, only data within the spheroid were considered in the present study. The goal was to provide an objective method for evaluating the necrotic diameter to compare with histological measurements.

THEORY

At steady state, diffusion of oxygen in a homogeneous sphere with angular symmetry is described by the simplified diffusion equation:

$$r^{-2}(d/dr(r^2 dP/dr)) - Q/K = 0 \qquad (1)$$

where P is the PO_2 (mmHg); r is the radial position (cm); and Q/K is the oxygen consumption rate divided by Krogh's diffusion constant (mmHg/cm^2). Integrating Equation (1) twice gives:

$$P = (Q/6K)r^2 - (a/r) + b \qquad (2)$$

where a and b are constants.

Two solutions to Equation (2) were successfully applied to PO_2 profiles from multicellular tumour spheroids (Jurman, 1986). These were the solutions for a homogeneous spheroid and a spheroid with a necrotic core ($Q = 0$) surrounded by a single homogeneous viable shell (constant Q/K). Other regression analyses investigated included ones with two viable layers and with two viable layers plus a diffusion-depleted stagnant layer in the surrounding culture medium. These other solutions required two additional boundary conditions at the interfaces between regions, equal fluxes and equal PO_2 values on either side. The results with these other solutions, however, were not realistic, so those methods will not be discussed in detail.

For a homogeneous, viable spheroid without necrosis, Equation (2) becomes:

$$P = P_S - (R_S{}^2 - r^2)(Q/6K) \qquad (3)$$

where R_S and P_S are the spheroid radius (cm) and surface PO_2 (mmHg), respectively.

For a spheroid with a homogeneous viable region plus a necrotic core, Equation (2) for the viable region becomes:

$$P = P_S - (R_S{}^2 - r^2 - 2R_N{}^3(1/r-1/R_S))(Q/6K), \qquad (4)$$

$$R_N \leq r \leq R_S$$

where R_N is the necrotic core radius (cm). An alternative, equivalent expression may be obtained for this region because P_S is completely determined by R_N, R_S, P_N and Q/K. This expression is:

$$P = P_N + (r^2 + 2R_N{}^3/r - 3R_N{}^2)(Q/6K), \qquad (5)$$

$$R_N \leq r \leq R_S$$

where P_N is the necrotic core PO_2 (mmHg). Obviously, inside this necrotic region Equation (2) becomes:

$$P = P_N, \qquad\qquad r \leq R_N \qquad (6)$$

Either Equations (4) or (5), with Equation (6) plus the zero flux and $P=P_N$ boundary conditions at $r=R_N$, could have been used to fit data from a spheroid having a necrotic core. Because Equations (5) and (6) required fitting for only three parameters (P_N, R_N and Q/K), rather than the four (additionally P_S) required with Equations (4) and (6), Equations (5) and (6) were used to fit these data.

MATERIALS AND METHODS

As previously described (Freyer and Sutherland, 1980), EMT6/Ro spheroids were grown in spinner flasks in Eagle's basal medium supplemented with 10% (v/v) newborn calf serum and equilibrated with air containing 5% (v/v) CO_2. The PO_2 was measured at from nine to sixteen radial positions using an oxygen microelectrode inserted on a radial track into the immobilized spheroid. Measurements were made in a chamber where the circulation of the culture medium around the spheroid duplicated as closely as possible the convection in the spinner flask. The spheroids were evaluated histologically according to the method of Mueller-Klieser et al. (1985) to find the necrotic diameter.

Two methods were employed to fit the spheroid PO_2 profiles to the two possible solutions, with and without necrosis. A noniterative method was used in which the root mean square (R.M.S.) error (comparable to the standard deviation of a mean) was computed at intervals over a predetermined range of each parameter, and those parameter values giving the lowest R.M.S. error were selected (Jurman, 1986). A nonlinear regression library routine utilizing the Marquardt method (Robinson, 1982) was also used to fit all of the data to both solutions. With a few data sets where there appeared to be several local minima, results from the noniterative method only were usable. The analysis assumed that all radial positions (including R_S) were correct and any errors were in P. The analysis therefore effectively fitted for P_S, P_N, R_N (with necrosis), and the average Q/K in the viable region.

Forty-six profiles from spheroids of EMT6/Ro were analysed. Their diameters ranged from 260 to 984 μm. For these forty-six profiles, parameter values were found for forty-three, three failing to yield a unique set of parameters. Necrosis was considered to be present when the solution with necrosis gave a lower R.M.S. error than the one without it.

Parameter correlation analysis indicated that without necrosis the correlation coefficient between P_S and Q/K was typically 0.8, while with necrosis it was typically 0.95 between R_N and Q/K. Nevertheless, convergence was obtained in all but three of the 46 data sets analysed, relatively independently of the initial estimates used. Uniqueness was assumed if two or three different initial estimates gave the same final values. When they did not, the noniterative method was employed.

RESULTS

Representative PO_2 profiles fitted to experimental data are shown in Figure 1 for a small spheroid without necrosis and in Figure 2 for a larger spheroid with necrosis. For the latter, 790 μm spheroid, the fitted values were: R_N = 128.9 μm and viable rim thickness equals

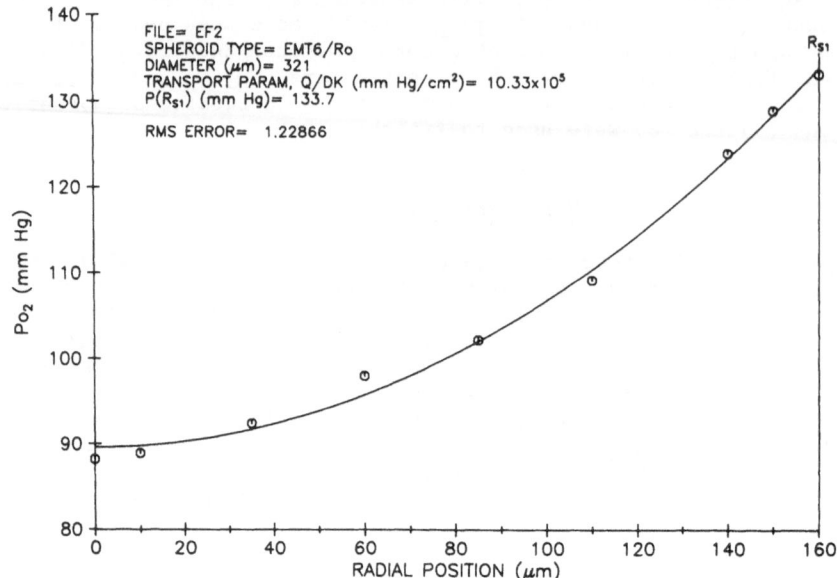

Figure 1. Typical computed PO₂ profile (solid line from Equation 3) fitted to experimental PO₂ values (points) in a small EMT6/Ro spheroid. R_{S1} = R_S, the measured spheroid radius.

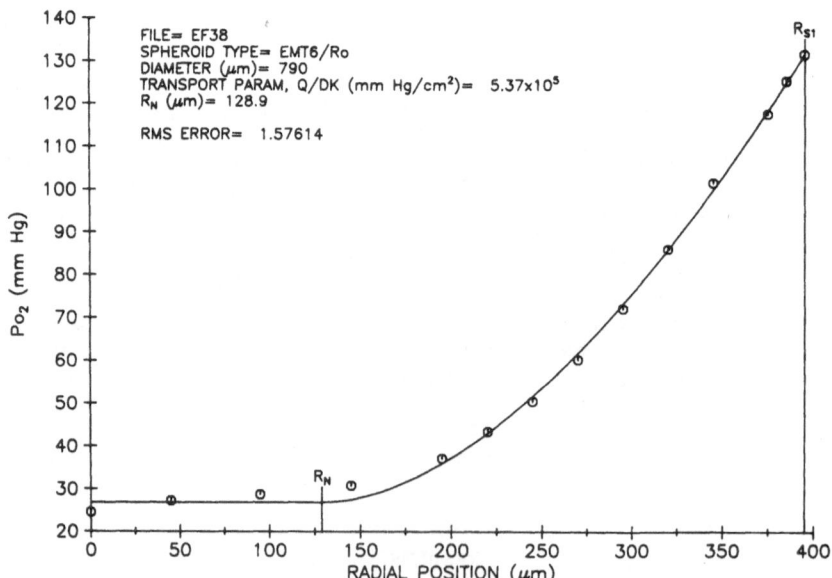

Figure 2. Typical computed PO₂ profile (solid line from Equations 5 and 6) fitted to experimental PO₂ values (points) in a large EMT6/Ro spheroid. R_{S1} = R_S, measured spheroid radius; R_N = fitted radius of necrotic centre.

384

266.1 μm. Except near the centre of the spheroid, the scatter was usually random. The R.M.S. errors of all fits ranged from $\pm$ 0.70 to $\pm$ 6.65 mmHg with a median of $\pm$ 2.54 mmHg. Near the centre, the points were almost always a couple of mmHg below the fitted curve, presumably because of the asymmetrical flow conditions around the spheroid (see Discussion).

Figure 3 shows that the fitted PO_2 at r=0 decreased with increasing spheroid size. The linear regression fit of the central PO_2 indicates that it would reach a PO_2 of zero on the average only at a diameter of 1002 μm. Also shown in Figure 3 are the identity lines (slope = 1) of the necrotic diameter (right-hand ordinate) based on histological evaluation, which gave an average viable rim thickness ($\pm$S.D.) of 170 $\pm$38 μm, and from nonlinear regression, which gave an average viable rim thickness of 280.5 $\pm$56.6 μm.

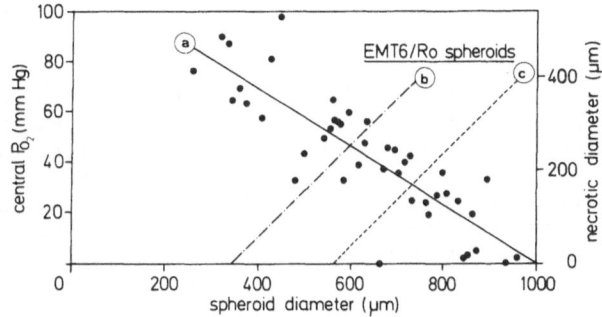

Figure 3.   The PO_2 at the centre of EMT6/Ro spheroids (left-hand ordinate) versus spheroid diameter. Linear regression line (a) has equation PO_2 = 115.5 − 2304R_S (r = −0.85; 130 $\leq$ R_S $\leq$ 492 μm). Dashed lines (b) and (c) show the necrotic diameter (right-hand ordinate) versus spheroid diameter for a constant viable rim thickness determined histologically (b) and by the nonlinear regression analysis (c).

DISCUSSION

Experimental PO_2 profiles measured in multicellular spheroids of EMT6/Ro cells could be fitted by the steady state solution to the angularly symmetrical diffusion equation with a constant value of Q/K in the consuming region. More complex approaches, such as two viable regions with different oxygen consumption rates (Jurman, 1986), failed to give meaningful results, possibly because of the low spatial resolution of the PO_2 measurements and their inherent experimental error. The median R.M.S. error, $\pm$2.5 mmHg, was approximately the precision of the entire measuring system. The rather good fit by the diffusion equation suggests that other transport phenomena, such as bulk flow, play no significant role in oxygen transport in spheroids.

A comparison of the average fitted viable shell thickness with its histological counterpart, 280 vs. 170 µm, suggests that there are regions which may appear necrotic but which in reality are oxygen consuming. A part of this difference may be related to systematic experimental error. Aside from the inevitable, probably random, experimental errors in PO_2, r, and the precise location of the spheroid surface (Groebe and Mueller-Klieser, 1986), there was a systematic error caused by nonuniform stirring. The point where the microelectrode entered the spheroid almost invariably had more vigorous stirring than the diametrically opposite point. The P_S therefore was not completely uniform over the entire spheroid surface and may have been as much as 10 mmHg higher at the point of entry than diametrically opposite. This had the effect of making the slope of the experimental PO_2 profile slightly positive at r = 0 (Figs 1 and 2). Nonetheless, the magnitude of the difference between histological and fitted viable shell thicknesses appears to be larger than could be explained by experimental error alone. Differences between functional and histological thicknesses have also been found in other studies with this and other cell types (Mueller-Klieser, Freyer and Sutherland, 1986; Sutherland et al., 1986).

As is characteristic of any modelling technique, systematic experimental errors tend to manifest themselves as changes in the model parameters. For example, asymmetrical boundary conditions over the spheroid surface tend to distort the fitted necrotic diameter. The PO_2 in the necrotic region decreased progressively as a result of the asymmetrical boundary conditions. This phenomenon was interpreted computationally in the nonlinear regression analysis as a smaller necrotic diameter and therefore a thicker viable shell.

CONCLUSION

Based on parameters derived from the computer fit, the following conclusions may be drawn:

1. The experimental PO_2 profiles follow the theoretical diffusion equation indicating that there is no significant bulk flow. Nonlinear regression can therefore be used to obtain relevant biological parameters.

2. The median R.M.S. error, ± 2.5 mmHg, was approximately the precision of the PO_2 measurements.

3. Central PO_2 decreased with increasing spheroid diameter reaching 0 mmHg at a diameter of ca. 1000 µm.

4. Necrosis, whether determined by histology or fitting, started long before the central PO_2 showed any evidence of hypoxia.

ACKNOWLEDGEMENTS

Part of the research reported herein was performed while T.K.G. was a Visiting Professor supported by the Deutscher Akademischer Austauschdienst, from June to August, 1986, in the Department of

Applied Physiology, University of Mainz.   In addition, the authors gratefully acknowledge the research support provided by grants no. Mu 576/2-2 from the Deutsche Forschungsgemeinschaft and no. EYO4085 from the National Eye Institute (U.S.A.). Partial graduate student support for L.A.J., was provided by Northwestern University. Peter J. Strouse conducted some of the very preliminary analyses of PO_2 profiles at the beginning stage of this research.   Rod D. Braun assisted with statistical analysis, discussion of results, and preparation of this manuscript.

REFERENCES

Busch, N.A., Bruley, D.F. and Bicher, H.I. (1982). Identification of viable regions in 'in vivo' spheroidal tumors: a mathematical investigation. In: Hyperthermia. Eds Bicher, H.I. and Bruley, D.F., Plenum Press, New York and London, (Adv. Exp. Med. Biol. 157, 1-7).

Carlsson, J., Stalnacke, C.G., Acker, H., Haji-Karim, M., Nilsson, S. and Larsson, B. (1979).   The influence of oxygen on viability and proliferation in cellular spheroids.   Int. J. Radiat. Oncol. Biol. Phys. 5, 2011-2020.

Franko, A.J. and Sutherland, R.M. (1979). Oxygen diffusion distance and development of necrosis and multicell spheroids. Radiat. Res. 79, 439-453.

Freyer, J.P. and Sutherland, R.M. (1980). Selective dissociation and characterization of cells from different regions of multicell tumor spheroids. Cancer Res. 40, 3956-3965.

Groebe, K. and Mueller-Klieser, W. (1986).   Non-linear regression analysis of PO_2-profiles in multicellular spheroids. Abstracts of the 24th Annual Mtg Radiation Research Soc., Las Vegas, p. 123.

Grossmann, U., Carlsson, J. and Acker, H. (1983). Oxygen consumption profiles inside cellular spheroids calculated from PO_2-profiles. In: Oxygen Transport to Tissue-IV. Eds Bicher, H.I. and Bruley, D.F., Plenum Press, New York and London, (Adv. Exp. Med. Biol. 159, 477-480).

Grossmann, U., Winkler, P., Carlsson, J. and Acker, H. (1984). Local variations of oxygen consumption within multicellular spheroids calculated from measured PO_2-profiles. In: Oxygen Transport to Tissue-V.   Eds Lubbers, D.W., Acker, H., Leniger-Follert, E. and Goldstick, T.K., Plenum Press, New York and London, (Adv. Exp. Med. Biol. 169, 719-728).

Jurman, L.A. (1986). Oxygen consumption and transport in multicellular tumor spheroids analyzed using a multiple compartment diffusion model. M.S. thesis, Chemical Engineering Department, Northwestern University, Evanston, IL 60201, U.S.A.

Kaufman, N., Bicher, H.I., Hetzel, F.W. and Brown, M. (1981). A system of determining the pharmacology of indirect radiation sensitizer drugs on multicellular spheroids. Cancer Clin. Trials, 4, 199-204.

Mueller-Klieser, W. (1984).   Method for determination of oxygen consumption rates and diffusion coefficients in multicellular spheroids. Biophys. J. 46, 343-348.

Mueller-Klieser, W., Bourrat, B., Gabbert, H. and Sutherland, R.M. (1985). Changes in O_2 consumption of multicellular spheroids during development of necrosis. In: Oxygen Transport to Tissue-VII. Eds Kreuzer, F., Cain, S.M. Turek, Z. and Goldstick, T.K., Plenum Press, New York and London, (Adv. Exp. Med. Biol. 191, 775-784).

Mueller-Klieser, W., Freyer, J.P. and Sutherland, R.M. (1986). Influence of glucose and oxygen supply conditions on the oxygenation of multicellular spheroids. Br.J. Cancer, 53, 345-353.

Mueller-Klieser, W. and Sutherland, R.M. (1982a). Influence of convection in the growth medium on oxygen tensions in multicellular tumor spheroids. Cancer Res. 42, 237-242.

Mueller-Klieser, W. and Sutherland, R.M. (1982b). Oxygen tensions in multicell spheroids of two cell lines. Br. J. Cancer, 45, 256-264.

Robinson, B. (1982). NLREG - Nonlinear regression subroutine package, VCC No. 328, Vogelback Computing Center, Northwestern University, Evanston, IL 60201, U.S.A.

Sutherland, R.M., McCredie, J.A. and Inch, W.R. (1971). Growth of multicellular spheroids in tissue culture as a model of nodular carcinomas. Jl Natl. Cancer Inst. 46, 113-120.

Sutherland, R.M., Sordat, B., Bamat, J., Gabbert, H., Bourrat, B. and Mueller-Klieser, W. (1986). Oxygenation and differentiation in multicell spheroids of human colon carcinoma. Cancer Res. (in press).

OXYGEN CONSUMPTION RATE OF TUMOUR CELLS AS A FUNCTION OF THEIR PROLIFERATIVE STATUS

S. WALENTA and W. MUELLER-KLIESER

Department of Applied Physiology, University of Mainz
Saarstrasse 21, D-6500 Mainz, F.R.G.

SUMMARY

The oxygen consumption rate ($\dot{Q}O_2$) of EMT6/Ro-cells cultured as monolayers was investigated as a function of their proliferative status. The transition of these cells from the exponential to the plateau growth phase was associated with a continuous decrease in $\dot{Q}O_2$ per single cell. This decrease can partially be attributed to a corresponding decline in cellular volume. In addition, the $\dot{Q}O_2$ per cell volume was also reduced during the passage of the cells through the growth phase described. The results lead to the conclusion that a reduction in cellular volume and factors which are still unknown may contribute to the metabolic changes observed.

INTRODUCTION

Investigations on multicellular spheroids (Freyer et al., 1984; Freyer and Sutherland, 1985) and on tumour ascites cells (Kallinowski, Schaefer and Vaupel, 1986) are indicative of an interrelationship between the oxygen consumption rate and the proliferative status of cancer cells. For systematic investigation of the time course of such a correlation between proliferation and metabolism, the oxygen consumption rate ($\dot{Q}O_2$) of tumour cells grown as monolayers was measured as a function of the time in culture.

MATERIALS AND METHODS

EMT6/Ro-cells which were derived from a mouse mammary sarcoma (Rockwell, Kallman and Fajardo, 1972) were used throughout this study. The cells were maintained in Petri dishes (10 x 20 mm^2) containing 10 ml of Eagle's Basal Medium with Earle's Salts (Flow Laboratories GmbH, Meckenheim, F.R.G.) and 10% Newborn Calf Serum (Gibco-Europe GmbH, Karlsruhe,F.R.G.). The cultures were equilibrated at 37.0 ± 0.2°C with air containing 5% (v/v) CO_2. The proliferative status of EMT6-cells in monolayer culture was assessed by recording the increase in cell number per dish against time in culture, for a period of 10 days. Prior to use in an experiment the cells were removed from the culture dishes by trypsinization and resuspended in fresh medium. Thus, cellular respiration was determined at physiological pH values under normoglycaemic conditions. The cell number and the cell volume distribution were recorded by an electronic particle counter (Coulter-Electronics GmbH, Krefeld, F.R.G.) interfaced to a pulse-height multi-channel analyser (Kuenze, Dolgesheim, F.R.G.). The mean cellular volume $\pm$ S.D. was determined by special software on a personal computer.

The respiration measurements were based on an injection procedure and a photometric determination of oxygen (Mueller-Klieser, Zander and Vaupel, 1986). In brief, the cell suspension was equilibrated in a thermostated tonometer with 95% (v/v) O_2, and 5% (v/v) CO_2. Samples of 500 μl were withdrawn from the tonometer into a thermostated, gas-tight precision syringe. The cell suspension in the syringe was stirred by a glass-coated magnet that fitted into the barrel of the syringe. For stirring, a large U-shaped-magnet was moved along the precision syringe by a special drive connected to a motor with adjustable speed. Aliquots of 100 μl were then injected from the syring into a gas-tight elution chamber at defined time intervals. By use of a special injection fluid, cells were immediately lysed and respiration completely stopped in the elution chambers. The oxygen introduced into the chamber with the injected sample, was washed out into nitrogen gas bubbling through the injection fluid. The amount of O_2 injected was measured by the colour reaction of pyrocatechol with oxygen, recorded with a photometer. Thus, each injection of a certain amount of O_2 into the measuring system led to a transient increase in the absorbance of the O_2-reaction solution and to a peak in the time course of the photometer reading. This reading was displayed on one chanel of a two-channel chart recorder. The area under this peak was proportional to the amount of oxygen injected. The photometer reading was integrated electronically to yield a sigmoidal curve with a plateau, the height of which was proportional to the respective quantity of oxygen. This signal was displayed on the second channel of the chart recorder. After calibration of the system by injecting given volumes of air, the oxygen consumption rate in the suspension and per single cell could be calculated from the cell concentration and the time-dependent decline of oxygen concentration in the precision syringe.

RESULTS

The typical growth behaviour of EMT6-monolayer cultures is characterized by an initial exponential increase of the cell number per dish and a transition to a plateau phase with a decrease in proliferation after 5 days in culture. Experiments were carried out on exponentially growing cells at days 2 and 3 and on plateau phase cells at days 5 through 9.

As shown in Table 1, · there was a significant reduction in $\dot{Q}O_2$ per cell from day 2 to day 9 in culture. The decline in the oxygen consumption per cell was associated with a drop in the mean cellular volume during growth (see Table 1). The results lead to the conclusion that the reduction in cellular respiration as a function of cellular proliferation can be attributed partially to a decrease in cellular volume. In addition to the changes in cellular respiration observed there was a distinct decrease in the oxygen consumption rate calculated per cell volume with increasing time in culture (see Table 1).

Table 1. Results for exponentially growing cells (2 days in culture) and cells in the plateau phase (9 days in culture)

Growth status	Cell volume $\bar{x} \pm$ S.E.M. μm^3	$\dot{Q}O_2$/single cell $\bar{x} \pm$ S.E.M. 10^{-17} mol O_2/s	$\dot{Q}O_2$/unit volume $\bar{x} \pm$ S.E.M. 10^{-8} molO_2/cm^3.s
Exponential	3813 ± 373	16.9 ± 0.83	6.5 ± 0.35
Plateau	1918 ± 191	9.6 ± 0.25	4.9 ± 0.16

DISCUSSION

The findings clearly show that the transition of tumour cells in monolayer culture from the exponential to the plateau growth phase is associated with a substantial decrease in both cellular oxygen consumption and mean cell volume. The concomitant decrease in $\dot{Q}O_2$ evaluated per cellular volume with time in culture implies that factors other than a reduction in cellular volume may also contribute to the metabolic changes observed. Furthermore, the results provide evidence for the biological significance of comparing the oxygen consumption rate on a cellular basis with that evaluated per cell volume. Such a comparison may contribute to a better understanding of the factors that determine the volume-related oxygen consumption of tissues in vivo. As a first approach to investigating the phenomenon described further, cytological investigations on the number, size and structure of mitochondria of tumour cells in the exponential and plateau growth phase will be performed.

REFERENCES

Freyer, J.P. and Sutherland, R.M. (1985). A reduction in the in situ rates of oxygen and glucose comsumption of cells in EMT6/Ro spheroids during growth. J. Cell. Physiol. 124, 516-524.

Freyer, J.P., Tustanoff, E., Franko, A.J. and Sutherland, R.M. (1984). In situ oxygen consumption rates of cells in V-79 multicellular spheroids during growth. J. Cell. Physiol. 118, 53-61.

Kallinowski, F., Schaefer, C. and Vaupel, P. (1986). Oxygen consumption rate of ascites tumor cells during growth. 34th Annual Meeting Radiation Research Society, Las Vegas, NV.

Mueller-Klieser, W., Zander, R. and Vaupel, P. (1986). A new photometric method for oxygen consumption measurements in cell suspensions. J. Appl. Physiol. 61, 449-455.

Rockwell, S.C., Kallman, R.F. and Fajardo, L.F. (1972). Characteristics of a serially transplanted mouse mammary tumor and its tissue-culture-adapted derivative. J. Natl. Cancer Inst. 49, 735-749.

BLOOD FLOW, VASCULAR RESISTANCE AND OXYGEN AVAILABILITY IN MALIGNANT

TUMOURS UPON INTRAVENOUS FLUNARIZINE

P. VAUPEL and H. MENKE

Department of Applied Physiology, University of Mainz
Saarstrasse 21, D-6500 Mainz, F.R.G.

SUMMARY

Tumour blood flow, an important determinant of the efficacy of presently available nonsurgical cancer treatments, significantly increased following a single I.V. injection of the calcium antagonist flunarizine. At a dose of 1 mg/kg, tumour blood flow increased approximately by 28% without a significant change in mean arterial blood pressure. The flow increase was paralleled by a similar improvement of the O_2 availability to the cancer cells. The data suggest that flunarizine may provide a means of improving delivery of antineoplastic agents to tumours. Furthermore, flunarizine may also enhance the effectiveness of irradiation by increasing tumour oxygenation.

INTRODUCTION

Calcium antagonists have been used successfully to enhance the cytotoxicity of several anticancer drugs under in vivo and in vitro conditions and are known to inhibit the respiration rate of cancer cells as well as the formation of metastases of spontaneous and experimental tumours (for a review see Vaupel and Mueller-Klieser, 1986). Besides these actions, which can be relevant for the therapy of neoplastic diseases, the main pharmacological properties of calcium antagonists are based on their actions in cardiac cells and on the arterial resistance vessels (Needleman, Corr and Johnson, 1985). Administration of calcium antagonists can result in a selective dilatation of arterioles and, thereby, can lead to an increase of peripheral blood flow as long as the mean arterial blood pressure does not decrease. If calcium antagonists could produce a similar vasodilatory effect on the arteriolar vessels feeding a malignant tumour, an eventual increase in tumour blood flow may potentially improve O_2 delivery to hypoxic cancer cells and/or improve local pharmacokinetics and may modulate pharmacodynamics of some antineoplastic agents. Because of these possible advantages of an increased tumour blood flow, we have initiated studies on the efficiency of flunarizine in raising tumour blood flow. Flunarizine was chosen because of its high specificity for peripheral arteriolar smooth muscle with little myocardial depressant activity (for a review see Flaim and Zelis, 1982).

MATERIALS AND METHODS

Animals and tumour system

The effect of flunarizine (1 mg/kg I.V., Janssen GmbH, Neuss, F.R.G.) on tumour blood flow (TBF) was studied in DS-Carcinosarcomas growing subcutaneously after injection of ascites cells (0.3 ml; about 10^4 cells/μl) into the hind foot dorsum of male and female Sprague-Dawley rats (mean body weight; 360 $\pm$ 41 g). Tumours were used in experiments when they reached a mean tumour net weight of 0.81 $\pm$ 0.05 g (mean tumour growth period: 7 $\pm$ 1 days).

Blood flow measurements

When the tumours had reached the desired size, the animals were anaesthetized with the neuroleptic-opioid combination droperidol (3.8 mg/kg I.P.) and fentanyl (0.075 mg/kg I.P.). Measurements of tumour blood flow (TBF) were performed by means of the 85-Krypton clearance technique (Vaupel, Ruppert and Hutten, 1977). The radioactive indicator was applied as a bolus injection through a PVC-catheter into the thoracic aorta (0.1 mCi = 3.7 MBq in 0.9% NaCl solution). A counting tube covered with a 6 μm thick Hostaphan membrane was placed over the tumour so that no compression of the tumour surface occurred. The registration of the washout process was performed as described earlier (Vaupel et al., 1977). For evaluation of TBF a special index was used that takes into account both the initial activities for each exponential function as well as the corresponding half times (the exponential functions served for approximation of the washout curves; for details see Vaupel et al., 1977).

Experimental protocol

Throughout all experiments the core temperature of the animal (which was placed on a thermostatically controlled heating pad), the mean arterial blood pressure (MABP) in the thoracic aorta, and the relevant respiratory gas parameters in the arterial blood were monitored. Heparin (350 USP-units/kg I.V.) was given as an anti-coagulant. The O_2 content in the arterial blood was obtained using actual O_2 and CO_2 partial pressures as well as pH values and consideration of the O_2 dissociation curve of Sprague-Dawley rats (Bork et al., 1975) and the respective haemoglobin concentration (Dave, Kallinowski and Vaupel, 1985). Oxygen availability was calculated from the equation:

$$O_2 \text{ availability} = TBF \times art.[O_2] \; (\mu l \; O_2 \cdot g^{-1} \cdot min^{-1} \tag{1}$$

Blood flow measurements were performed 30, 20 and 10 min before and 10, 20, 30, 45 and 60 min after injection of flunarizine into the external jugular vein. TBF data obtained before drug administration were pooled since there were no significant intra-individual differences.

Tumour vascular resistance (TVR) was calculated from the equation:

$$TVR = MABP/TBF \; (mmHg/ml \cdot g^{-1} \cdot min^{-1}) \tag{2}$$

RESULTS

Whereas the intravenous administration of a volume of saline equivalent to that of flunarizine did not itself affect tumour blood flow, the calcium antagonist, flunarizine, significantly increased TBF for 10 to 30 min from application, without a significant alteration of the mean arterial blood pressure. Tumour vascular resistance (TVR) is the mirror-image (see Fig. 1). At 10 min after I.V. application of flunarizine, mean TBF increased by about 21% (2p <0.02), at 20 min by 28% (2p <0.005), and at 30 min by approximately 24% (2p <0.05).

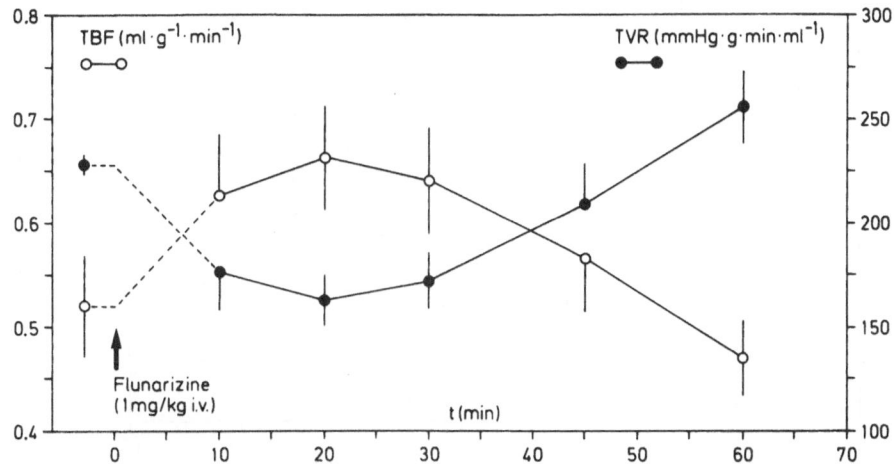

Figure 1. Tumour blood flow per unit weight (TBF) and tumour vascular resistance (TVR) before and after I.V. application of flunarizine (1 mg/kg). Values are means ± S.E.M.; open circles, TBF; filled circles, TVR.

These temporary increases in TBF were accompanied by similar increases in the O_2 availability to the tumour tissue (see Fig. 2). At 10 min after I.V. injection of flunarizine, the O_2 availability increased by about 17%, at 20 min by 24%, and at 30 min by about 20%. These improvements of the O_2 supply to the tumour tissue are all significant as compared with the control values before drug application.

DISCUSSION

Considering the lack of responsiveness or reaction of tumour microvessels to physiological or pharmacological stimuli, it is most likely that only host vessels incorporated into the tumour tissue will show some reactivity within a certain period of growth. From that one must conclude that the existence and the extent of a microcirculatory 'regulation' depend on the prevalence as well as on the proportion of residual normal host vessels in a growing tumour. Intratumour responsiveness can be mimicked by steal-phenomena and anti-steal-phenomena in conjunction with the vascular bed of the normal tissue adjacent to the tumour tissue. Therefore, in many instances the effect of vasoactive drugs on tumour microcirculation is indirect, i.e., changes in tumour blood flow result from changes in the resistance of the surrounding normal microvascular bed rather than from a direct effect on tumour microvessels themselves. Furthermore, very often alterations in systemic blood pressure upon drug application can affect tumour blood flow indirectly by changing perfusion pressures.

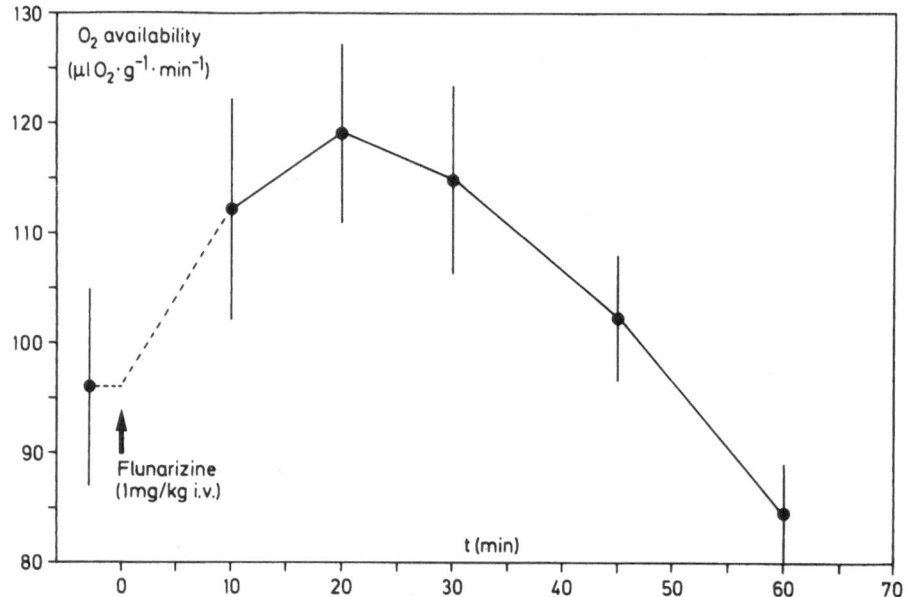

Figure 2. Oxygen availability to tumours before and after I.V. application of flunarizine (1 mg/kg). Values are means ± S.E.M.

Calcium antagonists are compounds that can significantly increase blood flow through certain tumours (see Table 1). These agents can decrease vascular resistance more in the tumour than in normal tissues (Kaelin et al., 1982; Kaelin, Shrivastav and Jirtle, 1984). Upon application of these drugs, not only arteriovenous shunt perfusion (Kaelin et al., 1984), but also nutritive blood flow through tumours seem to be increased, yielding a distinct reduction of the radiobiologically hypoxic cell fraction (Hill and Stirling, 1986).

Sites at which flunarizine may act to alter tumour blood flow are host tissue arterioles incorporated into the tumour mass and arterioles in the tissue surrounding the growing tumour. It has been postulated that the arterioles in the tumour periphery are the main resistance vessels influencing blood flow to more central regions of a tumour (Kaelin et al., 1984). In addition, inhibition of platelet aggregation and restoration of the red blood cell deformability within tumours are also thought to play a part in the flow increase observed after flunarizine application (Kaelin et al., 1984).

Based on these results one may speculate that calcium antagonists may be useful adjuvants to diagnostic procedures and anticancer treatment strategies. They may enhance local pharmacokinetics, modulate pharmacodynamics of cytotoxic substances, increase radiosensitivity, and improve delivery of both monoclonal antibodies and biological response modifiers.

Table 1.   Effect of flunarizine on tumour blood flow in rodents

Tumour (species)	Dosage (µg/g)	Flow change (%)	Duration of effect(min)	Reference
KHT sarcoma (mouse)	4 (I.P.)	+ 20-30	20-30	Hill and Stirling (1986)
C3H tumour (mouse)	4 (I.P.)	none	none	"
RIF-1 tumour (mouse)	4 (I.P.)	none	none	"
FS 13 fibro-sarcoma (mouse)	5 (I.P.)	none	none	Robinson et al. (1986)
SMT-2A tumour (rat)	1 (I.V.)	+ 50	45	Kaelin et al. (1984)
DS-CaSa (rat)	1 (I.V.)	+ 28	30	This paper

REFERENCES

Bork, R., Vaupel, P., Guenther, H. and Thews, G. (1975). Atemgas-pH-Nomogramme fuer das Rattenblut bei 37°C. Anaesthesist, 24, 84-90.

Dave, S., Kallinowski, F. and Vaupel, P. (1985). Blood flow and oxygen supply to human mammary carcinomas transplanted into nude rats. In: Oxygen Transport to Tissue-VII. Eds Kreuzer, F., Cain, S.M., Turek, Z. and Goldstick, T.K., Plenum Press, New York and London, (Adv. Exp. Med. Biol. 191, 753-762).

Flaim, S.F. and Zelis, R. (1982). Clinical use of calcium entry blockers. Fed. Proc. 40, 2877-2881.

Hill, R.P. and Stirling, D. (1986). The effect of flunarizine on blood flow and radiation response in the KHT sarcoma. 34th Annual Meeting, Radiation Research Society, Las Vegas, NV, Abstract Fh10.

Kaelin, W.G., Shrivastav, S. and Jirtle, R.L. (1984). Blood flow to primary tumors and lymph node metastases in SMT-2A tumor-bearing rats following intravenous flunarizine. Cancer Res. 44, 896-899.

Kaelin, W.G., Shrivastav, S., Shand, D.G. and Jirtle, R.L. (1982). Effect of verapamil on malignant tissue blood flow in SMT-2A tumor-bearing rats. Cancer Res. 42, 3944-3949.

Needleman, P., Corr, P.B. and Johnson, E.M., Jr (1985). Drugs used for the treatment of angina: organic nitrates, calcium channel blockers, and β-adrenergic antagonists. In: Goodman and Gilman's The Pharmacological Basis of Therapeutics. Eds Goodman Gilman, A., Goodman, L.S., Rall, T.W. and Murad, F., Macmillan Publ. Co., New York, pp. 806-826.

Robinson, B.A., Clutterbuck, R.D., Millar, J.L. and McElwain, T.J. (1986). Effects of verapamil and alcohol on blood flow, melphalan uptake and cytotoxicity, in murine fibrosarcomas and human melanoma xenografts. Brit. J. Cancer, 53, 607–614.

Vaupel, P., Ruppert, H. and Hutten, H. (1977). Splenic blood flow and intra-splenic flow distribution in rats. Pflugers Arch. 369, 193–201.

Vaupel, P. and Mueller-Klieser, W. (1986). Verapamil inhibits the respiration rate of cancer cells. In: Oxygen Transport to Tissue-VIII. Ed. Longmuir, I.S., Plenum Press, New York and London, (Adv. Exp. Med. Biol. 200, 645–648).

AUTHOR INDEX